The Invisible Invaders

Also by Peter Radetsky

Peak Condition
(with James Garrick, M.D.)

PaceWalking
(with Steven Jonas, M.D.)

Be Your Own Personal Trainer
(with James Garrick, M.D.)

The Invisible Invaders

. . . .

Viruses and the Scientists Who Pursue Them

. . . .

Peter Radetsky

BACK
BAY
BOOKS

Little, Brown and Company

Boston   New York   Toronto   London

FIRST PAPERBACK EDITION

Portions of this book appeared in somewhat altered form in *Discover,* April
1989, August 1989, September 1989, and November 1991.

LIBRARY OF CONGRESS CATALOGING-IN-PUBLICATION DATA

Radetsky, Peter.
The invisible invaders : viruses and the scientists who pursue them /
by Peter Radetsky. — 1st ed.
p.   cm.
Includes bibliographical references and index.
ISBN 0-316-73216-8 (hc) 0-316-73217-6 (pb)
1. Virology — History.   2. Viruses — Research — History.   I. Title.
QR359.R33   1991
616'.0194 — dc20   90-6612

10   9   8   7   6   5   4   3   2

MV-NY
Published simultaneously in Canada
by Little, Brown & Company (Canada) Limited

Printed in the United States of America

Once again, for Nancy
With even more love

Contents

Contents

. . . .

Illustrations

. . . .

. . . .

Preface

. . . .

I was talking with an old friend the other day. A fellow writer, he has known me since before I was published and has seen my family grow in numbers and years. I've watched him make one odyssey after another to distant corners of the globe as he tries to reconcile his strong sense of freedom and charity with the chaos he finds around him.

We talked about many things — children and the difficulty of growing up, the responsibility of writers, the turmoil in the world (he had just returned from Israel and Guatemala and was soon to take off to Nicaragua) — and he remarked that with so much political unrest all around us, he hoped my kids would be able to live full lives. I write much about science and medicine, and as he talked I was thinking of other kinds of turmoil. With the terrifying prospect of millions of deaths from AIDS by the early years of the twenty-first century, I replied, what did political considerations matter? Who was going to have energy left over for politics in a world of sick and dying people?

He then said something very interesting: "I'm not worried about AIDS. They'll come up with a vaccine in a couple of years, I'm sure of it."

That struck me. This worldly, sophisticated man was reflecting the thinking of the least aware of us — perhaps even most of us. After all, scientists have come up with vaccines against rabies, yellow fever, polio, measles, mumps, smallpox, and influenza; they'll undoubtedly come to the rescue once again. AIDS will simply be the next triumph on the list.

Maybe so — miracles can happen. But usually life isn't that easy. My friend's remark betrayed a startling lack of understanding about the amazing organisms that cause AIDS and many other diseases that plague us: viruses. And he's not alone. I realized that I knew almost nothing about them. "Virus" was a word I heard in the doctor's office when he was trying to explain that there was nothing he could do. "Bacterial infection" — now that meant a round of antibiotics and all would be well. But a virus? If you didn't stop it with a vaccine before it infected you, there wasn't much you could do. The doc would shrug his shoulders, give you a pat on the back, suggest a few aspirin, and announce that you'd just have to wait it out.

So as I was talking to my friend, it occurred to me that perhaps very few of us understand what a virus is and how it works. But until recently, except for the misery involved in waiting out periodic viral attacks, most of us saw no need to find out. We had medical scientists to make vaccines against the serious stuff, and were it not for the common cold — caused by viruses and still not preventable or curable — and those yearly miserable bouts with flu, and, perhaps, the seemingly indestructible herpesvirus, we could go merrily on our way, rarely having to think of the little creature and how it might affect us.

Those days are gone, maybe forever. It's no exaggeration to say that with AIDS a constant terror in our lives and new discoveries about the role of viruses in cancer and other diseases being made all the time, our very lives may depend on understanding what a virus is. So I began to think about this book. And when I looked into the world of viruses, an entirely unexpected thing happened: I lost myself. For it's a world teeming with interesting people doing interesting things, people who have fascinating stories to tell. And at its heart is a creature almost beyond imagination, the virus.

What I've found convinces me more than ever that my friend and the great, great majority of the rest of us have little idea of the impact viruses have on our lives. For they are not simply dangerous enemies, the only organisms besides ourselves that pose a threat to our survival; they're our co-travelers in life, our most intimate fellow workers. Viruses are literally everywhere — inside us, outside us, constantly permeating the boundaries of self.

And they're more. Rather than simply being along for the ride, viruses may be prime movers. They may swap our genes around, rearrange our destinies, act as agents of the ecosystem. In their admirable simplicity and appalling efficiency, they may be the most successful life-form of all — that is, if they can be said to be alive in the first place. For good or ill, how we deal with viruses, these tiny creatures at the heart of medical research, will have profound impact on our health and continued survival.

I hope that *The Invisible Invaders* provides an entrée into the important and exciting world of viruses. I've tried to present the emerging age of viral research in a way that will both inform and intrigue readers. I've tried to tell interesting stories and convey important information. And, above all, I've tried to be absolutely accurate in terms of the workings of these invisible invaders and the people who spend their lives dealing with them.

These past years researching and writing this book have been a rich journey for me. As I've traveled around the country, talking to people involved in the world of viruses and observing them at work, and as I've immersed myself in the literature, I've felt privileged to be a close observer at a particularly momentous time in human history. For we have in the past few decades discovered some of the fundamental secrets of life and are close to detecting more. Viruses have to a large extent enabled us to come so far and so near. The story in these pages, therefore, is not simply one of discovering the world around us — it is of discovering ourselves.

Santa Cruz, California
August 1990

. . . .

Acknowledgments

. . . .

Thanks to the many scientists who talked with me, not all of whom appear in these pages. And special thanks to those who were kind enough to review sections of the manuscript for accuracy. It is a better book for their efforts.

Thanks also to my agent, Kris Dahl, who supported this project from its inception; to Bill Phillips, who brought me to Little, Brown; and to my editor, Debbie Roth, for her enthusiasm, clarity, and sensitivity.

Finally, thanks to family and friends for their encouragement and forbearance.

The Invisible Invaders

Introduction

· · · · · · · · · · · · · · · ·

The Invisible Invaders

· · · · · · · · · · · · · · ·

It was not so long ago — no more than a couple of decades, really — that medical science stood on the brink of preventing infectious disease. Or so it seemed. We had antibiotics to battle bacteria-caused killers such as pneumonia and tuberculosis, and vaccines to prevent the onset of virus-caused illnesses such as polio and smallpox. In some parts of the world, improvements in the way people live — in hygiene, the quality of food, the disposing of wastes — further reduced the incidence of disease. Cancer and heart disease still attacked and killed millions, but these maladies were not considered the result of infections. They were more likely the result of heredity, or environmental factors, or the luck of the draw, or God only knows what. Yet it seemed only a matter of time before they too would succumb to the onrush of science. As far as being in control of our own medical destiny, it seemed the best of all possible times to be alive.

Well, it wasn't so long ago, but it might as well have been ages. Those heady, innocent days are gone forever. Thanks to medical science, we know more than we might have thought possible about the workings of life and the processes of disease and death — much more than we did only a few scant years ago. And that knowledge undoubtedly will increase exponentially over the next few years. We live in an explosion of knowledge. Yet it's fair to say that the essential insight afforded us by all this knowledge is the realization of how much we don't know.

Nowhere is the dichotomy between our comparatively deep understanding of the nature of life and our vast simultaneous ignorance of it more clear than in the case of the virus. Indeed, the virus — this tiny, not-quite-alive-yet-not-entirely-dead substance; this *thing* that possesses an awesome power and, it almost seems, a sly intelligence; this organism that is so helpless and vulnerable as to be laughable — the virus is the primary reason why our earlier notions of conquering disease now seem naive and faraway.

For the onset of terrifying virus-caused diseases such as AIDS, the discovery that viruses such as hepatitis B and the human leukemia viruses can cause cancer, and the emergence of seemingly new and deadly kinds of viruses such as retroviruses have dashed for the time being any grandiose hopes of conquering disease. These viruses have brought home the reality of disease and the evanescence of life. No longer can we say, "Not me." These viruses can attack anyone, and when they do, they are devastating. They remind us of something we do our best to forget: our vulnerability.

Yet despite the power of the metaphor, *attack* may be the wrong term to use with respect to viruses. Viruses are our co-travelers in life. They may play as great a role in shaping our lives, for good or ill, as does our own personal will.

Ours is the age of viruses. With respect to medical science, viruses constitute a primary focus, but viruses are of great interest to more than medical researchers alone. It was viruses that gave impetus to the science of molecular biology, and it is viruses that continue to provide means to manipulate the elements of that invisible world. In that sense, viruses provide great hope in our effort to extend and improve our lives.

But for most of us, most of the time, viruses are nothing more or less than a great menace. And there's simply no escaping them. Viruses cause more sickness than anything else on earth. In the United States alone, viruses are to blame for almost 80 percent of all acute illnesses. The list of virus-caused diseases, past and present, is daunting. Epidemics of yellow fever, polio, influenza, and smallpox immediately come to mind. So do less immediately devastating diseases such as hepatitis, viral pneumonia, mononucleosis, rabies, and herpes. Viruses are behind the common childhood maladies — mumps, measles, rubella, and chickenpox — and cause the most familiar illness of all, the common cold.

Exotic — and lethal — afflictions such as Lassa fever, encephalitis, dengue, and aseptic meningitis are the work of viruses. Lately, as noted, it's been shown that viruses also cause certain kinds of cancer and may be involved with diabetes, multiple sclerosis, chronic fatigue syndrome, and a host of unexplained neurological diseases. And, of course, it is a virus, a particularly frail and mutable virus, that is at the heart of the most notorious disease of our day, AIDS.

Why should something that has been around as long as we have, if not longer, cause us so many problems? No one knows where or how viruses began, but they've probably been on earth since the beginning. Some scientists regard them as the earliest form of life; others consider them a backward step in evolution; others see them as having sprung from the genetic material of the earliest cells. In any case, one aspect of the historical record is clear: viruses have been causing disease for at least the last three thousand years. The pockmarked, mummified face of Ramses V is mute testimony to the virulence of smallpox as long ago as ancient Egypt, and hieroglyphs from the period depict people with shriveled legs, as though having suffered through a bout with polio. With all that history behind us, shouldn't we have learned by now?

The answer is that it has only been during the last century that people have begun to be aware of the existence of viruses, much less how to deal with them. For example, influenza, one of the most common and dangerous of viral diseases, gets its name from the Italian, *influenza,* because the malady was blamed on the "influence" of the heavenly bodies. From the early fifteenth century, when the term was coined, to the late sixteenth century, the time of the first recorded epidemic, to the early twentieth century, when the world suffered the worst influenza plague in history, blaming the heavenly bodies was as effective a response as any known.

Viruses have affected the course of history many times over. It was smallpox, unwittingly imported by the Spaniards, that allowed them to conquer the Indians of Mexico in the sixteenth century. The disease killed several million Aztecs, a feat the vastly outnumbered Spaniards could never have accomplished on their own. Later, in other New World conflicts, European and English intruders introduced the disease deliberately — the first recorded instances of biological warfare. But they received their just deserts: smallpox went on to kill sixty million Europeans and New World settlers in the eighteenth century alone.

Yellow fever was another dreaded viral plague. "Yellow jack" — British soldiers called it that because yellow quarantine flags were used to warn of its danger — decimated the native peoples of the Caribbean. In Haiti, it killed most of Napoléon's twenty-five thousand men in less than a month. In Panama, twenty thousand workers died of the disease between 1881 and 1889; along with malaria, it forced the French to forsake building the canal.

And viruses have caused other kinds of panic, as well. In flower-growing Holland during the seventeenth century, men paid fantastic prices for tulip bulbs that produced blossoms with colorful streaks. Wealthy men offered their daughters in marriage, along with house and farm, in exchange for a single bulb. Wild speculation broke out, so much so that the government had to step in to restore order. The phenomenon became known as "tulipomania." The reason for the unusual color pattern: a viral infection.

We know all this now, but our knowledge has been hard in coming. It wasn't until the turn of the century that Dutch botanist Martinus Beijerinck (pronounced "*buy*-er-ink") called the world's attention to a virus — something that was not quite like anything that had ever been conceived of before. As Beijerinck tried to track down a mysterious disease-causing substance that would not show itself in the finest of microscopes — microscopes that could display even the tiniest of bacteria — a strange thought came to him: there was something loose in the world, something that was so small it might be no larger than a molecule, something that was alive, able to reproduce, and very, very dangerous. He called this invisible substance a virus, from the Latin for "poison" or "slime."

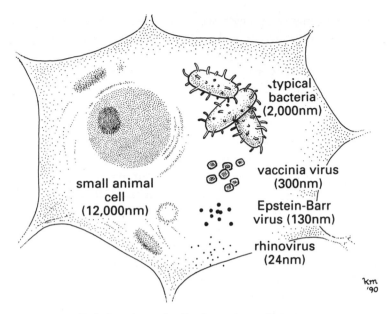

Relative sizes of cells, bacteria, and viruses

Viruses are among the smallest of all living things. An animal cell dwarfs even the largest of them, vaccinia, while the tiny rhinovirus appears as a mere pinpoint in comparison.

Today we know a good deal more about viruses than Beijerinck did, but his reaction — that they were "positively unnatural" — is no less apropos. Imagine a thing somewhat bigger than a molecule but considerably smaller than most living cells. A single virus ranges from about .01 to .3 microns in diameter — that is, as little as two millionths of an inch across, a size so small it is virtually impossible to conceive of it. For example, it would take ten thousand rhinoviruses (the viruses that cause most cases of the common cold) lined up side by side to span the space between words on this page.

Imagine further that this thing inhabits a shadowy borderland between life and death, becoming animate only when making its way into the living cells of a susceptible host. And imagine that, once inside, this thing reproduces by commandeering and recasting the now doomed cell to make replicas of its own viral form. A virus is

7

nothing less than a submicroscopic missile, armored with protein, at the heart of which is a clear jelly of nucleic acid, the essential genetic information of life itself. The target of the missile: all living things. (Or, in the words of Sir Peter Medawar, a 1960 Nobel laureate in medicine: "A virus is a piece of bad news wrapped in protein.")

Some viruses are so resilient that they can survive a Fahrenheit temperature of two hundred degrees below freezing. Others are so tough that a force a hundred thousand times greater than gravity cannot crush them. Certain viruses can survive outside our bodies for hundreds of years, in a kind of suspended animation, waiting until the opportunity presents itself to infiltrate and take over congenial living cells. Others — the AIDS virus is one — are so fragile that a few minutes of exposure to air or water will kill them off. And, to cap it all, viruses are uncommonly beautiful. Electron microscopes reveal a gallery of subminiature geodesic domes and icosahedrons, helixes and cylinders, spaceships and lunar landers. There's simply no figuring them.

But take heart. Although the list of diseases caused by this strange substance can seem endless, the fact is that we are surrounded by viruses, inside and out, and most of them are quite willing to leave us alone. We usually don't even notice their presence (it's when we do that problems arise). And unlike other organisms — even the tiniest and most basic, which can reproduce on their own if by no method more sophisticated than splitting in two — viruses are impotent. They have neither the ability to produce their own energy nor to make new viruses. They are the most helpless of parasites, for to survive they need to appropriate the vitality of living cells.

The process works like this. Viruses are literally everywhere. They float on wind and water, hitch rides on animals, insects, and birds, bide their time on desk tops, fingertips, the pages of this book — until they encounter their target: a specific kind of cell in a specific kind of organism. Viruses are very picky in this way. For example, a hepatitis virus is interested in the cells of your liver. It'll have nothing to do with your brain cells, say — that's the province of polio viruses — or the cells of your nose, which belong to the common cold. In the same way, other viruses have their preferred targets: a rabies virus infects nerve cells, and an AIDS virus attacks white blood cells, part of the armamentarium of your immune system.

But viruses have no way of seeking out these cells. Having no vitality of their own, they must depend entirely on chance. Without appropriating the life energy of our own cells, viruses are powerless to do so much as move. They simply float, directionless, buffeted this way and that by our bodies' interior movement, until they actually make contact with a congenial cell. It's then that the tiny missile attacks.

Different viruses wage their campaign in different ways. For example, the AIDS virus pulls itself close to a white blood cell with a built-in grappling hook, fuses its outer skin to that of the cell, and jettisons its viral contents inside. Other viruses, such as the influenza virus, depend on the cell itself to invite them in — which it does as a matter of course. The cell's intent, however, is not hospitality: it means to destroy the intruder. It surrounds the virus and draws it into its interior. That's its first mistake.

The cell's second mistake is to dissolve the virus's armor coat of protein. In so doing, it destroys part of the intruder, but it also exposes the virus's deadly core, its nucleic acid. These coiled strands of RNA or DNA immediately take over the metabolic machinery of the cell. In a submicroscopic coup d'etat, the virus usurps the cell's old commanding general, its own DNA, and issues new orders: stop making cells and start making viruses.

Thoroughly demoralized, the cell does as it's told. Responding to the manufacturing blueprint supplied by the virus's nucleic acid, the cell begins to churn out viruses. The process can be astonishingly quick. For example, a bacteriophage, a virus that invades bacteria, can cause its defeated host to produce as many as a hundred new, fully grown viruses within half an hour.

Only one task still remains: unleashing the young viruses. This the newly formed viruses do themselves by budding out of the cell, in some cases tearing away a bit of the cell's protective membrane skin to function as their own viral overcoats. Then, escaped into the bloodstream, the new viruses are free to seek out new cells and begin the process all over again. The original cell, badly wounded, usually dies.

But some viruses — the AIDS virus, for instance — may not immediately commandeer the cell's genetic machinery to make new viruses. They bide their time — actually slipping their genes into those of the cell and remaining there for years, if need be — until the time is right for them to explode into action. They are, in effect, time bombs within us. Other viruses, such as the human leukemia virus and those that cause cancer, transform, rather than destroy, the cells they infect, causing them to multiply out of control.

Another characteristic of viral attacks is that there may be little if anything we can do about them. For the fact is that once a virus has successfully invaded a cell, it's relatively safe from the body's defenses. It's also protected from most outside help the body may receive. For example, antibiotics and similar drugs do nothing against viruses. Because viruses set up shop *within* infected cells, to kill a successful virus directly a drug would have to destroy the host cell itself — a task the invading virus may do quite nicely all by itself.

If the virus's target is a tiny, one-celled creature such as a bacterium, the death of that cell means the death of the entire organism. If the target cell is part of a large, many-celled animal such as a human being, its loss may mean nothing. Who will miss a single cell? The problem is that the released viruses promptly invade and infect new cells, and the process begins again, geometrically increased.

It's then that our immune system is faced with its greatest challenge. If the virus has slipped by our initial defenses and safely ensconced itself inside the body's own cells, we must actually kill parts of ourselves to reach the invaders. And that's exactly what the body does, by sending agents of the immune system such as "killer cells" to destroy the infected cells and the viruses within them. If the infection has become widespread, the body must fight back in kind, killing infected cell after infected cell. In that case, we get sick, for it's often the body's violent reaction against intruders that causes the symptoms of disease. In most cases, the symptoms will abate as the body gradually gains the upper hand, slowing and then stopping the viral invasion. But if for one reason or another the virus continues to spread and the body continues to react, the result may no longer be just a period of sickness leading to recovery — it may be death.

So by far the best way to stop a viral attack is to meet it *before* the virus actually invades the individual cells. This the body does by means of other agents of the immune system called antibodies. These are special proteins in the blood or on the surface of certain cells that recognize the approaching viruses and stick to them, rendering them harmless, fit to be gobbled up and discarded. Vaccines work by exposing the body to a seemingly dangerous virus that has been artificially stripped of its infectivity; this stimulates the production of antibodies against the real thing, so that if a legitimate attack should come, the body will be ready.

But viruses are not helpless in the face of such strategies. They have ingenious ways of dealing with these defenses. If one great characteristic of viruses is their ability to infect, another is their ability to change. Faced with an immune system, or a vaccine, that successfully prevents an invasion, certain viruses simply mutate: they change their form so that the body's antibodies no longer recognize them as a danger. Flu viruses are notorious for their ability to mutate, which is why every few years bring a new variety of flu. The body is never quite ready for the latest version of the virus. Cold viruses operate in much the same way, further augmented by the fact that there are so many of them. Were there just one cold virus, there might be a vaccine against it. But there isn't just one cold virus — there are a couple hundred different kinds. Fight off one today, there are enough waiting around the corner to keep you busy the rest of your life. And perhaps the most striking example of a virus's ability to mutate is the AIDS virus, which has changed more often and in a shorter time than any other virus ever recorded. It's this mutability (and the fact that AIDS viruses attack the immune system itself) that causes such consternation in scientific circles about the likelihood of controlling the disease.

The fact is, however, that when it comes to AIDS, there's more optimism than ever before. In general, with respect to all disease, we find ourselves poised on the threshold of unprecedented discoveries. For example, viruses themselves are now being groomed to fight disease. By using viruses to make cells produce what *we* want them to — antibodies against a variety of dangerous intruders, say, or altered genes that may prevent genetic disease — researchers are in effect tapping viral technology for our own benefit. Turnabout is fair play. It's a new day, truly an ongoing revolution, the ramifications of which can hardly be imagined.

What follows is an exploration of this emerging age of viruses. We'll relive the discovery of these amazing creatures and experience some of the high — and low — points along the way, as scientists struggled to understand their inner workings and how to manage their outer effects. We'll follow some of the leading viral detectives of our day as they track down invisible invaders and devise ingenious methods to deal with them. We'll see how viruses provided the key to unlock the fantastic world of molecular biology and genetic engineering, and how researchers and physicians learned to manipulate viruses, in effect enlisting the help of these ancient enemies to fight the viruses' own consequences as well as other diseases. Finally, we'll consider the origins of viruses and their possible role in the grand scheme of life.

It's now clearer than it has ever been that previous advances in our understanding of viruses have been no more than a preamble. The anticipated discoveries in virology and microbiology, and those who will make them, loom larger than ever before. Medical scientists know that only too well and are running as hard as they can to keep up with the incredible little creatures. We now have illnesses for which there are no known causes and viruses for which there are no known illnesses. We have discovered viruses that in their eccentric and surprising form hardly seem viruslike at all. We have discovered viruses whose function is still a mystery. An unknown and provocative submicroscopic world beckons. We're just at the threshold — and that's where the excitement is. It behooves us not to be left behind.

I

Beginnings

One

. . . .

The World of Humors

. . . .

On the morning of the twenty-seventh of April, 1774, in the fifty-ninth year of his reign, Louis XV of France awoke dizzy and shivering, with a headache and pains in his back. It had not been the best of times for the aging king. Once affectionately called Louis the Well-Beloved, the divine inheritor of the throne of his great-grandfather, the Sun King, Louis XIV, he was now referred to as Louis the Well-Hated. He had subjected his country to a series of disastrous wars that divested France of much of its empire, and his personal excesses had done much to deplete the royal treasury, as he had been notorious for his love of women and drink. But now, at sixty-four, Louis was growing fat and old and increasingly subject to sieges of depression. He felt himself wearing down. "I see that I am no longer young," he remarked to his physician, "and I must begin to rein in the horses." "Sire," replied the doctor, "you would be better advised to unhorse them." The terrors of mortality filled his mind, intensified by the recent deaths of two courtiers who had suddenly collapsed at his feet. So he might have expected it when, at his country château in the village of Trianon on the outskirts of Versailles, Louis awoke ill.

His doctor's remedy for such ailments had always been exercise, so Louis set out for a day's hunting in the fresh country air. But he was too weak to mount his horse. He was reduced to following the chase from a carriage, and by half past five in the afternoon when the party returned to the château, he was worse. His head ached more painfully than before, and he was exhausted. He fell into bed early.

By the next day, the news of Louis's illness had traveled the few miles to Versailles. In the afternoon the king's first surgeon arrived at the country château. He demanded that the monarch be transported back to the palace. "It is at Versailles, Sire, where one must be ill." For an illness of the king of France, no matter how minor, was an occurrence with far-reaching ramifications. It required the full mobilization of the machinery of state. That meant the ministrations of no less than fourteen royal medical experts and the constant presence of a mind-boggling number of attendants and courtiers. Futures and fortunes could rise and fall on the fluctuations of the king's fever; announcements of the royal condition were the most important events at court. Weak and worried, Louis acquiesced.

Never again would Louis XV see his château at Trianon. Nor would he ever again see the great garden at Versailles. In fact, he would never again see anywhere but the inside of his bedchamber. He didn't yet know it, but he was dying, and, although it wasn't yet apparent, the killer would turn out to be the scourge of his times, smallpox.

Smallpox. For us it is a thing of the past. As the first disease to be eliminated by human effort, it was the target of one of the greatest triumphs of medical science. Many of us recall it — if we recall it at all — through memories of the gentle scratch of a smallpox vaccination or the resulting shallow crater upon our skin. But in Europe during Louis XV's time, smallpox killed as many as one person in ten; more than half of those were children. Of those who were fortunate enough to survive into adulthood, about 95 percent suffered smallpox at some time in their lives.

The disease erupted suddenly. You might feel perfectly well one day and the next find yourself miserable with headache and back pains, chills and dizziness. The following day came the fever, perhaps as high as 104 degrees Fahrenheit, and often delirium or even a coma. On the third and fourth day, the symptoms often abated, causing you to feel as though the worst were over. But then a rash appeared and, soon after, the first pox sores. At first these red splotches showed up around the neck and forearms, but soon they spread to the upper arms and torso, particularly the back, and then to the legs. They commonly swelled to about a third of an inch in diameter, then broke open and eventually scabbed over.

It was at this point, when the patient's suffering was greatest, with excruciatingly painful sores on the face and in the mouth, that a physician might be able to predict the chances of survival. If the sores did not touch each other, there was real hope. Less than 10 percent of such cases ended in death. If the sores ran into each other, there was a fifty-fifty chance of surviving. But if bleeding erupted beneath the skin, death was certain. And it was a death of the most terrible sort, as the pox-covered body literally rotted away, accompanied by great stench and horrible suffering.

And the lucky ones, those who didn't die? Even then the smallpox often left a terrifying memento. At worst, it destroyed sight; at its height, smallpox was one of the world's leading causes of blindness. And at best, it left disfigured or pockmarked skin, most apparently on the face.

And if smallpox was the terror of the European world, it was even worse in unexplored territory. New World peoples who had never known smallpox were ravaged when the illness was introduced by the European invaders. Historians have wondered, for example, how the Spanish conquistadors under Hernando Cortes, with only about five hundred men, could so quickly and thoroughly destroy an Aztec civilization of millions. A large part of the answer was a secret weapon that not even Cortes knew he possessed: smallpox. Just one Spanish soldier carrying the disease was enough to infect the entire Aztec population. Estimates are that by 1522, just two years after the Spanish invasion, between three million and four million Aztecs were dead of the disease, and the Spaniards had expanded their empire.

So Louis XV, without knowing it, was embarked on a great and terrible adventure. It wouldn't last long, but it would prove to be symptomatic of life at that time. Like a hero of ancient tragedy — although Louis was certainly anything but a tragic figure — he exemplified in the manner of his dying the state of living as he knew it.

It was a privilege he would have gladly forgone. Back in the gilt and glitter of his own bedroom in the main palace at Versailles, Louis felt the fever intensify as the pains in his head became worse. He couldn't sleep, and at times his mind wandered. The next morning, the first physician and first surgeon consulted with three other royal physicians and decided that the only course left to them was clear: His Majesty must be bled.

Bled! How curious the practice seems to us now — now that we know that blood circulates the disease-fighting armamentarium of the immune system, not to mention nutrients, while at the same time carrying away wastes and poisons. In virtually all instances, it's the strategy of modern medicine to enhance or replenish blood rather than deplete it. But in Louis's time, bleeding a patient was de rigueur.

The question that springs to mind to a person of the twentieth century is, why? For an eighteenth-century European, however, it was a question that was beside the point. Although the purported benefits might have been somewhat difficult to display, the reasoning behind the practice was well established and virtually incontrovertible. Bleeding was based on a theory that was more than two thousand years old — the theory of "humors" or "temperaments."

No one knows exactly where the notion of humors came from, but its first articulation in the West was in the teachings of Hippocrates in the fifth century B.C. Long before the seventeenth-century discovery that blood circulates through the body, Hippocrates considered disease the result of an imbalance in what were thought to be the body's four basic humors, or bodily fluids: blood, phlegm, white bile, and black bile. These humors, in turn, were representative of the four elements that comprised the world — water, earth, fire, and air. As all food and drink consisted of these elements, we therefore consumed them as we ate — except for air, which was directly provided by breathing. In the process of digestion, they became the bodily fluids, which nourished the body but never lost the characteristics of the elements themselves. Water and blood were moist; earth and phlegm were cold; fire and white bile were hot; and air and black bile (a nonexistent fluid invented to complete the hypothesis — reality must follow theory, after all) were dry. So the body was, in effect, a microcosm of the universe. The forces in play in the cosmic setting were no less active in the flesh.

As the theory passed through the centuries, it became subtly altered and embellished. Well before Louis's time, each of these bodily humors became further related to certain temperaments, or personality traits. One particular humor tended to dominate, so the theory went, providing people their temperamental makeup. Dry black bile depressed the spirit, producing melancholy, pensive, solitary individuals. On the other hand, the fire of hot white bile made for a choleric personality, quick to anger (we use the terminology still: "You've got your bile up"). Cold phlegm dulled the temperament, leading to sluggish or indifferent behavior — a phlegmatic personality. And moist blood brought about sanguine (or "bloodlike") behavior — robustness, hopefulness, cheerfulness.

It was the classical notion of harmony taken to the extreme. Just as the four elements worked in the natural universe, they worked, in the form of humors, in the bodily universe. And, just as an imbalance on the cosmic level could bring about great natural disasters — fires, earthquakes, hurricanes, floods — so an imbalance on the bodily level, that is, too much of any one humor, might bring about disease. Disease was, then, for the most part the result of a "plethora" of what in balance was a natural condition. When things went awry, the body struggled, often successfully, to restore its natural sense of balance, but in some cases it simply needed help. It was the physician's task to help restore that balance.

So, Louis XV was bled. No specific diagnosis had yet been made — his doctors were only beginning to suspect smallpox — but Louis was bled. It was to no avail. The king's fever continued to increase. There was no longer any doubt that he was seriously ill.

By the afternoon of the third day, the entire royal faculty of medicine was involved — six physicians, five surgeons, and three apothecaries. They crowded around the king's bed, flanked by scores of courtiers and attendants. Louis, by this time beside himself with pain, worry, and weakness now caused not only by his illness but by the loss of blood, called upon each specialist in turn to feel his pulse, describe his symptoms, and declare the nature of his illness. None was able to do so, undoubtedly hindered by the fact that Louis had complained so much about the brightness of the room that great pains were taken to keep light out of his eyes. It's likely that none of the learned doctors could see well enough to perform a competent examination. In any case, after deliberating for some time, and conferring in whispers, the committee of doctors decided that it was necessary to bleed Louis a second time and, if that was unsuccessful, a third.

Imagine the scene: there, propped up in the shadows of the royal bed, flanked by two large carved palm trees and protected by a low balustrade from the attendants straining for a glimpse, Louis lay surrounded by royal physicians with his lifeblood streaming out of his arms, the entire scene reflected in the huge mirrors and gilt carved paneling of what was considered one of the most dazzling rooms at Versailles. By this time, the king may have been the only one of those congregated in the royal bedchamber not to suspect that he was dying of smallpox, but he was now so severely weakened by lack of blood that it hardly mattered what his disease really was. The common practice was to bleed the patient until he fell unconscious — only then could you be certain that enough blood had been removed. And how much blood was that? It varied. "Generally speaking," wrote a medical researcher a few decades later, "as long as bloodletting is required, it can be borne; and as long as it can be borne, it is required." In Louis's case, the amount came to "four large basin-fuls" of blood.

So, as Louis lapsed into unconsciousness, imagine the physicians of the faculty of medicine, having just drained away what little vitality still remained in the king, bobbing their heads in solemn discussion, most likely suspecting the worst but unable to do anything about it, continually called to stand at the king's side, "as if he imagined that, surrounded by so many satellites, no harm could happen to His Majesty." And envision the multitudes of courtiers and attendants, buzzing, bustling, and wondering what was going on, trying to anticipate the effect of the royal illness on their own futures one way or another. (Their proximity as it turned out, was not entirely a privilege — fifty courtiers soon came down with smallpox; ten died.)

Finally, imagine the chagrin of the poor attendant who, later that evening, while the king was being moved from his state bed to a small camp bed in the corner of the room, happened to shine a light near the royal forehead. What he saw prompted him to summon the first surgeon, who in turn summoned the rest of the royal medical assembly, each of whom took perhaps his first good look at the king. On Louis's forehead was an unmistakable rash marked by red pustules standing out against his pale skin.

Smallpox. The news soon spread throughout Paris. Prayers were ordered for the king's recovery, but with little effect — the churches and chapels became deserted. In 1744, when the king previously had been seriously ill, over six thousand masses were said on his behalf in Notre Dame Cathedral. In 1757, when he was ill again, the number went down to six hundred people. Now, three persons were willing to pay for a private mass for the king. Louis's own family displayed more concern, however. While his male heirs and their wives quietly left the sickroom, his three daughters, at great risk to themselves, stayed behind. Even his mistress, Madame du Barry, who had never had smallpox and so lacked immunity, remained near the dying monarch.

The next morning, Louis was so weak the consensus was that he wouldn't last more than another couple of days. It was decided that the archbishop of Paris would come to prepare the king for his last confession. But His Excellency foresaw a difficult problem. For political reasons, he and the church had favored Louis's dalliance with Madame du Barry. To administer a last confession now would be to force Louis to send her away. The old man was therefore faced with a dilemma: should he do what his religious calling required of him and administer the sacraments, or should he continue to serve the church in more secular ways?

The practical side won out. The old prelate made his way into the sickroom, greeted the prostrate Louis, spent the next few minutes chatting about commonplace matters, then wished the king a speedy recovery and departed.

Louis was transformed. He had expected — dreaded, that is —

the Last Sacrament. What he received instead was some friendly chitchat and best wishes — all of which must mean that he wasn't as sick as he had thought. Overjoyed, he called for du Barry and "wept with joy, and covered her hands with kisses." He called for his fourteen medical attendants and each hour had them in turn take his pulse and examine his tongue. All of them manifested, "each in [his] own way," according to an observer, "the satisfaction they found in the beauty and colour of this portion of the King's Most Excellent Majesty."

And so it went until the third of May, the seventh day of the illness, when Louis's condition took a dramatic turn for the worse. His face became swollen and yellow, he no longer made discernible movements, and he spoke so softly that du Barry had to bend over him to hear his words. They were not the ones she anticipated. The king ordered du Barry to leave Versailles. "I belong to God and my people," he said, and then, in du Barry's words, the king "seized my hand, which he tenderly kissed, and then whispered an affectionate 'Adieu.' "

With Madame du Barry banished from court and his family no longer allowed into the sickroom, the full hopelessness of his condition engulfed Louis. In terror, he had holy water sprinkled on his bed to exorcise demons and ordered that money be sent to every church in Paris so that masses would be said for him. It was far too late. Louis XV received the Last Sacrament on the seventh of May, the eleventh day of his illness. With his family, the clergy of the realm, the grand officers of the Crown, the ministers and secretaries of state, and nearly everyone at court in attendance, all with lighted candles in their hands, the king heard his repentance read to the assembly.

"Messieurs," announced the cardinal, "the King charges me to inform you that he asks pardon of God for having offended Him and for the scandal he has given his people; that if God restores him to health, he will occupy himself with the maintenance of religion and the welfare of his people."

"I should have wished for sufficient strength to say it myself," murmured Louis from his bed.

From that moment, the public life of the king of France was ended. Everyone who wasn't compelled to stay quickly abandoned the sickroom, leaving Louis to die — which he did soon enough. The following day, he became delirious; and by the next day, according to one historian, "the body of the King was falling to pieces, in a state of living putrescence, and the smell was horribly fetid." Not even incense and open windows could overcome the stench. On the next day, the tenth of May, 1774, shortly after three in the afternoon, the ordeal ended. Louis XV was dead.

It remained only to bury the king. And here, too, prudence prevailed. The funeral ceremonies and lying-in-state of his predecessor Louis XIV had lasted seven weeks; now the elaborate protocol was ignored. Tradition required that Louis's body be autopsied and embalmed, but when the king's chamberlain directed the chief surgeon to proceed, the doctor quickly explained to the courtier that during the operation "your duties oblige you to hold the head of the defunct. I declare to you, if the body is opened, neither you nor I nor anyone present at the autopsy will be alive a week later."

No autopsy was performed. Louis's body was not embalmed, nor was his heart cut out and placed in a special coffer, as was the custom with Bourbon kings. Instead, two protesting grave diggers were persuaded to load the corpse into double lead coffins and then soak the remains in wine and quicklime. Two days later, in the dead of night, the king of France was conveyed to the royal crypt at Saint-Denis, "the funeral resembling rather the removal of a load one is anxious to get rid of than the last duties rendered to a monarch." His cortege consisted of forty of his bodyguards and a few pages carrying torches. And there is at least one report that one of the grave keepers who helped inter Louis in the royal crypt died of violent vomiting brought on by the intolerable stench.

Louis XV wrote in his will: "I have governed and administered badly, because I have little talent and I have been badly advised." It is as accurate a summation of his reign as any, and a judgment that was dramatically reaffirmed nineteen years later in the early days of the Revolution, when rioters broke into the royal crypts at Saint-Denis and removed the body, as well as those of Louis's predecessors, and unceremoniously threw them into a ditch. Indeed, Louis may be more charitably remembered for the nature of his death than for his life.

Another of his remarks, however, his notorious *"Après moi, le déluge"* ("After me, the deluge"), which is famous both for its cynicism and its accuracy — the French Revolution followed his death by a mere fifteen years — may now be seen as an ironic harbinger of better things. The life Louis knew, with respect to medicine at least, was only slightly discernible from that of an inhabitant of the early Renaissance, the Dark Ages, or, for that matter, ancient Greece or Rome. In fact, the science of medicine was no science at all in Louis's time. Rather, it was a jumble of often competing strategies and philosophies, most of which stemmed from ages long past; in any case, they did not require what we now refer to as empirical proof to establish their utility. These philosophies were a result of hunches concerning the workings of life rather than experiments designed to discover it. Louis was bled because theory demanded it, not because the practice showed any particular merit. And if a treatment should fail, as did many, it was the fault of the constitution of the patient rather than the approach of the physician.

Testing a procedure to determine its efficacy, an approach that seems second nature to us today, was not yet a part of the world of Louis XV. That was yet to come. For in this prescientific world, a person lived by the grace of God or the luck of the draw — a determination not even the privilege of birthright could influence — and the role of medical practice was largely irrelevant. The workings of the physical universe were a mystery that reflected themselves in the capricious nature of the body's health.

But by the end of Louis's eighteenth century, and no farther away than across the English Channel, an event would arise that helped to change all that. It was the beginning of an awareness of the nature of life and disease that is still evolving today, an awareness that has led to the discovery of viruses and to the elimination of the virus-caused scourge called smallpox. *"Après moi, le déluge,"* indeed. For an obscure country doctor in England would soon loose a deluge that only now, in the last years of the twentieth century, is beginning to crest. Today, we ride that wave, with all the possibilities and excitement it affords. Louis XV was not so lucky. But for all its significance, this deluge grew quietly and haltingly at first — and it began with a milkmaid.

Two

. . . .

It Began with a Milkmaid

. . . .

Sometime between 1763 and 1770, just before Louis XV came down with smallpox, a young apprentice surgeon named Edward Jenner heard the words that were to change his life and, eventually, the lives of millions of people around the world. They would prompt him to take the first effective step against viruses — creatures he didn't even know existed.

Jenner hailed from the village of Berkeley, in Gloucestershire, near the southwestern coast of England. This was dairy land, and the broad-faced, solidly built Jenner reflected the place of his upbringing. Just a teenager, he was serving a seven-year apprenticeship with a local surgeon as the first step in becoming a physician in his own right. So it was a familiar duty for him to accompany his mentor on his rounds, which, in this case, included treating a milkmaid during an outbreak of cowpox.

Cowpox was something of an occupational hazard in dairy country. It caused sores on the udders of dairy cows and was highly contagious, not only to other animals but to people as well, especially those who tended the cows. In humans, painful, pustular sores appeared on the hands and wrists, and the disease caused a few days of fever and nausea besides. In fact, the sores might be mistaken for those of smallpox if you didn't know any better.

It was then, so the story goes, as Jenner was examining the milkmaid's newly erupted case of cowpox, that the young woman made the statement that was to change his life and the practice of medicine ever since. "Now I'll never take the smallpox," she announced, "for I have had the cowpox."

Her attitude wasn't surprising. The notion that exposure to cowpox brought about immunity to smallpox was widely held in this part of the country, at least by common, uneducated people. It was just as widely dismissed by most physicians. So, the milkmaid's statement might well have had little impact on Jenner; but for some reason, it stayed with him. It stayed with him through a subsequent two-year apprenticeship in London with the famed surgeon John Hunter, and it stayed with him over the more than twenty years he spent as an increasingly successful and prosperous country doctor back home in Berkeley — until finally he was able to do something about it.

Why was Jenner so taken with what most doctors considered an old wives' tale? There must have been something in his experience that made him particularly receptive to such an untested notion. Perhaps that something was his own bout with smallpox — or, rather, his bout with the only known method of effectively dealing with the disease, a procedure called inoculation.

If the milkmaid's idea that exposure to cowpox would immunize her against smallpox was questionable, there was a truth that no one disputed: if you survived one battle with smallpox itself, you'd never have to fight another — you were immunized. That being the case, it was only a matter of time before someone tried to take advantage of the fact by consciously infecting others with the disease in the hopes of causing a mild, survivable case that would then protect against more severe attacks later on. The procedure came to be known as *variolation* (from the Latin term for smallpox, *variola*) or, more commonly, *inoculation*.

The earliest attempts to inoculate against smallpox probably took place in ancient China. The Chinese "sowed the pox" by using powdered scabs from mild cases of the disease and inhaling the infectious matter up their noses through ivory straws. They called the procedure "heavenly flowers." Other Eastern societies inserted smallpox scabs directly into incisions. But there was no way of controlling the dose. A supposedly mild exposure to the disease in effect might not be so mild after all, and could kill or mutilate rather than simply immunize. Besides that, smallpox, no matter how mild, is highly contagious. Infect one person and you could easily cause an outbreak of the disease.

But despite the uncertainty and danger of the procedure, it was more successful than not, and by the beginning of the eighteenth century inoculation had become a popular folk practice among some European peasants, who called it "buying the smallpox." Travelers reported seeing inoculations in Poland, Denmark, Scotland, and Greece, and the practice was well established in England as well. It hadn't caught on, though, with the ruling and wealthy classes — not, that is, until Lady Mary Wortley Montagu, the wife of the British ambassador to Turkey, decided to change all that.

Lady Mary was a wellborn beauty, one of the most popular hostesses in London. In 1715 she was struck by smallpox, which had killed her brother just two years earlier. Lady Mary didn't die, but the attack ruined her beauty, destroying her eyelashes and leaving her with a badly pockmarked face. By the time her husband was posted to Constantinople in 1716, Lady Mary had a vested interest in smallpox. Within two weeks she had written her now famous letter to her friend Sarah Chiswell (who was to die of smallpox nine years later).

"*Apropos* of distempers," Lady Mary wrote,

I am going to tell you a thing that will make you wish yourself here. The small-pox, so fatal, and so general amongst us, is here entirely harmless, by the invention of *ingrafting*, which is the term they give it. There is a set of old women, who make it their business to perform the operation, every autumn in the month of September, when the great heat is abated.

People send to one another to know if any of their family has a mind to have the small-pox; they make parties for this purpose, and when they are met (commonly fifteen or sixteen together), the old woman comes with a nut-shell full of the matter of the best sort of small-pox, and asks what vein you please to have opened.

She immediately rips open that you offer her with a large needle (which gives you no more pain than a common scratch) and puts into the vein, as much matter as can lie upon the head of her needle, and after that binds up the little wound with a hollow bit of shell; and in this manner opens four or five veins.

. . . The children or young patients play together all the rest of the day, and are in perfect health to the eighth. Then the fever begins to seize them, and they keep their beds two days, very seldom three. They have very rarely above twenty or thirty [sores] in their faces, which never mark, and in eight days they are as well as before their illness. Where they are wounded, there remain running sores during the distemper, which I don't doubt is a great relief to it.

Every year thousands undergo this operation: and the French Ambassador says pleasantly that they take the small-pox here by way of diversion, as they take the waters in other countries. There is no example of anyone that had died in it: and you may believe I am well satisfied of the safety of this experiment, since I intend to try it on my dear little son. I am patriot enough to take pains to bring this useful invention into fashion in England. . . .

Within a year, without informing her husband, and against the bitter opposition of the embassy chaplain, who called the practice an unchristian operation that could succeed only in the infidel, Lady Mary had her six-year-old son inoculated. There were no complications. A week later, when Lady Mary finally told her husband what she had done, the boy was out of danger, "singing and playing and very impatient for his supper."

In 1721, three years later, with the Montagu family again in London and facing a deadly smallpox epidemic, Lady Mary had her four-year-old daughter inoculated, thereby beginning the practice among many of the royal and aristocratic families of England and Europe. In France, half a century later, the family of Louis XV quickly had themselves inoculated following the death of the king, as did the royal family of Austria. And inoculation flourished across the Atlantic in the British colonies, as well.

But inoculation was not the answer to preventing smallpox. Contagion was a continual problem, as was the question of how to perform the practice on the people who most needed it. For if by Jenner's time many of the wealthy and aristocratic classes had taken pains to become protected against the disease, and if inoculation had made some headway among country peasants, poor people in the cities still suffered smallpox as much as ever. Living closely together, often in appalling conditions, they made up the vast majority of the population in the cities and were the incubators of disease. Suspicious of new approaches, resentful of the practices of their social superiors, and inured to their situation by a fatalistic outlook that considered disease the natural condition of life, these people resisted even the infrequent feeble attempts to introduce inoculation into their midst. The result was that despite the existence of a treatment that may not have been perfect but certainly offered hope for something better, the rate of smallpox infection in England actually rose during the eighteenth century. Inoculation offered a foundation upon which to build, but clearly something else was necessary if this horrible disease was to be curtailed.

Jenner himself was inoculated at the age of eight. It was this experience that may have motivated him to pursue the cowpox connection, for it was the custom in England to prepare patients for inoculation by an elaborate and brutal method. Jenner's preparation lasted six weeks. He was bled until pale, then purged and fasted repeatedly, until he wasted to a skeleton. He was denied solid food in favor of a vegetable drink that was supposed to sweeten the blood; after the inoculation itself — the least traumatic event of the entire experience — he was removed to an "inoculation stable," and, according to an early biographer, "haltered up with others in a terrible state of disease, although none died." His recovery, amidst the moaning and crying of other inoculated children, took another three weeks, and he was sickly for some time thereafter.

If this experience, and the words of the milkmaid, did indeed haunt Jenner's mind, he seemed in no hurry to do anything about them. His first years as a physician in Berkeley were marked more by his carefree style of living than by any breakthroughs in the battle against smallpox. He rode the countryside with a group of bachelor friends, frequented the taverns and medical-society meetings, and even composed light verse. (Concerning a patient to whom he had made a gift of two ducks, Jenner wrote,

> I've dispatched, my dear madam, this scrap of letter
> To say that the patient is very much better.
>
> A regular Doctor no longer she lacks,
> And therefore I've sent her a couple of Quacks.)

Still, he hadn't forgotten the problem of cowpox and its relationship to smallpox. He even made a presentation of his interests to fellow physicians in his local medical club, but with little effect. Talk did no good. If Jenner was to make any headway, he must prove the idea one way or another.

In 1789, the year the American colonies became a nation, and a year after Jenner married a local woman named Katherine Kingscote, there was an outbreak of swinepox in Gloucestershire. Swinepox was a disease similar to cowpox that attacked pigs rather than cows (horsepox was yet another similar disease). Up until this point in his life, Jenner had been anything but rash in his investigations, but now, at the relatively wise old age of forty, he inoculated his ten-month-old son, Edward, Jr., with matter taken from a pustule of the baby's nurse, who had caught the swinepox infection.

After doing so little for so long, what in the world made him take such an extreme step? And how, one wonders, did his wife of little more than a year feel about it? In any case, eight days later several small pustules appeared on baby Edward's skin, and he took sick. Sometime after that, Jenner inoculated him with smallpox itself, not once but *five* times. The infant showed no reaction whatsoever. Two years later, Jenner inoculated the child with smallpox again. This time there was a reaction, and a severe one, but not from the smallpox. The inoculation material turned out to be contaminated — a constant danger that later threatened to undermine Jenner's work altogether. Young Edward contracted a fever and his arm swelled all the way to the armpit. But he quickly recovered, and a year later Jenner inoculated him with smallpox once again. And once more there was no reaction. Apparently, the swinepox protected against smallpox.

Armed with the encouraging results of the experiment, Jenner began traveling from farm to farm, researching cases of milkers who had been protected from contracting smallpox by previous cowpox infections. And in 1796, cowpox again invaded the countryside, providing Jenner with the wherewithal to do what he had been wanting to do for years: actually test the immunizing power of the disease.

That meant he must infect someone with cowpox and then, later, with smallpox, much as he had done with his own son. But whom could Jenner find for such a dangerous test? It would have to be someone who had never had smallpox or cowpox — a tough order in his place and time — and, moreover, who would consent to putting his life in jeopardy. That person turned out to be an eight-year-old boy named James Phipps, whose parents consented to the procedure "once they had considered the alternative." For the cowpox material itself, Jenner turned to a young milkmaid named Sarah Nelmes, who had contracted the disease after milking an infected cow. Jenner made a detailed drawing of the sores on her hand and wrist. They were just what he was looking for — large and approaching their greatest infectivity. He described them later as looking "like the section of a pearl on a rose leaf."

On the fourteenth of May, 1796, a day famous in medical history, Jenner pierced a sore on Nelmes's wrist with a lancet and then, with the tip of the instrument now covered with cowpox matter, made two scratches on young Phipps's arm. It was the first recorded vaccination — a term that Louis Pasteur later popularized as a tribute to Jenner after adapting it from Jenner's term *vaccinae*, "of the cow."

"On the seventh day," Jenner wrote, Phipps "complained of uneasiness in the axilla [armpit] and on the ninth he became a little chilly, lost his appetite, and had a headache. During the whole of this day he was perceptibly indisposed, and spent the night with some degree of restlessness, but on the day following he was perfectly well."

Phipps developed a single pustule on his arm, but on the tenth day it scabbed over, and a few weeks later the scab fell off, leaving the tiny crater that in the years to come became the well-known mark of a vaccination that "took."

On the first of July, seven weeks after the experiment commenced, Jenner inoculated Phipps with smallpox. Nothing happened. Several months later Jenner inoculated him again, still with
no effect. In the next twenty-five years Jenner inoculated "poor
Phipps," as he called him, some twenty times, and twenty times there
was no reaction.

Jenner was elated. Jubilantly he wrote a friend: "I shall now pursue my experiments with redoubled ardor." But again he ran into
an insurmountable obstacle: once more, cowpox disappeared from
the countryside. So Jenner waited. And he waited. For two years he
waited, a delay that must have been agonizing, until finally cowpox
reappeared in Gloucestershire. Quickly Jenner returned to the attack, infecting a five-year-old boy named John Baker with material
from a servant's horsepox sores and another boy of the same age,
William Summers, with pus directly from the sore on a cow's udder.

Baker took sick on the sixth day with cowpox-like symptoms and
by the eighth day appeared better; so far, so good. But then, mysteriously, he went rapidly downhill. "The boy," Jenner wrote, "was
rendered unfit for inoculation from having felt the effects of a contagious fever in a workhouse soon after this experiment was made."
He might have written, a later critic suggested, "The child, it appears, was rendered unfit for inoculation by unhappily becoming a
corpse." For John Baker soon died of what probably was an unre-
lated infection.

William Summers, on the other hand, presented no such setback.
He soon developed a single pustule, much as had James Phipps, and
in short order Jenner took the matter from this sore and vaccinated
an eight-year-old boy, William Pead, and then from Pead's sore a
seven-year-old girl, Hannah Excell. She, in turn, provided material
that Jenner used to vaccinate four others, including his youngest son,
Robert, just eleven months old. This process of serial passage —
passing vaccine from person to person — was Jenner's second great
contribution, for it showed that cowpox could be passed from person to person indefinitely, thus freeing him from dependence on an
infected animal.

Jenner then proceeded to test the vaccinations by inoculating the
first in the line, William Summers, with smallpox; his nephew and

assistant Henry Jenner inoculated the last in the series, a Mr. Barge. As Jenner expected, the results were positive — the inoculations had no effect.

For Jenner that proved the case. He felt no need to test the others in the series; it was abundantly clear to him: cowpox provided immunization against the horrors of smallpox. "I presume," he wrote, "it may be unnecessary to produce further testimony in support of my assertion that the cow-pox protects the human constitution from the infection of the small pox." And, the matter now settled in his own mind, he proceeded to write a paper describing his efforts.

The result — *An Inquiry into the Causes and Effects of the Variolae Vaccinae, a Disease Discovered in some of the Western Counties of England, particularly Gloucestershire, and known by the name of the Cowpox* ("By Edward Jenner, M.D. F.R.S. & C.") — is one of the most well known works in medicine. In it he described a hodgepodge of medical histories — naturally occurring cases of cowpox and horsepox, and, of course, his own vaccination series — all of which he found to confer resistance to smallpox. Upon this relatively meager evidence, he rested his case.

It was enough. Jenner promised protection from smallpox "for ever after"; one can only imagine the effect on a world that lived in constant dread of the horrors of disease but had little experience with the possibility of medical intervention. The paper caused a sensation and made Jenner instantly famous. Within two years it was translated into several languages and reprinted around the world. But though Jenner was suddenly famous, he was neither universally admired nor appreciated. For every grateful response to his work, there was equally forceful criticism. For example, the eminent London physician Henry Cline used the infectious material from Jenner's series of vaccinations and wrote: "I think the substituting of the cowpox poison for the smallpox promises to be one of the greatest improvements that has ever been made in medicine. . . . The more I think on the subject the more I am impressed with its importance." On the other hand, many newspapers and magazines printed accusations that being vaccinated caused people to take on cowlike characteristics. Some people sprouted horns, the accounts stated; others began to moo. A cartoon by the artist James Gillray, "The Cowpock — or the Wonderful Effects of the New Inoculation," depicted an aloof, disinterested-looking physician resembling Jenner injecting cowpox vaccine into a woman's arm, while around him cluster a group of previously vaccinated patients, all of whom are grotesquely disfigured in various bovine ways — cows bursting out of arms and thighs, cows popping out of noses and mouths, faces sprouting horns and fur. To us, today, the cartoon is humorous; to Jenner, it must have seemed anything but that.

Even less humorous were the efforts of the first independent promoters of vaccination. One of them, a Dr. William Woodville, took it upon himself to identify a suitable cowpox sore and then, without bothering to consult with Jenner about proper technique, began a series of vaccinations that eventually ran to some six hundred cases. Unfortunately, knowing neither the optimum time to use the cowpox material nor when to inoculate with smallpox later on, Woodville contaminated his vaccine to the extent of causing the death of a nursing baby and bringing about a number of severe complications in others. Rather than causing the one familiar, shallow crater indicating a successful vaccination, Woodville caused cases that looked like smallpox itself. Some of his patients broke out with up to a thousand pustules. The experience so confused Woodville that he wrote a paper claiming that cowpox was a very severe affliction and that inoculation was still the best way of preventing smallpox. Jenner angrily made his way to London to set Woodville right.

But that was not the end of it. More opposition followed. Groups called "Anti-Vacks," for anti-vaccinationists, tracked down an old farmer named Benjamin Jesty, who in 1774, over twenty years before Jenner's first vaccination, had given cowpox material to his wife and two children in an attempt to prevent smallpox. Jesty, they said, was the real originator of vaccination — Jenner was nothing more than a fraud. It was soon generally acknowledged, however, that although Jesty may indeed have beaten Jenner (who knew nothing of Jesty's prior actions) to the punch, he didn't follow through by testing and developing the vaccination procedure.

Despite the multitudes of naysayers, whom Jenner characterized as "snarling fellows and so ignorant withal that they know no more of the disease they write about than the animals which generate it,"

Jenner had unleashed a flood that would not be held back. The practice of vaccination became commonplace throughout the world. By 1800, at least a hundred thousand people had been vaccinated, a number that just two years later more than doubled. Vaccination quickly became a family custom in Italy, and Germany and Sweden soon established compulsory-vaccination laws. President Thomas Jefferson of the infant United States of America, observing that "every friend of humanity must look with pleasure on this discovery, by which one evil more is withdrawn from the condition of man," wrote Jenner that his gift was greater than that of the discovery of the circulation of blood, and he established the White House as the distributor of vaccine to the American South. The House of Commons awarded Jenner grants totaling thirty thousand pounds, and tribute poured in from all over the globe. Jenner had indeed become "vaccine clerk to the world." When he died of a stroke in 1823, he was the most honored man of his time.

POSTSCRIPT: In the years after his father inoculated him, Jenner's son Edward became a sickly child and exhibited signs of mild mental retardation. At least one historian speculates that the secondary infection resulting from the inoculation might have caused the boy's problems. No one knows. Edward Jr. died from consumption at the age of twenty-one, and for long afterward Jenner would burst into tears while talking of his dead son.

The cow from which Sarah Nelmes contracted cowpox was named Blossom. The animal's hide subsequently hung for years on Jenner's coach house wall.

Jenner built a cottage for young James Phipps and looked after him until Jenner himself died, twenty-seven years later. The house is now the Jenner Museum.

In his *Inquiry,* Jenner described the pustule caused by vaccination as "the vesicle formed by the virus." *Virus.* It's the first use of the word in our story so far. We know now that smallpox is indeed caused by the submicroscopic, infectious organism we call a virus, but Jenner meant something very different when he used the term.

To him, *virus* was a commonplace expression that signified a poison, a pestilence — the *result* of some unhealthy condition rather than its cause. It would be another century and more before the term came to mean something new, something virtually no one had ever conceived of before, something so small and powerful that today it still elicits disbelief and awe. In using the word, Jenner was simply reflecting the practice of his time.

More significantly, however, he was revealing his ignorance of the real processes at the heart of his revolutionary technique. For although Jenner discovered that exposure to cowpox does indeed confer immunity to smallpox, he had not the faintest idea why; it was enough that it worked at all. But just what was it in the cowpox material that protected against subsequent exposure to smallpox? What processes did the cowpox inoculation set into motion to cause the body to be able to throw off a smallpox attack? And what caused smallpox in the first place? Although there were many attempts to answer these questions, they were no more than theories. Neither Jenner nor anyone else had answers.

Jenner's contribution, certainly, was enormous. It was the first effective, safe, and systematic attack against serious disease (natural cowpox infection may have protected against smallpox, but it may not have — before Jenner, no one knew, because there were no accurate records). Moreover, it heralded an approach to disease that would eventually lead to the discovery of viruses. At the same time, however, it was a bare beginning. It can be argued that results are what count: whether or not Jenner understood his vaccine is not to the point; it *worked*. And surely uncounted millions of people owe him thanks for keeping the scourge of smallpox from their door. But it was impossible to generalize from Jenner's discovery. Vaccination with material generated by a like disease served to prevent smallpox, yes, but the same method, for unknown reasons, would not halt tuberculosis — "consumption," as it was called then — or scurvy, or pneumonia, or syphilis. On the other hand, Jenner's method was many years later adapted to prevent diseases such as yellow fever, measles, mumps, polio, and the myriad of others for which we today have vaccinations.

How could you know which diseases were amenable to vaccination and which were not — other than by trial and error, that is? Without *knowing* how the vaccine worked, and *knowing* what caused the ailment, how could you *know* what to do about it? The answer was, you couldn't. For all his impact, Jenner and the rest of those who lived in his times were still blinded by a veil of misapprehension and ignorance.

That veil was about to be lifted. Jenner's discovery heralded not only the coming age of disease prevention, but an age that looked for causes and processes — not only of disease, but of the workings of life itself. Jenner stands out not only for discovering the means to rid the world of smallpox but also for having the courage to challenge the prevailing dogma of his times, to confront the ignorant resistance of his people, to go ahead and investigate possibilities. He had the courage to try something new. In that, Jenner seemed to stand alone.

Not for long. A new age was coming. A year before Jenner's death in 1823, "the most perfect man who ever entered the kingdom of science" was born. "Science" — the choice of words itself indicates the beginnings of a new world. This man's greatest contribution in a lifetime of great contributions would be to strike to the heart of the cause of things — diseases, life processes — and in doing so to exemplify a credo that would make possible the discovery of viruses. And perhaps the most spectacular manifestation of that effort was his battle against another virus-caused disease, a disease that was truly one of the most horrible of his or any other day: rabies.

Three

. . . .

Chance Favors the Prepared Mind

. . . .

On July 4, 1885, a mad dog attacked a nine-year-old boy named Joseph Meister as he was walking to school in the town of Meissengott, in the northeastern province of Alsace, in France. Too small to defend himself, the boy could only cover his face with his hands as the dog knocked him down and mauled him savagely, and he might have been killed had it not been for a bricklayer who rushed over and beat off the animal with an iron bar. The man then picked up the terrified boy, who was covered with blood and slobber, and carried him home.

Meanwhile, the dog ran back to his master, a grocer named Theodore Vone, and bit him on the arm. Vone shot the animal dead. Later on, the body was opened to find a stomach full of hay, straw, and fragments of wood, evidence that the dog had been running amok. There was no doubt that it was rabid. Young Joseph Meister suddenly faced the horrible prospect of a death from rabies.

Rabies. The French call the disease *la rage* — "the rage" — a well-chosen designation, because rabies, then and now, produces one of the most terrible deaths imaginable. First comes fever, then depression, then restlessness that turns into uncontrollable excitement. The muscles of the throat convulse, and saliva froths and runs down the chin, causing great thirst. But even the smallest sip of water may produce convulsions and more thirst. In the later stages of the disease, the mere sight of water can bring on convulsions and paralysis — thus the archaic name for the disease, *hydrophobia*, "fear of water." After four to five days of this kind of suffering, death can be a blessing.

Today we know that rabies is caused by a virus that attacks the nervous system and makes its way along the neural pathway to the brain. But no one knew this in 1885. It seemed clear only that the disease was caused by the bite of a rabid animal, usually a dog. If the bite didn't break the skin, you were probably safe. If it drew blood, you might well face death.

Hysteria surrounded the disease — a circumstance exacerbated by the fact that there was next to nothing that could be done about it. Treatments, such as they were, tended to be desperate and bizarre. In ancient times, certain notables recommended eating the livers of mad dogs as a cure; others prescribed a paste made of crayfish eyes. Closer to Joseph Meister's day, some doctors suggested inoculating victims with a poison — the venom of a viper, say — to counteract the poison of rabies, and another equally unsuccessful approach consisted of "dipping" the infected person into a lake or river, on the premise that forcibly being immersed in dreaded water could cause such shock that the afflicted person's system might reverse itself and expel the disease. Edward Jenner was familiar with the technique (although there's no record that he ever used it). He once asked a "dipping doctor" how long he held his patients underwater. The man replied that he pushed the patients under three times, each time holding them until "they have done kicking."

The most effective — although no less brutal — treatment was to cauterize the bite wound, thereby killing the tissues contacted by the rabid animal's saliva and reducing the likelihood of infection. The cauterizing agent might be a red-hot iron, or gunpowder sprinkled over the wound and then ignited, or a variety of caustic acids. If the wounds were particularly deep, the physician was supposed to use long, sharp cauterizing needles and push them well into the flesh. One can only imagine the suffering caused by the procedure, but it was the only treatment that had even the remotest possibility of success.

So in the evening, twelve hours after the mad dog's attack, Joseph Meister's mother took her little son to a local doctor, who proceeded to use carbolic acid to cauterize the worst of the boy's fourteen wounds on his hands, legs, and thighs (he had at least been successful in protecting his face). The doctor then advised Madame Meister to take her son to see the great Louis Pasteur in Paris. If anyone would know what to do, the doctor suggested, it would be Pasteur — high tribute, as Pasteur was a chemist rather than a physician. But such was the renown that he enjoyed in France. No matter that he was now sixty-three years old, partially paralyzed from a stroke, with most of his great discoveries behind him — he was still Pasteur.

Two days later, Joseph and his mother, accompanied by Theodore Vone, appeared unannounced at the doorway of Pasteur's laboratory in Paris. By this time, the little boy was so stiff and sore he could hardly walk, and he was frightened nearly out of his wits. Madame Meister threw herself at Pasteur's mercy, begging him to save her son.

Pasteur was moved by the little boy's plight. He examined Joseph's wounds; then he looked at Vone's bite, determined that it hadn't broken the skin (and so was unlikely to be infectious), and sent him back to Meissengott. But he arranged for Joseph and his mother to remain in Paris, and asked them to return at five that afternoon. That done, he turned his mind to a considerable problem: what could he do for this young boy?

As it turned out, he could do a great deal, for Pasteur had been studying rabies for the past five years. He saw in the affliction a chance to demonstrate, in a particularly useful and dramatic way, what he considered a fundamental truth of life, a truth that he had been advancing against entrenched opposition for much of his career. For a century after Jenner performed the first vaccination, and a century before our day, Pasteur's world was to a large extent still mired in the old ways of looking at life, the ways of Hippocrates, and the humoralists, and the bleeders, the way expressed by a contemporary Paris clinician: "Disease is in us, of us, by us." To this world Pasteur offered a new reality, one that was so striking as to seem absurd, so complex that it seemed impossible, and so simple that it could not be denied. It was the theory that germs cause disease, and soon it would lead to the discovery of viruses.

Two centuries earlier a Dutch dry-goods dealer named Antony van Leeuwenhoek (*"lay*-vuhn-hook") had discovered a new world. As a hobby, something to do in his town of Delft after he put down his linens and drapes, he taught himself to grind lenses and smelt silver, gold, and copper to produce the first precision microscopes the world had known. He then trained these lenses, some of which could magnify to the 270th power, on rainwater, and the tartar from between his own teeth, and even his own feces (on those occasions when he "was troubled with a looseness"). What he saw made him look again in disbelief. There under his lenses swarmed a menagerie of tiny creatures. One leapt about in a drop of water "like the fish called a pike." Another squirmed this way and that, then tumbled forward in somersaults. Some of these strange beings were corkscrew-shaped and whirled crazily under his lenses; others were rod-shaped things that seemed sluggish and inert. There were more creatures in his mouth alone, he declared, all "moving in the most delightful manner," than there were people in the Netherlands.

Van Leeuwenhoek had discovered the world of microbes. What he called "animalcules," tiny animals, others would later call microorganisms, and, still later, protozoa and bacteria (but not viruses — not yet; they were much too small to be seen through van Leeuwenhoek's magic lenses). After writing the Royal Society of London about his discoveries (and confirming them by allowing emissaries from the society to view firsthand the marvels that existed under his lenses), van Leeuwenhoek became well known throughout Europe and through the years received a stream of prominent visitors, including Queen Mary II of England and Peter the Great of Russia, eager to view for themselves the microscopic wonders he described.

But van Leeuwenhoek never suspected that these tiny creatures might be the cause of his loose bowels or rotting teeth. Nor did he wonder where they came from. For him, they were "exciting novelties." His gift was in describing rather than investigating. In the years to come, an occasional insightful observer reinforced and expanded on van Leeuwenhoek's discoveries. Still, by the middle of the nineteenth century the reality of teeming microscopic life thriving in the heart of the world apparent to the senses had made few if any inroads into the prevailing views concerning the nature of things. The role of these tiny creatures, if any, in the processes of life and death was unknown. More than that, in most quarters the microscopic realm was disregarded. Other than occasionally ruminating on the origin of this miniature world, most influential thinkers simply ignored it.

Louis Pasteur, while influential, was not a thinker, at least not in the sense of theorizing about life. In fact, he distrusted the value of theory alone. Theorizing without investigating was for people he despised, the people who *thought* that life was one way or another without *knowing* it to be so. For Pasteur, the utility of theorizing was as prelude to examining the nature of things, to proving them one way or another by experiment.

It fell largely to Pasteur, then, to discover and demonstrate the impact of microorganisms and, at the same time, to determine where these unbelievable creatures actually came from. In fact, it was with Pasteur that the science of microbiology — and with it modern medicine — began. Systematic scientific research really began with Pasteur. For him, discovering *why* was as important as determining *how*. That had been his approach since at least 1856, almost thirty years before young Joseph Meister appeared at his door, when Pasteur found himself thrown squarely into van Leeuwenhoek's world of tiny creatures. It was a world from which he would never extricate himself.

Pasteur, a professor of chemistry, had answered a call of distress from a distiller in the north of France whose sugar beets simply weren't fermenting into alcohol. While investigating the problem, Pasteur found that the process of fermentation, which had been considered chemical and inorganic, was actually caused by living microorganisms, or "germs." The distiller's problem, it turned out, was that he had been plagued with the wrong germs. His vats were replete with microscopic rodlike organisms that caused a ruinous fermentation into lactic acid rather than into profitable alcohol; conversion to alcohol, Pasteur affirmed, was the work of yeast globules. Pasteur determined that by their own life processes — that is, simply by being — these microorganisms affected the world around them. The yeast globules brought about the fermentation of sugar into alcohol by the process of their own reproduction. Similarly, the rods produced lactic fermentation through their own multiplication. It was then that Pasteur made the conceptual leap that initiated a new era. "Everything indicates," he wrote, "that contagious diseases owe their existence to similar causes."

Like most notable events in science, it wasn't an entirely unprecedented leap. As early as the first century B.C., the possibility had been raised that invisible living beings — *animalia minuta* — breathed in with air or eaten, might be the cause of disease. Much later, in the middle of the sixteenth century, an epidemic of syphilis gave impetus to an even stronger expression of the same idea. A Veronese nobleman named Girolamo Fracastoro suggested that communicable diseases, which syphilis certainly was, were transmitted by a *contagium vivum*, a living agent. These "seeds of disease which multiply rapidly" might be spread through direct contact with sick people, contaminated objects, or through the air. And identical seeds caused identical diseases. Disease, therefore, was a particular affair with particular causes.

But as insightful as these hypotheses were, they were only that — hypotheses. You couldn't demonstrate their truth, after all. Who had ever seen a *contagium vivum?*

Discovering and demonstrating the role of microbes to a world of previously deaf ears was what Pasteur spent the rest of his career doing. He had found his calling, and it took over his life. Whenever possible, Pasteur quartered his assistants near the laboratory so as to have them constantly on hand. The working day lasted from eight in the morning to six in the evening, with few holidays. "I would consider it a bad deed," Pasteur said, "to let one day go without working."

At the end of the day, Pasteur brought his preoccupations home to the dinner table. Stubby and thick-bodied, with a high forehead, trim black beard, and an intense gaze behind wire-rimmed glasses, he exuberantly drew his wife and children into the events at the laboratory, all the while wiping glasses, plates, and silverware in the hopes of removing contamination. "He minutely inspected the bread that was served to him," recalled his assistant Adrien Loir, "and placed on the tablecloth everything he found in it: small fragments of wool, of roaches, of flour worms. Often I tried to find in my own piece of bread from the same loaf the objects found by Pasteur, but could not discover anything. All the others ate the same bread without finding anything in it."

His was a contentious nature that simply wouldn't take no for an answer — unless that no was well proved — and he enjoyed putting in their place all those he considered misinformed or badly motivated. The most famous of his confrontations concerned a fundamental question about microorganisms: where do these microscopic creatures come from?

As far as the scientific establishment was concerned, the question had been answered long before. These microorganisms — these animalcules — simply sprang whole from the material they inhabited. If they were present in the souring of milk, for example, they represented an alteration in the vital matter of the milk itself. If they participated in the fermentation of sugar to alcohol, they were the essence of sugar transformed. The doctrine was called "spontaneous generation," and it had been the answer to such questions of origins for a long, long time.

From antiquity it had been believed that plants and animals sprang to life from the fundamental matter of inanimate objects, and that the life principle was set free by the death of living things. Eels arose from the muck of rivers, for example. Bees flew perfectly formed from the entrails of dead bulls, and maggots crawled to life from the depths of decaying meat. As late as the sixteenth century, a celebrated physiologist could state that you could create a mouse whenever you cared to simply by placing some dirty linen in a container and adding a piece of cheese or some grains of wheat.

If such recipes for the generation of familiar animals were no longer taken seriously in Pasteur's time, it remained easy enough to affirm the doctrine with respect to microorganisms. So small that they were invisible to the naked eye and virtually unmeasurable (we now know that bacteria measure from about .5 to 2 microns in diameter, or about 1/25,000 of an inch — and viruses, of course, are even smaller), and, as one-celled creatures, so uncomplicated in form that they undoubtedly represented some kind of transitional life, these creatures, it was thought, must simply spring into being from inanimate matter.

For Pasteur the notion was ridiculous. Microorganisms came from other microorganisms, just as all living things came from their own predecessors, and in the early 1860s he devised a series of experiments that proved his contention. To begin with, he sucked air through a tube whose opening was plugged by cotton, thereby trapping any dust or other microscopic hitchhikers floating by. Under the microscope he could see in the dust the spores of minute plants and other microorganisms, and when he placed them in a warm nutrient solution they multiplied rapidly — proof that these invisible travelers needed little else to reproduce but themselves. No mud bogs, no putrefying meat, no dirty linen and scraps of cheese here.

Next Pasteur made flasks with long necks and small openings that easily could be sealed off by melting the glass closed, and into them he poured a nutrient solution of sugar and extract of yeast. Then he boiled the flasks to kill any living organisms inside and to expel most of the air, which escaped as a current of steam through the long neck. He next sealed the necks with an alcohol lamp while the steam was still escaping, thereby sterilizing the contents.

For days afterward, he heated the flasks in an incubator and, as he predicted, no discoloring took place — nothing grew. The flasks were indeed pure — there were no germs inside. Now came the test. Pasteur took some of the flasks to the laboratory and snapped off the ends with a pair of pliers. There was a sharp hissing sound as air rushed back inside, and immediately Pasteur sealed the flasks closed again. Then he replaced them in the incubator and once again commenced to wait. This time the results were different. The liquid in most of the vessels soon became cloudy. Microorganisms were starting to ferment the liquid — proof that these germs had rushed in with the air rather than grown spontaneously from the liquid itself. If they had generated spontaneously, the solution would have begun to ferment long before Pasteur let air inside.

But some of the exposed flasks remained clear. It must be, Pasteur reasoned, that the distribution of microbes in the air was uneven. Air in some locales was less contaminated than elsewhere. And it must be, he suspected, that where there was abundant life — that is, food for these creatures — there were more germs than in more isolated, inhospitable places. But, once again, suspecting it so didn't prove it so. Pasteur decided to take his microbe show on the road.

In a series of tests beginning in 1860, he tested his flasks in environments as varied as the calm air and unchanging temperature of the cellars of the Observatory of Paris and the peaks of the French Alps. Again, the results were clear: air carried microscopic germs, the amount of which varied depending on locale and temperature.

In 1864 Pasteur clinched the matter. At the Sorbonne, before the *crème* of Paris society and influence — Alexandre Dumas, Princess Mathilde, George Sand — Pasteur demonstrated perhaps his most elegant experiment. He held up a flask with an especially long and curved neck, much like that of a swan. Although air flowed into the flask, nothing grew in the liquid it contained, because microbe-containing dust and other microscopic matter could not reach the liquid — it settled out in the swanlike neck. Only when Pasteur tipped the flask, so that the liquid came into contact with the settled material and its hitchhiking microbes, did the liquid become cloudy with microbial growth.

"No," he thundered, "there is now no circumstance known in which it can be affirmed that microscopic beings came into the world, without germs, without parents similar to themselves. Those who affirm it have been duped by illusions, by ill-conducted experiments, spoiled by errors that they either did not perceive or did not know how to avoid." For better or for worse, a world consisting of generation after generation of dynamic, living microorganisms was with us for good.

*   *   *

Sometimes there breaks out in the poultry-yard a disastrous disease, commonly known as chicken cholera. The animal which is a prey to this infection is without strength, trembles and has drooping wings. The feathers of the body are ruffled giving it the form of a ball; an unconquerable drowsiness overpowers it; if forced to open its eyes it appears to waken from profound slumber; soon the eyes close again, and in most cases, the animal does not change its position until death comes, after a dumb agony. At most it sometimes shakes its wings for a few seconds.

So read a report of the Paris Academy of Sciences. The cause of the disease, and its cure, were unknown.

They would not be unknown for long, for in 1876 an obscure German doctor named Robert Koch proved that what Pasteur and isolated theorists before him had suspected was true: microbes did indeed cause specific diseases. Koch identified the bacterium that caused anthrax, an often fatal disease of animals as well as humans.

Given Koch's lead, others quickly followed; and in 1878 an ambitious young professor named Toussaint at the Toulouse veterinary school identified a virulent microbe that he suspected caused chicken cholera. He was not, however, able to grow the microbe in the laboratory, as he couldn't find a congenial medium in which it could survive. If this germ was to be studied, it would somehow have to be grown. Toussaint sent it to Pasteur.

Pasteur was now fifty-six. He had been through much. He had discovered that some microbes need air to survive and some don't, and had coined the terms *aerobic* and *anaerobic* to describe these characteristics. He was credited with single-handedly rescuing the wine-making and brewing industries of France, for he had determined that bacteria were spoiling the wine and beer, and that the microbes could be killed by applying controlled heat. This technique, which came to be called pasteurization, may be the most famous of all his contributions. Pasteur also saved France's silk industry by discovering and suggesting how to control the microbe causing a disease that was killing the country's silkworms.

Overseas, an English surgeon, Joseph Lister, incorporated Pasteur's teachings about the function and spread of microbes to introduce sterile, antiseptic surgery, an approach that has saved uncountable lives. But Pasteur was unable to honor Lister's invitation to visit his hospital and observe his techniques; for in 1868, when he was forty-six, Pasteur had suffered a stroke that left him permanently paralyzed on his left side, with a stiff hand and a dragging leg. Fortunately, his mind was not affected. If anything, the stroke focused Pasteur's priorities even more strongly. During the attack he had complained, "I am sorry to die; I wanted to do much more for my country." Finding himself still alive, he began to marshal all his efforts in the direction his earlier work had led him — toward the goal of determining the role of microorganisms in disease.

So when the chicken-cholera microbe arrived in his laboratory, Pasteur took on the challenge eagerly. He soon found an environment in which to grow it — a broth of chicken gristle — and now, able to experiment with the microorganism, he quickly discovered other interesting things. For one, the microbe was so virulent that the smallest drop of the culture on a crumb of bread was enough to kill the unlucky chicken that ate it. And when the afflicted bird excreted the microbe, it quickly infected other birds; such was the cause of chicken-cholera epidemics.

For another, although rabbits succumbed to the disease as quickly as chickens, guinea pigs rarely took sick. When Pasteur gave the culture to guinea pigs, usually all that resulted was an abscess at the point of inoculation that eventually opened and then healed over. Within the pus in the sore appeared the same chicken-cholera microbe, as though it had passed through the body of the guinea pig untouched and was now preserved as in a test tube. When applied to chickens once more, it had lost none of its virulence: the birds soon died.

It was an evocative finding. "Chickens or rabbits living in contact with a guinea pig suffering from such abscesses might suddenly become sick without any apparent change in the health of the guinea pig itself," Pasteur explained. "It would be sufficient that the abscesses open and spread some of their contents onto the food of the chickens or rabbits."

Not only might this phenomenon explain an occasional chicken-cholera epidemic, it had enormous implications for the spread of disease in general. It was possible, Pasteur had found, for one animal species to harbor and spread the disease of another species — perhaps even humans. It was the first articulation of the fact that there are "carriers" of disease, who, while not affected themselves, can spread the disease to others. "How many mysteries in the history of contagions will one day be solved as simply as this!" Pasteur exclaimed.

Yet the most noteworthy, and surprising, of Pasteur's discoveries was yet to come. After writing of his chicken-cholera findings to the academy in 1880, Pasteur hurriedly left Paris for a few weeks' holiday. Upon his return, he noticed a beaker of cholera culture that he had forgotten to dispose of before his departure. It had been resting on his workbench, untended, all that time. Since they had not been fed with fresh nutrients, the germs in the culture had stopped reproducing some time earlier; they were therefore no longer of use. Pasteur was on the verge of emptying out the flask when he suddenly changed his mind. He decided to inoculate a few hens with the old culture. To his astonishment, the birds became mildly sick and then recovered. For some reason, in this case the killer germ didn't kill.

People discussing what happened next sometimes attribute it to luck, and it's true, Pasteur was uncommonly lucky in his scientific life. Others call it an accident, the happy result of pure chance — and it's true, it was an accident that the culture had been left unattended. But what Pasteur did after this accident — that's what made possible the discovery that followed. Pasteur himself recognized the role chance plays in science (it's a theme we'll encounter again in this story). He also realized that what counts is what you do with your chances. "In the field of experimentation," he wrote, "chance favors only the prepared mind."

So, Pasteur decided to test this old, accidentally neglected culture in yet another way. He obtained some fresh, virulent chicken-cholera broth and injected it into some fresh chickens. The birds promptly died. Then he inoculated the chickens who had already survived a dose of the old, neglected culture with the same new solution. Instead of dying, these chickens just kept on pecking.

Pasteur was stunned. What in the world had gone wrong? These chickens should be dead. They should have been dead long ago, when he had given them a dose of the old culture, and they certainly should be dead now, having received a batch of proved killer germs. But there they were, clucking as incessantly as before. Only one thing could have happened: somehow the old culture had made them immune to the disease.

Immune! Pasteur immediately thought of Jenner. But whereas Jenner's cowpox inoculations hadn't paved the way to inoculations for other diseases, Pasteur saw in this accidental result the possibility of a generalized procedure. If he could discover *why* the old culture had lost its virulence, perhaps he could repeat the accident. While he didn't understand the process by which the culture conferred immunity, perhaps he could make other cultures that similarly didn't kill but immunized. And here, paying respect to what Jenner had accomplished, he took Jenner's term *vaccinae*, "of the cow," and made it apply to a general procedure. *Vaccination* would no longer describe only what Jenner had done; now it would mean any inoculation that conferred immunity to disease. And if vaccination could be done for smallpox, and now chicken cholera, why couldn't it be done for other diseases?

Soon Pasteur concluded that it was the exposure to air — specifically oxygen — that had weakened, or "attenuated," the cholera bacteria to the point where they could no longer kill. And he showed that the attenuation could be controlled over time. A fresh culture would kill ten out of ten chickens. Expose it to air for a period of days, and it would kill eight of the ten; for a few more days, five; and so on, until after three months it would kill no longer. It could then be used as a vaccine.

Pasteur had turned the corner. During the next three years, in an astonishing burst of creative work, Pasteur discovered two more ways to attenuate disease-causing microorganisms and consequently came up with two more vaccines. For swine erysipelas he found that the offending microbes attenuate when passed from rabbit to rabbit; the bacterium that caused anthrax weakened and became suitable for use as a vaccine when exposed to high temperature. All of Pasteur's vaccines proved to be effective — in some cases, spectacularly so, especially when introduced by his characteristically dramatic demonstrations. But no matter how great the economic and agricultural impact of Pasteur's vaccines up to this point, their chief use was in controlling animal diseases. It remained to apply the principle of immunization to the afflictions of humans. For that, Pasteur chose rabies. And in doing so he moved into yet another realm. He was about to take on a type of microbe as yet unknown, the virus.

In 1880, when Pasteur turned his attention to the disease, only three things could be said about rabies with any certainty at all: whatever caused it was contained in the saliva of the rabid animal; it was transmitted by a bite that broke the skin; and after transmission it took anywhere from a few days to months before the symptoms appeared in its victims. Beyond that, all was mystery. Pasteur had his work cut out for him.

His first task was to try to isolate the cause, which he suspected to be a microbe. He tried directly inoculating animals with rabid saliva, which, of course, had to be collected by hand — a hazardous occupation. On one occasion, Pasteur himself drew saliva from a mad bulldog, which was dragged from its cage by a lasso around the neck, then held down on a table by two courageous attendants while the semiparalyzed Pasteur bent over the animal's foaming mouth and carefully sucked a few drops of the lethal saliva through a glass tube between his lips. One slip might have meant a fatal bite on the face.

But these inoculations proved to be inconclusive. Sometimes the rabbits, guinea pigs, and other animals came down with rabies, but all too often nothing would happen, or another disease would appear. And in any case, the rabies simply took too long to show up for the purposes of manageable laboratory study.

So Pasteur tried a different tack. He obtained mucus from the mouth of a five-year-old child who had died of rabies just four hours earlier. After mixing the mucus with water, he injected it into laboratory rabbits. This time all the animals died within thirty-six hours, and when blood from the dead rabbits was injected into other rabbits, they too died. It seemed possible that Pasteur had indeed found the virulent germ.

But what was it? Pasteur put some of the rabbit blood on a slide under the microscope. He saw something very interesting: "an extremely short rod, somewhat constricted in its center, in other words shaped like an 8." Pasteur cultivated the microbe in the laboratory, injected it into rabbits and dogs, and examined the blood of the dead animals to find the same microbe once again. Could this be the cause of rabies? For a time, he was tempted to think so. But then Pasteur began to find the same microbe in the saliva of healthy children, and children suffering other diseases, and healthy adults. No, it couldn't be the rabies germ. Refusing to be sidetracked, he began to search elsewhere — and so inadvertently passed by the cause of a far wider killer than rabies. Years later, it was determined that his figure-8 microbe was the pneumococcus bacterium, the cause of pneumonia. No wonder the rabbits had died quickly.

Pasteur pushed on. Having decided to look elsewhere than in saliva, he turned his attention to the nervous system. The early symptoms of rabies — the difficulty in swallowing, the creeping sensations on the skin — and the later paralysis all pointed to an involvement of the nerves. With the help of his assistant Emile Roux, Pasteur injected rabbits with a solution containing the ground-up tip of a rabid dog's spinal cord. This time a greater proportion of the rabbits contracted rabies than when injected with saliva, but again there were some that kept hopping as though nothing had happened. And again it simply took too long to see a result one way or another. Pasteur decided to inject the rabid nerve material directly into the brain of a laboratory dog.

That is, he decided that it should be done; he was entirely incapable of doing it himself. A prime target of antivivisectionists (in this, as in so many other ways, Pasteur's experience foreshadowed our own times), Pasteur was notoriously solicitous of his laboratory animals, preferring to leave the actual laying on of hands to others. So, Roux made the injection, and in two weeks the dog began howling and snarling madly, attacked and destroyed its bed, then stiffened and in another five days died. When the researchers repeated the experiment on other dogs, they obtained much the same results. They had finally found a reliable means of transmitting the disease and ensuring a manageably short incubation period. But the next step, cultivating and studying the infective agent, as they had done with the chicken-cholera microbe, seemed insurmountable. Their usual method of growing the agent in a nutrient solution in the lab just wouldn't work, for they had no idea what that agent might be. Whatever it was, it didn't show up under the microscope. Perhaps it simply was too small to be seen. Yet it existed — there had to be a way to work with it.

"Since this unknown thing is living," Pasteur said, "we must cultivate it; failing an artificial medium, let us try the brain of living rabbits; it would indeed be an experimental feat!" The experiment was a great success. He passed the rabid solution from one rabbit brain to another, over and over, with the result that the agent actually became more virulent. It killed in a progressively shorter length of time, until by the hundredth passage incubation time was down to six days — and there it stayed. They now had a constant supply of a "fixed" infective agent, even if they didn't know what that agent was. They could cause rabies at will; now they could try to see if there was a way to prevent it.

The inspiration for the final breakthrough came from Emile Roux. While Pasteur was busy experimenting with passing the rabies infective agent from rabbit to rabbit, Roux was studying how long the agent could survive in the spinal cord. He had hung part of the infected spinal cord of a rabbit in a flask by suspending it from a stopper wedged in the upper opening. One day Pasteur happened to walk into the incubator where Roux had placed the flask. Another assistant, Adrien Loir, later described how Pasteur fell into a characteristically impenetrable spell of concentration.

> At the sight of this flask, Pasteur became so absorbed in his thoughts that I did not dare to disturb him, and closed the door of the incubator behind us. After remaining silent and motionless a long time, Pasteur took the flask outside, looked at it, then returned it to its place without saying a word.
>
> Once back in the main laboratory, he ordered me to obtain a number of similar flasks from the glass blower. The sight of Roux's flasks had given him the idea of keeping the spinal cord in a container with caustic potash to prevent putrefaction, and allowing penetration of oxygen to attenuate the [microbe].

Pasteur had remembered how the chicken-cholera bacteria had been weakened by exposure to oxygen; perhaps a similar exposure would work for the agent of rabies. If so, he might then be able to use the attenuated agent in a vaccine. Whereas the chicken-cholera vaccine had been an accidental discovery, however, this time he would experimentally diminish the virulence of the virus. Every day, Pasteur cut off a piece of the cord, crushed and mixed it with water, and injected a few drops into a fresh rabbit. As the days went by, the virulent solution took longer and longer to infect each rabbit, until at the end of two weeks it had virtually no effect at all. The agent, whatever it was, had been weakened to the point where it was too feeble to harm the animal.

Now came the crucial step. Could he use the diminished virus to provide immunity to the disease? Always profoundly empirical, Pasteur decided to reverse the experimental weakening process rather than try any one particular dose. As it had taken two weeks to attenuate the virus, he would see if reversing the process would gradually build up immunity. He began by injecting into a dog a solution of the fourteen-day-old, nonvirulent cord. On the next day, he injected a thirteen-day-old-cord solution, and on the day after, one that had aged twelve days, and so on, increasing the virulence until on the fourteenth day he gave the dog a fresh cord that could otherwise kill within ten days. Nothing happened. The dog remained as healthy as before the injections had begun. Then, as a final test, Pasteur injected the strongest dose of rabies he could find directly into the brain of the dog. No effect. He could even put the dog into a kennel with a rabid animal, and although the mutt would emerge battered and bitten and generally the worse for wear, he would not emerge with rabies. The vaccine was a success.

It was now May of 1884. By this time, Pasteur was working with 125 dogs, not to mention a whole courtyard of rabbits, guinea pigs, and chickens. It had taken four years, and he still hadn't identified the cause of rabies, but he had produced a vaccine against the disease. Now the question was, what should he do with it?

At first he thought about vaccinating every dog in France, some 2.5 million of them. He quickly gave up the notion as impossible. Soon, however, he began receiving requests to vaccinate people. Pasteur was hesitant. The vaccination process was long and arduous, fourteen days of continual injections and monitoring. And although he was confident of his vaccine, he had used it only on animals. The emotional and ethical distance from an animal experiment to a human one was long, indeed.

Such was the state of affairs as the year turned to 1885. The populace endured the rabies season of late winter and spring. The "dog days" of summer arrived. It was then that Joseph Meister knocked at Pasteur's laboratory door.

"In . . . experimentation, chance favors only the prepared mind." Yes, it was true, but this might have been a chance too harrowing to take advantage of. Should Pasteur administer his animal-tested vaccine and risk killing or paralyzing the little boy? If such were the outcome, his enemies would have a field day. If he failed, they would accuse Pasteur of scientific hubris, of unforgivable, inhumane presumption. They would brand him a murderer. And if he succeeded, they would claim that the child might have survived anyway, that Pasteur was just using young Joseph for his own self-aggrandizing purposes.

Pasteur felt the crushing weight of decision. And there was an added consideration: young Joseph already had been bitten. He almost certainly was infected. To Pasteur's other worries was added the possibility that his vaccine might be useless against an already-infected individual, or might even augment the virulence of the infection.

So while Joseph and his mother waited in a room Pasteur arranged for them, he went for advice. He sought the counsel of two men, Edme Vulpian, a physiologist who had impressed Pasteur through his work on a commission that was investigating the rabies vaccine, and Jacques Grancher, a physician and bacteriologist. Both of them immediately urged Pasteur to give his vaccine to little Joseph — it was his duty, he could do no less. And when they examined the boy later in the evening, they advised Pasteur that because of the number and severity of the wounds, there was no time to lose: it was almost certain that if nothing were done, Joseph would die. They would share the responsibility if need be, but treatment must begin at once.

So at eight in the evening on July 6, 1885, sixty hours after the child had been bitten, Pasteur, still plagued by grave misgivings and acute anxiety, administered the second human vaccine the world had seen, and the first that had been manufactured, through laboratory experimentation, from the same agent that caused the disease. And while the gradual administration of increasingly virulent virus might be hard on the boy, there would be no shortening of the process. This was a human life; Pasteur would stick to what he knew. He injected into Joseph Meister's hip a solution made up of the spinal cord of a rabbit that had died of rabies two weeks earlier.

The boy cried a great deal before the inoculation but quickly calmed down when he realized it was no more than a slight prick of the skin. And when he returned to his room, which was located near the pens of rabbits and guinea pigs and chickens, he became happier than he had been since before his attack. Pasteur followed the first inoculation with two more each day for the next two days, then one each day from then on. By July 11, he could report in a letter that "all is going well, the child sleeps well, has a good appetite, and the inoculated matter is absorbed into the system from one day to another without leaving a trace."

But his cheery written tone did not mirror his behavior. As the inoculations became more and more virulent, Pasteur grew more and more subject to worry and attacks of anxiety. "My dear children," wrote his wife, "your father has had another bad night; he is dreading the last inoculations on the child. And yet there can be no drawing back now! The boy continues in perfect health."

In his letters, at least, Pasteur kept up his tone of optimism:

> I think great things are coming to pass. Joseph Meister has just left the laboratory. The three last inoculations have left some pink marks under the skin, gradually widening and not at all tender. There is some action, which is becoming more intense as we approach the final inoculation, which will take place on Thursday, July 16. . . . Perhaps one of the great medical facts of the century is going to take place; you would regret not having seen it!

Yet at night the demons returned. Pasteur had dreams of Joseph suffocating in his own saliva and, as the days counted down, found it increasingly hard to work.

Finally, the last day arrived: July 16. Pasteur gave the boy a dose of one-day-old spinal cord, strong enough to kill a rabbit in seven days. In the evening, after playing with the animals, Joseph demanded a kiss from "Dear Monsieur Pasteur" and went merrily to bed. Pasteur suffered a night of insomnia, punctuated by visions that the boy would die.

But Joseph Meister didn't die. The boy flourished. In ten days, he was home in Alsace, none the worse for wear. The vaccine had been a success.

Three months later, with young Joseph again healthy, Pasteur described his treatment for the Academy of Sciences in Paris. After he was finished, the chairman of the academy rose. "We are entitled to say that the date of the present meeting will remain for ever memorable in the history of medicine," he proclaimed, "and glorious for French science: for it is . . . one of the greatest steps ever accomplished in the medical order of things — a progress realized by the discovery of an efficacious means of preventive treatment for a disease, the incurable nature of which was a legacy handed down by one century to another."

The public agreed. Within fifteen months after Joseph Meister's treatment, 2,490 people received Pasteur's rabies vaccine. Pasteur saw his treatment quickly become a common practice for people who had been bitten by animals known or presumed to be rabid. But he also saw it become the center of controversy and hostility. Although of the 1,726 French nationals treated during that period only 10 died — a mortality rate of less than 0.6 percent — for some people that was too many. The attacks were spearheaded by a clinician, Dr. Michel Peter, whose doctrine, "Disease is in us, of us, by us," Pasteur had systematically demolished with his discoveries in the microbial world. The rabies treatment "has become outright dangerous," Peter stated. "Pasteur confers on the persons whom he inoculates the hydrophobia of rabbits — 'laboratory rabies.' " And this reaction was comparatively restrained. Some were calling Pasteur no less than "an assassin."

Nevertheless, as with Jenner's smallpox inoculations, the rabies vaccine was a gift too dear to be thrown away. Pasteur's treatment endured, and by 1935, 51,107 people had been vaccinated against rabies by the Institut Pasteur alone, with only 151 deaths — a mortality rate of less than 0.3 percent. Today in the United States, a onetime rabies vaccination is routinely given to dogs and cats, with periodic boosters to keep up immunity. The result has been that the country's domestic animals no longer carry the threat of rabies. That threat has passed to wild animals such as foxes, raccoons, skunks, and bats. In humans, the incidence of rabies has dropped to virtually nothing in the United States. It is, for the most part, a disease of the past.

POSTSCRIPT: Pasteur lived for another ten years after establishing his rabies vaccine. In 1887 he suffered another stroke and thereafter was unable to participate personally in experiments. The next year, the Institut Pasteur officially opened, funded with the help of subscriptions from all over the world. And in 1892, on his seventieth birthday, Pasteur was honored by a jubilee at the great amphitheater of the Sorbonne. He entered on the arm of the president of the Republic of France to a prolonged ovation, then sat to hear eulogies from representatives of the great cities of the world.

It's likely, though, that no remarks were appreciated more by Pasteur than those of Joseph Lister of England. "You have," he said, "raised the veil which for centuries had covered infectious diseases; you have discovered and demonstrated their microbian nature."

Young Joseph Meister grew up to become the gatekeeper of the Institut Pasteur in Paris. In 1940, fifty-five years after his vaccination, when Meister was sixty-four, he was guarding the gate as the Nazis invaded Paris. The soldiers ordered Meister to open Pasteur's crypt. Rather than do so, he committed suicide.

With Pasteur's life began the science of microbiology. Two centuries after van Leeuwenhoek saw amazing creatures tumbling about beneath his lenses, and a century after Jenner followed his hunches to employ a microscopic world for the benefit of humankind, Pasteur offered compelling evidence that life does indeed involve the work of tiny organisms, invisible to the naked eye, that infiltrate the world around us and within us and leave their mark in our comings and goings, our living and dying. With Pasteur, the microbial world became real.

He thereby liberated the world of biological science. Long shackled by the customs and philosophies of ages far gone, biologists now produced an explosion of new findings, using their microscopes to travel the invisible world as enthusiastically and profitably as explorers of previous centuries sailed their tall ships through the apparent world. The few decades that followed Pasteur's first studies of microorganisms have been called the golden age of bacteriology. During these remarkable years, scientists discovered numerous disease-causing bacteria and established basic laboratory practices still commonly in use. Tuberculosis, cholera, diphtheria, typhoid, pneumonia, meningitis, tetanus, bubonic plague, syphilis, dysentery — researchers had discovered the microbial causes of all these diseases by the turn of the century.

But as widely as they traveled, and as much as they accomplished, there was more to find. The microscopic world Pasteur had illuminated, as vast and rich as it turned out to be, contained a deeper layer. It was a fact that was not lost on Pasteur. His inability to find the cause of rabies had prepared him for the possibility that there were infectious organisms too small to be seen through a microscope. Yet he conceived of these organisms simply as ultramicroscopic microbes that differed only in size from the microorganisms he already knew. He did not imagine the nature of what we know today as a virus — his newly identified microbes offered enough of a challenge to the imagination and will — but ironically, with his rabies experiments, Pasteur came face to face with the very creature itself. For the cause of rabies is indeed a virus. It was finally discovered in 1903 at the Pasteur Institute in Constantinople. Pasteur was confronted with trying to prevent a disease caused by an agent he could not see and had no precedent for understanding. It's remarkable that he did so well.

In 1887, however, just two years after Pasteur treated Joseph Meister, a virus was actually seen. That feat was accomplished at the University of Edinburgh, in Scotland, by a surgeon and bacteriologist named John Buist. But as momentous as the event was, it passed virtually unnoticed. For the sad fact was that poor Dr. Buist had no idea what it was that he saw. Following the techniques established by Pasteur and Koch, Buist drew fluid from a smallpox vaccination sore on the arm of a patient, smeared it onto a glass slide, dried it, and stained it with a dye that made its components visible under the microscope. When he looked through the lenses, Buist saw some reddish, round dots. He made measurements of the dots and then tried to figure out what they were.

The dots didn't look like any bacteria he knew of, but they did resemble spores, the tiny seeds some microorganisms produce for reproduction. Buist thought that these spores must be the cause of cowpox and smallpox, but as he was unable to grow them in culture and then cause the disease in the lab, he couldn't prove his suspicion. There the matter ended.

As it turns out, Buist was at least partially right — what he saw does cause cowpox. He was the first person to look upon a cowpox virus. The reason he was able to see it is that it is one of the largest of viruses, measuring a whopping 1/100,000 of an inch on a side (about a quarter the size of a bacterium). Thanks to the electron microscope, we know now that what looked like a dot to Buist is really more nearly rectangular, with rounded corners. It is almost identical to the virus that causes smallpox — which is why it confers immunity to that deadly disease.

But why couldn't Buist grow the virus in culture? The reason, as virologists were to learn to their frustration for years afterward, is that viruses don't follow the same rules as bacteria: they simply won't survive and multiply in a lifeless nutrient solution. That Pasteur and Roux conceived of cultivating their rabies virus, without knowing what it was, by passing it from rabbit brain to rabbit brain seems an even more impressive achievement in this light.

So although Buist did gaze upon a virus, he was less able to deal with it than was Pasteur, who never saw one. If the world of viruses was ever to become recognized, another leap of the imagination was necessary — a new conceptual basis almost as shocking as the original idea of microorganisms. It may be that given a longer life, Pasteur himself might have made that leap; certainly, he was capable of it. As it turned out, however, the legacy of the pioneers of microbiology was well served, for it was the ingenuity of one of Pasteur's assistants that made the discovery of viruses possible, and it was a spiritual descendant of old Antony van Leeuwenhoek, a fellow citizen of his native city of Delft, who claimed the distinction of making the final conceptual leap into this unknown new world.

Four

. . . .

Obscure if Not Positively Unnatural

. . . .

Back in 1876, while disproving yet another claim that microorganisms could burst to life spontaneously, Pasteur demonstrated that virtually all water, even distilled water housed in a washed receptacle, contains germs. The exception was deep-well water — water that had slowly percolated down through soil which acted as a filter, trapping microorganisms so that the water passed through pure.

The discovery set Pasteur's assistant Charles Chamberland to thinking: if sandy soil filtered microbes out of water, might there not be use for a laboratory filter that accomplished much the same thing? The result was his invention, in 1884, of what came to be known as the Chamberland filter. This was a candle-shaped filter made of unglazed porcelain that functioned much as did the soil surrounding a well. When liquid of any kind was poured into the top of the filter, the tiny pores in the porcelain trapped foreign bodies, allowing the liquid to emerge pure from the spigot at the bottom. (For years afterward, Chamberland filters were used in households to filter pure drinking water.)

Pasteur, as well as every other microbiologist, used Chamberland filters to show that it was the bacteria held back by the porcelain that caused disease, not the clear, innocuous fluid that passed through. But in some cases, fluids were able to pass through and still bring about disease. Emile Roux demonstrated that the bacterium that caused diphtheria, for example, produced a toxin that easily passed through the Chamberland filter. Inoculate with this toxin, and the result was diphtheria. But while such toxins could cause disease in their immediate victim, they couldn't transmit it beyond that. They didn't reproduce and multiply, and thereby spread infection. These were onetime infections — the bacteria that produced the toxins were the real culprits. And the filter readily trapped these bacteria.

When it came to rabies, however, it was a different story. Whatever it was that caused the disease wasn't caught behind by the fine porcelain pores. The filtered fluid itself was infectious. Something odd was happening here. Pasteur suspected that an ultramicroscopic microbe was passing through. But, of course, he never figured out what it was.

In 1892, a Russian botanist named Dmitry Ivanovsky used a Chamberland filter to investigate a troublesome problem called tobacco mosaic disease. The affliction threatened the production of tobacco throughout Russia and Europe, as it caused infected plants to become dwarfed and their leaves blistered and mottled, which gave them a mosaiclike appearance. In an attempt to discover the cause of the disease, Ivanovsky crushed the leaves of infected plants and passed the sap through the filter. He expected to obtain a neutral fluid, while the bacteria at fault would be left behind. But when Ivanovsky anointed new tobacco plants with the filtered fluid, the characteristic mottling began again. The cause of the disease remained in the fluid; it had no trouble passing through the filter.

Ivanovsky was enough impressed to italicize his findings in his report to the Academy of Sciences in Saint Petersburg: *"The sap of leaves infected with tobacco mosaic disease retains its infectious properties even after filtration through Chamberland filter candles."* Either something had gone wrong — perhaps the filter was faulty — or, like diphtheria, tobacco mosaic disease was the result of a toxin produced by bacteria.

The only way to find out one way or another, of course, was to find the offending bacteria. But this, despite efforts demanding "a great deal of time and trouble," Ivanovsky was unable to do. And even though he claimed that he had been able to see "the growth of the tobacco microbe and then to determine its presence in the tissues of infected plants" (under "certain special conditions" he did not describe), Ivanovsky let the matter stand. He closed his report with the hope that he would be able to conduct more experiments "to clarify these questions" and went off to other things.

For six years that's where the matter continued to stand. Then, in 1898, another botanist, the Dutchman Martinus Beijerinck, decided to investigate the same problem. Unaware of Ivanovsky's work (for the Russian had published his results in relatively obscure Russian-language journals), Beijerinck performed similar experiments with the Chamberland filter and obtained similar results: the filtered sap of the crushed leaves readily infected other leaves. And, like Ivanovsky, when he set out to isolate the cause of the infection, he could find nothing. But there the similarities ended. For what Martinus Beijerinck accomplished after that ushered in a new way of looking at life and disease. He was about to introduce the world to viruses.

Beijerinck was the epitome of the temperamental scientist. He would burst into his lab at the Delft Polytechnic School, a tall, striking figure in dark coat and high collar, with a full mustache that accentuated his angular, even rakish features. Around the rooms he would prowl, shutting all windows, disdainfully sniffing for the faintest remnant of cigarette smoke, and inspecting the benches for as little as a drop of spilled water. Then he would confront his handful of students. "One has to have witnessed the high tension found there," reported an eyewitness,

> and to have heard the conversations sometimes lasting for hours, with one of the experimenters, where usually Beijerinck did the talking and the other the listening — fascinated by the stream of surprising and new remarks with thousands of suggestions for new experiments which the professor poured out over his unfortunate head. The student tried to take it all in, but at last was almost in

despair, because his head was unable to contain the overpowering amount . . . while Beijerinck, as fresh as if nothing had happened, went on to another student to lose himself entirely in the latter's subject.

And woe to the student who didn't follow through on these suggestions. "Sir," Beijerinck addressed one such young man, "there are two types of monkeys. One type is interested if one shows a coin and will hold it firmly, the other type will at once drop it. The first type can be trained, the second type cannot. If you were a monkey, you would belong to the second type!"

For women, he had little regard and less time, although — or perhaps because — he is said to have had one great disappointment in love. He never married (he lived with his two sisters) and did not approve of his collaborators' marrying, either. "A man of science does not marry," he said, and he always pointedly opened his lectures, "*Gentlemen* and Ladies."

Yet those who were able to endure the man found the effort well worth it. He provided constant support for those students who measured up to his standards, helping a number of them go on to prosperous careers, standing up for them in the face of criticism from other professors, and even inviting a particularly faithful assistant and his family to live in the laboratory while their fire-damaged house was being repaired.

"Beijerinck was like a mighty building," wrote one of his assistants. "Wandering through its unfamiliar courts and archways, a visitor might sometimes knock against, and be hurt by, protruding stones, but after leaving the building and contemplating at a distance its superb architecture, the former visitor would be lost in rapture."

By temperament and ability, Beijerinck was perfectly suited to pioneer the world of viruses. It may be that a less intransigent man would have permanently turned away from the matter of the mysterious infectious juice of tobacco mosaic disease, much as Ivanovsky had, and a less confident man, having failed before, might have been wary of further investigations, but Beijerinck cared for nothing more than knowledge. "An almost Dionysian joy often came over him when his experiments were successful," wrote his assistant.

"Then his brown eyes would glitter, and, with a staring look and uplifted left forefinger, he would explain the significance of his discoveries to his disciples."

The significance of *this* discovery, though, would take some unique explaining. In fact, it's possible that Beijerinck himself didn't immediately comprehend the extent to which he was cultivating new ground. All he knew in 1898 when he took up the study of tobacco mosaic disease was that it was extremely odd that the strange sap passed through the Chamberland filter and still retained its ability to infect other tobacco plants. Immediately, he made his first break with what had been accomplished before. Rather than suspecting, as had Ivanovsky, that the Chamberland filter was defective and so allowed bacteria to pass through, or, like Pasteur, surmising that there might be ultramicroscopic microbes that the filter simply didn't hold back, Beijerinck took his experiment at face value and declared that the filtered sap contained no microorganisms at all.

Then he went another step further than Ivanovsky. Rather than simply spreading the sap on one series of tobacco plants, concluding that the cause of the infection might be a toxin, and leaving the matter at that, Beijerinck continued to inoculate plants. He ground up the leaves from recently infected plants, passed the juice through the filter, spread it on healthy plants, then ground up those leaves, filtered them, and inoculated another series of healthy plants, and so on. And what he found was that each successive series of plants similarly showed the characteristic mottling and blistering of tobacco mosaic disease. This sap not only infected once, it infected again and again.

How could that happen? There was only one answer: it must be that somehow the infectious principle in the sap was reproducing, multiplying in the tobacco leaves, so that each successive series became as infectious as the last. But as far as the science of Beijerinck's day was concerned, only bacteria and other microorganisms were able to reproduce in such a fashion. Toxins certainly couldn't. Yet there were no microorganisms in the filtered sap, he was sure of that. It must be that whatever caused the disease was unlike anything conceived of previously.

Beijerinck backtracked and double-checked. He repeated his experiments, with the same results. He let the filtered sap sit for three months before applying it to tobacco leaves; it infected once again. He exposed it to alcohol and formalin, enough to kill microorganisms. Still the fluid infected. He knew that some bacteria contain spores tiny enough to pass through filters, so he heated the fluid to ninety degrees centigrade (194° F), ten degrees less than the temperature needed to kill spores. If there were bacterial spores in the fluid, it would still infect, but he would have killed off whatever else might be there. Instead the sap lost its infectivity altogether. Beijerinck had indeed killed off whatever else might be there — but *that* included the infective principle; the culprit here certainly was not spores.

So, he was back to square one. If he was going to crack this one, he would have to go beyond established microbial thought and techniques. For now the question became not simply what kind of microbe it was, but whether this elusive cause of tobacco mosaic disease was a microbe at all. For Beijerinck, in 1898, living in a world less than twenty-five years removed from the profound shock and enormous influence of Pasteur's and Koch's demonstrations that microbes can cause disease — a citizen in good standing of the burgeoning golden age of bacteriology — to consider that something other than a standard microbe was involved was a kind of heresy. After all, Pasteur and Koch had had the testimony of van Leeuwenhoek. They at least knew there existed a microscopic world of amazing creatures; it was up to them to discover their function. What precedent did Beijerinck have? None. What could he do? The evidence he had uncovered was totally contradictory. This unknown agent infected as microbes did; it even seemed to multiply in similar fashion; but it didn't grow in culture medium in the lab, as did microbes, and it didn't show up under the microscope. Moreover, it didn't remain behind in the Chamberland filter, and alcohol and formalin didn't kill it; these were characteristics of nonliving material like toxins, but this thing was not a toxin. Beijerinck had to determine if it was a *thing* at all — that is, was it indeed a particle, a cell? Did it have discrete boundaries, was it an enclosed, living organism? Or rather was it diffused, dissolved? Unbelievable as it may sound, might it be dead?

To find out, Beijerinck came up with an ingenious test. He

smeared some of the infectious sap on a Jell-O-like block of solidified agar, a medium made of seaweed. If, he reasoned, the infective agent penetrated into the block of agar, it could not be a microbe, for bacteria would simply remain on the surface, unable to diffuse inside. If, on the other hand, he did indeed find evidence of the substance within the block, he would be up against something no one had ever seen before.

Beijerinck allowed the inoculated agar to sit for ten days. Then he carefully sliced off the surface layer of the block, exposing the fresh agar beneath. He took a sample and applied it to a healthy tobacco plant. Soon the leaves of the plant became mottled and blistered — the familiar pattern of tobacco mosaic disease. This amazing agent had indeed penetrated inside the agar. It was no conventional microbe. It was something else altogether.

So Beijerinck sat down to report his puzzling findings. The conclusions he arrived at introduced concepts up to then unknown. He titled his paper "Concerning a *contagium vivum fluidum* as a cause of the spot-disease of tobacco leaves." *Contagium vivum fluidum* — the Latin phrase means "contagious living fluid." This infectious sap was indeed a contagious fluid, *not* a microbe-containing solution; that much was clear. But "living"? How could a fluid be living? It was a difficult notion for Beijerinck, but such was the direction his experiments had taken him. "It seems to me," he wrote, "that the reproduction or growth of a dissolved particle is not absolutely unthinkable, but it is very hard to accept, and the idea of a self-supporting molecule, which is a corollary of this view, seems to me obscure if not positively unnatural."

With this famous passage, Beijerinck had inched his way into the unknown, toward what we know today as a virus. His contagious living fluid he described as "a dissolved particle," as he recognized that this mysterious agent must surely be a particle of some kind. That notion led to the next concept, that of "a self-supporting molecule," which today, almost a century later, is no less hard to accept than it was for Beijerinck. But that's pretty much what a virus is — a collection of molecules, actually, which, in and of itself, is whole, complete, self-supporting. Unlike a bacterium, it is not a living cell, with the independent capability of growth and reproduction (Beijerinck would tackle that problem too), but it doesn't need to be. While not exactly alive, it certainly is not dead. It simply exists as it is — "obscure if not positively unnatural."

Beijerinck then went on: "Hence it might conceivably serve as an explanation that the contagium, in order to reproduce, must be incorporated into the living protoplasm of the cell, into whose reproduction it is, in a manner of speaking, passively drawn."

It was a prophetic remark, and an astonishing one. To be able to assert that something existed which relied for its own reproduction upon the reproductive capabilities of living cells, a process that was not itself understood at the time, displayed amazing insight. For although viruses aren't exactly "passively drawn" into the reproductive processes of cells — they force the issue themselves — they are indeed incorporated into the cells' "living protoplasm." In a manner of speaking, they become one with the cell in order to ensure the propagation of future viruses. Sitting in his turn-of-the-century laboratory, having never seen nor heard of such a business, crusty, acerbic Beijerinck suspected all this. Yet, still a man of his time, he found it all a bit bewildering.

"This would at least reduce two riddles to only one," he wrote, "since the incorporation of a virus into the living protoplasm, even if well-documented, cannot by any means be considered a thoroughly understandable process." (If he felt embarrassed by his inability to explain further, he needn't have — it would take at least another fifty years before people began to understand the process.)

Finally, he used the term *virus*. There it was. The old word was now being used in a new context. What had been general and vague was now specific. The terminology stuck. Although it would take some time to become accepted in all circles, it would eventually refer to this kind of microbe and this alone. And because the virus had been able to pass through Beijerinck's Chamberland filter, it became known in these early years as a "filterable virus." Once more, the world's idea of the nature of life had changed.

The immediate reactions to Beijerinck's concept of a virus were varied. Pasteur's old pupil Emile Roux admitted that it was "very original" but suggested that before accepting it the possibility of a spore-bearing microorganism should be considered (as we've seen, Beijerinck had already done just that). Another bacteriologist, the German Ernst Joest, even more strongly upheld the prevailing idea that all life was made up of living cells. "A contagium vivum fluidum can only exist in the form of organized cell individuals," he maintained. Therefore Beijerinck's virus was "unthinkable" as the cause of tobacco mosaic disease, as well as any other infectious diseases.

However, another German scientist, Carl Oppenheimer, saw intriguing possibilities in Beijerinck's findings. He suggested that certain human diseases, such as smallpox, scarlet fever, and syphilis, might well be caused by viruses. If so, he wrote, "our concept of the 'cell' may undergo such a radical change as to admit that vital forces can actually act in unorganized [that is, noncellular] media." (While mostly wrong in specifics — only smallpox is caused by a virus, the others by other microorganisms — he was right in the larger sense. One of the profound realizations inspired by Beijerinck's discovery of viruses was that life can exist in other than cellular form.)

And there was yet another reaction. It was from Dmitry Ivanovsky.

Predictably, perhaps, once Beijerinck's paper appeared Ivanovsky claimed the discovery for himself. He went on to criticize Beijerinck's agar block experiment and then stated that subsequent to his early research he himself had transmitted the disease serially from plant to plant and now concluded, with Beijerinck, that the virus multiplied in the living plant itself. He made no mention of the fact that seven years earlier he had considered the infectious sap to be a bacterial toxin.

Beijerinck responded immediately, and in a manner that might not have been anticipated. Contentious though he may have been, he was above all a seeker of knowledge. He "willingly acknowledged" Ivanovsky's prior claim and explained that he had not known of the Russian scientist's work. And with that he was off to other research, forever leaving behind the problem of tobacco mosaic disease.

Ivanovsky, on the other hand, spurred by Beijerinck's investigations, returned to the disease and in 1902 published a doctoral thesis on the subject. In it he declared (correctly, as it turned out) that "the most likely conclusion is that the contagium is contained in the sap in the form of solid particles." He also proposed (incorrectly) that the cause of the disease might be a microbial spore. But he was never able to identify or cultivate it, and he concluded his thesis by stating, "On the whole the problem of the artificial cultivation of the microbe of the mosaic disease must be reserved for future investigations."

For the Russian scientific community, it was enough. Though he never suggested that the mysterious infectious agent was other than a derivative of some microorganism whose nature was already understood, Ivanovsky, who died in 1920, is now hailed in his country as the founder of virology. The Institute of Virology of the Academy of Medical Sciences of the USSR is named in his honor, and its award for the year's best work in virology is called the D. I. Ivanovsky Prize.

Beijerinck continued his work at Delft until his mandatory retirement in 1921, when he was seventy years old. Widely praised and honored, he remained himself, indignant that he should be forced out of what he always considered his real home, the laboratory. When, during the celebration that marked his seventieth birthday, his staff proposed to raise the national flag in his honor, Beijerinck vetoed the idea. "One does not hoist flags on the day of one's funeral," he said.

Nine years later, on New Year's Day, 1931, after a quiet country retirement in the company of his one remaining sister, Beijerinck died of an excruciating disease, cancer of the rectum. During these last years, he devoted himself to his first love, botany. He cultivated an extensive garden, which he delighted in showing to visitors, and about which he was no less exacting than ever — as his long-suffering gardeners could testify. At the age of seventy-five, about halfway through his retirement, he closed a letter to his successor at Delft with the words, "Fortunate are those who now start." They have since been inscribed on the wall of his laboratory.

Beijerinck knew he was on to something important when he discerned the existence of the tobacco mosaic virus. Although he never returned to this research, what he learned changed his notion of the nature of life. It even changed his mode of speech. Who would have suspected that the terse, understated Beijerinck of his earlier days — the man who could declare, "A discovery is great when one can communicate it in passing" — would be capable of the kind of dramatic and poetic address he delivered in 1913? "The existence of these contagia proves that the concept of life — if one considers metabolism and proliferation as its essential characters — is not inseparably linked up with that of structure," he suggested.

> In its most primitive form, life is, therefore, no longer bound to the cell, the cell which possesses structure and which can be compared to a complex wheel-work, such as a watch which ceases to exist if it is stamped down in a mortar.
>
> No; in its primitive form life is like fire, like a flame borne by the living substance — like a flame which appears in endless diversity and yet has specificity within it ... which can be large and which can be small: a molecule can be aflame ... which acts as a catalyst that brings about in its environment changes all out of proportion to its own size — which does not originate by spontaneous generation, but is propagated by another flame.

Beijerinck himself acted as a flame for others investigating this new world of viruses. Just as Pasteur's and Koch's demonstrations that bacteria caused disease sparked an explosion of discoveries of disease-causing bacteria, so Beijerinck's conclusions led to an explosion of interest in filtering infectious material in the hopes of finding new viruses. The research didn't take long to bear results. In the same year that Beijerinck first published his findings, 1898, two German scientists, Friedrich Löffler and Paul Frosch, discovered that material passed through a Chamberland filter could cause foot-and-mouth disease in animals. Three years later, in 1901, the Americans Walter Reed and James Carroll found the first filterable agent known to cause human disease: the yellow fever virus. And soon, dozens more plant, animal, and human diseases were found to be caused by filterable agents.

Perhaps the most interesting of those early discoveries involved a Rockefeller Institute pathologist named Peyton Rous. In 1909, at the age of thirty, he began studying a tumor in a chicken that had been brought to him by a poultry breeder. Rous determined that the tumor was a sarcoma, one of the most virulent forms of cancer. He then decided to see if he could spread the tumor to other chickens by applying material from the cancerous cells that he had passed through a filter.

It was an innovative approach, because at the time, cancer had only recently been recognized as a disease involving cells reproducing out of control — a normal process run amok. It was therefore nothing like infectious diseases such as yellow fever, smallpox, or rabies. It certainly was not caused by an infectious agent, filtered or no — that much the nineteenth-century bacteriologists had decided. Cancer tumors were considered "spontaneous," without a demonstrable cause. But Rous persevered. He ground up part of the tumor, passed it through a filter, then injected the resulting clear fluid into other chickens. A number of them developed tumors exactly like the first one. Rous then took material from the new tumors, passed it through a filter, and similarly injected it into other chickens, with much the same results. In his report in 1911, he avoided the word *virus* — he called his infectious material the "tumor agent" — but the results were clear. Now cancer, too, seemed to be caused by a virus.

(Rous's discovery was largely ignored, however. The tumors appeared in only about half the chickens, and, influenced by an assertion of Koch's that to be considered a legitimate cause of disease a microbe must bring about the problem every time, on demand, scientists simply weren't convinced by Rous's findings. In 1915, Rous gave up his investigation, never again to return to it.)

Yet, despite this spate of activity, and despite the seemingly unassailable persuasiveness of these discoveries, the very concept of viruses, much less their actual existence, continued to provoke controversy. It simply contradicted the hard-won ascendency enjoyed by the cell biologists. It had been difficult enough to convince a world mired in such ancient dogmas as the humoral origin of disease and the idea of spontaneous generation that there existed microscopic, one-celled creatures called microbes that had a hand in all life processes. Now here were some people trying to advance a new, contradictory concept. The fact that there might be microbes smaller than any yet discovered — perhaps that might be accepted. But to suggest that these ultramicrobes, these "viruses," were fundamentally different in structure and function, that they were made up of something other than the fundamental form of all life, the cell — now, that was too much.

It's easy, then, to understand the regard these fledgling "virologists" had for Beijerinck. He had given voice to a new field. In 1925, four years after Beijerinck's retirement, a Pasteur Institute microbiologist named Félix d'Hérelle eulogized him before the Amsterdam Academy of Science: "People have discussed the concept of Beijerinck a great deal, but I do not think that they have grasped all its profundity." D'Hérelle then asked to pay his respects to Beijerinck in person, with no success, as the great man entertained few visitors during his last years. It wasn't that Beijerinck didn't know who d'Hérelle was. On the contrary, the old man was well aware of the young microbiologist and followed his career with great enthusiasm. Beijerinck considered d'Hérelle's work a confirmation of his own early findings concerning the existence of viruses, and wrote so in several letters to former colleagues.

By the time d'Hérelle was through, the entire scientific world, and much of the public as well, would be aware of his work and of the existence of viruses — perhaps all too aware. It wouldn't be easy, and it wasn't always pleasant, but because of Félix d'Hérelle, viruses were about to invade the public consciousness for good. And it may be that a large part of the reason was his choice of a term to describe his discovery: *bacteria eaters.*

Five

· · · · · · · · · · · · ·

The Virus That Eats Bacteria

· · · · · · · · · · · · · · ·

"TINY AND DEADLY BACILLUS HAS ENEMIES STILL SMALLER." So read a headline in the September 27, 1925, issue of the *New York Times*. "What is the smallest living being?" the article began.

The microbe? Microbes have lesser microbes that prey upon them. Science is far from having determined the dimensions of these tiny beings which owe their acquaintance with the medical world to M. d'Hérelle of the Pasteur Institute of Paris. They are much smaller than the little flea's "lesser fleas." A little flea would always be able to see its parasitic "lesser" flea . . . but a microbe would no more be able to see its "lesser microbe," which enters into its interior and causes it to explode, than man with the naked eye can see the microbe that is responsible for his illness.

It's likely that Félix d'Hérelle very much enjoyed seeing such an article in print, for ever since he had announced in 1917 that he had discovered a mysterious agent that attacked and killed bacteria, he himself had been the target of attacks. A little-known, forty-four-year-old, Canadian-born bacteriologist, who was he to challenge the accepted notions concerning the microscopic world? Microbes existed, of course — of that there was no doubt. But to claim that something smaller might be preying upon these almost infinitesimal creatures . . . well, it might have been possible, but it was not likely. And then to state categorically that the ultramicroscopic invaders were not simply smaller microbes but actually viruses — that was simply too much. Beijerinck's concept of viruses was still too new and as yet unproved. At the least, d'Hérelle's theory should be presented as no more than that — a theory. But that wasn't the man's style. In fact, not content to stop there, he went even further. He suggested that these organisms were a special kind of virus, one that preyed only upon other microbes, and he came up with the term that through its evocative, descriptive power made scientists have to contend with his claims whether they wanted to or not: *bacteriophage,* "bacteria eater."

And what a discovery it was! Imagine: a typical bacterium, invisible except through the microscope, measures about one micron across (a micron is a thousandth of a millimeter — or $1/25,000$ of an inch); and a large virus — a smallpox virus, say — for all intents and purposes visible only to the electron microscope, is about one-quarter that large; but a small bacteriophage, able to prey on bacteria much as the bacteria themselves prey on a cell, is only about *twenty-five thousandths* of a micron, or twenty-five millimicrons — that is, a millionth of an inch. It would take forty bacteriophages placed side by side to span the diameter of a typical bacterium. With bacteriophages, we descend into the submicroscopic world to a dizzying degree.

The *Times* article provided welcome support to d'Hérelle. As noted, he needed lots of it, for he and his theory were continually under siege — and not only because of the controversial nature of his contentions. The man himself was controversial. Colorful, outspoken, and self-serving, d'Hérelle offered a tempting target for all those dissatisfied with or threatened by the seemingly outlandish nature of his discovery. But in time, all that might have blown over were it not for another circumstance — another reason why he was so forcibly attacked and widely distrusted. It seems that in d'Hérelle's initial report of the discovery and in all subsequent publications he somehow neglected to mention a crucial fact: Félix d'Hérelle was not really the first person to find the viruses that eat bacteria.

The story begins in England in the early 1910s, in a research facility of the University of London called the Brown Institution. The superintendent of the Brown, which was a hospital "for the care and treatment of Quadrupeds or Birds useful to man," was Frederick W. Twort. Born in 1877, the son of a hardworking country doctor of modest means, Twort went into medicine at least in part because "it was imperative that I should earn at least sufficient to be able to pay for lodgings and food." Given a mandate to do any research work he wanted "provided expenditure could be kept within the limits of the small income available," Twort embarked upon a project that had little to do with quadrupeds or birds. Like so many other bacteriologists, he was trying to disprove Beijerinck. The irascible Dutchman had theorized that viruses were subcellular, submicroscopic infectious agents that not only couldn't be filtered but couldn't be grown in the laboratory. Twort, like others, suspected otherwise. He figured that viruses were probably nothing more than very small bacteria and most likely would grow in the lab if only the appropriate culture conditions could be found.

Twort went about trying to find the appropriate conditions in an interesting way. He theorized that all pathogenic viruses — that is, all viruses that cause disease — were the descendants of safe, nonpathogenic viruses. With bacteria, anyway, it was true that for every disease-producing variety there were other, neutral types. So if viruses were only an ultrasmall variety of bacteria, why shouldn't they behave similarly? And because bacteria grew in culture media, why wouldn't the relatively benevolent ancestors of known viruses also readily make their home in the lab?

To find out, he turned to the discovery that had started it all, Jenner's smallpox vaccine. Finding in the vaccine a reliable source of virus — in this case, vaccinia virus, as the cowpox virus had come to be called — Twort smeared smallpox vaccine on a base of agar, hoping that a mild, ancestral version of the virus might begin to grow on the nutrient jelly. It didn't, but other things did. Those things were colonies of bacteria that had contaminated the vaccine.

Here, faced with a failed experiment, Twort demonstrated his version of the truth of Pasteur's dictum that "chance favors only the prepared mind." Instead of simply throwing out the whole business, chalking it up to experience, and trying again, he noticed that something unusual was going on. Some of the colonies of bacteria were undergoing what he called a "glassy transformation": they had stopped growing and were becoming watery and transparent. He then found that if he touched a healthy spread of bacteria with a bit of the glassy material, it too became transformed: the watery, transparent spot spread outward from the point of initial contact. Moreover, when Twort passed the glassy material through a Chamberland filter and inoculated a fresh batch of bacteria, the new colonies stopped growing and likewise became transparent and watery. Even when he diluted the material a *millionfold*, it caused the same transformation, and it could transmit this "disease" from colony to colony indefinitely. After being stored for six months, even, it caused the transformation. But — and this was a big "but" — Twort could not induce the mysterious material to grow and multiply by itself on any laboratory medium. It showed signs of life only when in contact with the bacteria. It behaved, in other words, just like a virus.

Twort published his unusual findings in 1915. His article, brief and to the point, described the mysterious changes he had observed and then suggested that "the transparent dissolving material" might have three possible causes. First, it simply might be a stage in the life cycle of the bacteria — nothing more, nothing less. Second, it might be caused by an enzyme, a catalyst released by the bacteria in the last stages of life. Finally, the agent of this strange transformation might be a virus that infects and destroys these particular bacteria. That was it. Given the incomplete nature of the evidence, the cautious, thoughtful Twort simply would not commit himself any further. "From these results it is difficult to draw definite conclusions," he wrote. "In any case, whatever explanation is accepted, the possibility of its being an ultra-microscopic virus has not been DEFINITELY disapproved because we do not know for certain the nature of such a virus." It was an entirely reasonable conclusion. He gave the unknown culprit no name other than "the Bacteriolytic

Agent," and concluded with a characteristic apology: "I regret that financial considerations have prevented my carrying these researches to a definite conclusion." He hoped that "others more fortunately situated" might follow his lead and continue similar investigations.

The article elicited little if any reaction. Twort's invitation to continue his investigations produced no takers. And soon thereafter, something happened that effectively dashed any meager hopes he still might have had of continuing his research: with World War I approaching, Twort, a captain in the Royal Army Medical Corps, found himself posted to the base laboratory at Salonika, in Greece. He was thus reduced by circumstances to nothing more than talking about his discovery, and while serving in Greece he gave a lecture to his fellow medical officers explaining the strange transformation he had observed. It may be that in the audience was a French medical officer named Félix d'Hérelle.

Reasonable is one term that could never be applied to d'Hérelle. Born in Montreal in 1873 to a French-Canadian father and a Dutch mother, he learned his medicine in France and Holland and during his life worked in Guatemala, Mexico, South America, Egypt, Algeria, Tunisia, Indochina, and Russia. For five years he was a professor at Yale, until he left for reasons that remain clouded. An associate during those years remembers that everywhere he went there were fireworks. So when in 1917, two years after Twort's publication and lecture, d'Hérelle published his own account of the phenomenon — in which he coined the term *bacteriophage,* claimed the discovery for himself, and failed to mention Twort — he was acting entirely in character.

But no one knows for sure if d'Hérelle actually did hear Twort speak. D'Hérelle was in Tunisia shortly before Twort's presentation, that much is certain; and it's possible that on the way back to Paris he stopped in Salonika. Even if he did hear Twort, however, the significance may be moot — d'Hérelle may have already begun his own independent study of the bacteriophage. Yet here, too, is confusion.

There is no confusion, however, regarding the impact of his 1917 article. Entitled "On an Invisible Microbe, an Antagonist of the Dysentery Bacillus," it described an unusual phenomenon that d'Hérelle noticed while investigating an outbreak of dysentery in Paris in 1916, after he returned to France from Tunisia, well after Twort's early work. While examining stool samples from stricken people, d'Hérelle found material that did strange things to the bacteria that caused the disease. When he passed the material through a Chamberland filter and then added the resulting fluid to test tubes filled with cloudy dysentery bacteria, the cultures suddenly became clear. And when he put a drop of the filtered fluid on a "lawn" of bacteria growing on an agar block, the bacteria simply disappeared, leaving a clear spot.

Many years later he recalled the incident in detail.

At this time we often got cases of bacillary dysentery in the hospital of the Institut Pasteur in Paris. I resolved to follow one of these patients through from the moment of admission to the end of convalescence, to see at what time the Principle causing the appearance of clear patches first appeared. This is what I did with the first case which was available.

The first day I isolated from the bloody stools a Shiga dysentery bacillus, but the spreading on agar of a broth culture, to which had been added a filtrate from the faeces of the same sick man, gave a normal growth.

The same experiment, repeated on the second and third days, was equally negative. The fourth day, as on the preceding days, I made an emulsion with a few drops of the still bloody stools, and filtered it through a Chamberland candle; to a broth culture of the dysentery bacillus isolated the first day, I added a drop of the filtrate; then I spread a drop of this mixture on agar. I placed the tube of broth culture and the agar plate in an incubator at $37°$. It was the end of the afternoon, in what was then the mortuary, where I had my laboratory.

The next morning, on opening the incubator, I experienced one of those rare moments of intense emotion which reward the research worker for all his pains: at the first glance I saw that the broth culture, which the night before had been very turbid, was perfectly clear: all the bacteria had vanished, they had dissolved away like sugar in water. As for the agar spread, it was devoid of all growth and what caused my emotion was that in a flash I had understood: what caused my clear spots was in fact an invisible microbe, a filtrable virus, but a virus parasitic on bacteria.

For d'Hérelle, unencumbered by the hesitations and qualifications that plagued Twort, there was no question as to what was happening. These clear patches and the abrupt clearing of the bacterial fluid could mean only that a virus was killing the bacteria. It was a "living germ." It was a bacteriophage.

But that was only part of it. Although the mere discovery of such fascinating creatures was significant enough, d'Hérelle went even further: far from simply being an evocative phenomenon of nature, perhaps these bacteriophages, these bacteria eaters, might actually function as our allies. Perhaps they might be the means of eliminating bacterial disease.

> Another thought came to me also: . . . the same thing has probably occurred during the night in the sick man, who yesterday was in a serious condition. In his intestine, as in my test-tube, the dysentery bacilli will have dissolved away under the action of their parasite. He should now be cured.
>
> I dashed to the hospital. In fact, during the night, his general condition had greatly improved and convalescence was beginning.

It was entirely likely, therefore, that the bacteriophage had something to do with his recovery. By attacking and killing disease-causing bacteria, it might aid the body's own disease-fighting capabilities. In fact, when d'Hérelle injected the virus into rabbits, he discovered that it immunized the animals against a later, otherwise lethal dose of the dysentery bacteria.

Further research even more strongly convinced d'Hérelle of the enormous potential of his discovery. He took case histories of thirty-four dysentery patients and found that the disease waxed when there were few bacteriophages present in their digestive tracts and waned when the virus became active there. Therefore, if you could cultivate large quantities of the virus and give it to people showing the first symptoms of dysentery, it should be possible to induce immunity to the disease. Moreover, that immunity should be as contagious as the disease itself, since bacteriophages spread from person to person as readily as bacteria. Bacteriophages, therefore, might be the key to preventing dysentery in vast populations afflicted with the disease.

And if this particular bacteriophage could provide immunity against this particular bacteria, it stood to reason that other types would attack other pathogenic bacteria. He had discovered, d'Hérelle announced, nothing less than the agent that might do away with all bacterial disease. He had found the long-sought magic bullet.

From that moment on, d'Hérelle's life changed. In contrast to the lack of response to Twort's paper, which in the space of just two years had been forgotten — if, indeed, it had ever been noticed in the first place — the reaction to d'Hérelle's announcement was strong and lasting. Suddenly bacteriologists in all parts of the world started looking for, and finding, bacteriophages. In the next few years, d'Hérelle and others discovered viruses that attacked the bacteria that caused many diseases — anthrax, bronchitis, diarrhea, scarlet fever, typhoid fever, typhus, cholera, diphtheria, gonorrhea, bubonic plague, osteomyelitis — and an avalanche of articles dealing with bacteriophage therapy turned up in the medical literature. D'Hérelle quickly became famous as the prophet of this exciting new possibility of combating disease, a process that many researchers termed the "d'Hérelle phenomenon."

The enthusiasm prompted by the discovery was not confined only to the scientific world. Antibiotics were still far in the future, and people were desperate for ways to cure often fatal bacterial diseases such as diphtheria, cholera, pneumonia, and tuberculosis. D'Hérelle's finding gave new hope. The promise of bacteriophages was so great that the good and gray *New York Times* was moved to gush: "The bacteriophage increases the range of the science of bacteriology considerably. Its discovery has explained many facts that until the present have remained obscure. . . . the life or death of a patient may depend on the outcome of this fierce battle waged by beings too small to be seen by the most powerful of microscopes."

But while few argued with d'Hérelle's suggestion that these bacterial agents might have important medical applications, his unequivocal insistence that the agents were viruses — that and his abrasive personality — raised the ire of many bacteriologists. For the first few years following his discovery, there was little his opponents could do — the ground swell of interest, fueled by the rapid discovery of many new bacteriophages, was simply too strong. Then, in 1921, new ammunition for his foes appeared. Researchers at the Pasteur Institute in Brussels not only found fault with the heart of d'Hérelle's 1917 announcement that the attacking agents were viruses, they also dropped an unexpected bombshell.

In the first place, said the leader of the group, Jules Bordet, the director of the Belgian branch of the institute and a world-famous Nobel Prize winner and expert in the field of immunology, the cause of the bacteria's death is not a virus at all but an enzyme that stimulates its own production. It was a familiar objection and by itself might have caused d'Hérelle to lose little sleep. But there was more:

> It is currently admitted that d'Hérelle has been the first to observe the lysis [destruction] which he attributes to a bacteriophage . . . but the burden of an exact history makes it necessary for us to cite a previous work which d'Hérelle has not known and that we ourselves have been ignorant of until now that contains the observations that d'Hérelle has made. This remarkable work by F. W. Twort appeared in Lancet in 1915, that is to say, two years before the research of d'Hérelle.

There it was. The Belgian researchers had resurrected the all but forgotten Frederick Twort. In this light, d'Hérelle's contribution was really little more than coining the word *bacteriophage*. Somewhat disingenuously, the Belgians wrote, "Without wanting to diminish the interest of the observations of d'Hérelle we believe that it is a duty to recognize the incontestable priority of Twort in the study of this question."

The gauntlet had been thrown. Twenty-two years earlier, when faced with a similar challenge to his precedence, Beijerinck had graciously acceded to Ivanovsky. But that was not d'Hérelle's style. Besides, his combative nature could not have been soothed by the fact that, armed with the Belgian claims, many bacteriologists were now referring to bacteriophage as the "Twort-d'Hérelle phenomenon." D'Hérelle decided to fight.

While admitting that Twort had indeed made a prior discovery, he immediately denied that it was identical to his own. Then, in a symposium attended by Bordet, as well as by Twort himself, d'Hérelle charged his opponents with neglecting the entire range of facts in the matter — a remark little suited to mending fences. He proceeded systematically to eliminate every possible explanation concerning the nature of the bacteriophage except his own, which "I have held since my first publication."

When it was Twort's turn to speak, he was characteristically cautious and, presumably gratified to have been rescued from obscurity, tried to walk the thin line between d'Hérelle on the one hand and Bordet and his colleagues on the other. He reiterated the views expressed in his 1915 article and stated that he could totally support neither side.

If his stance was intended to make peace, it failed. In the years to come both sides escalated their attacks, with d'Hérelle repeating his contention that what Twort had observed was different from bacteriophages. Soon after the meeting, moreover, d'Hérelle presented a very interesting new piece of information: contrary to what he had expressed earlier, his work with the dysentery epidemic in 1916 wasn't the first time he had encountered bacteriophages. He had actually begun to notice the phenomenon as far back as 1910, well before Twort's investigations, while studying, of all things, diarrhea of locusts. He had noticed in the diarrhea clear spots, which he concluded must be caused by some unknown, filterable agent. But, he noted, "I could not reproduce the phenomenon at will; I therefore could not study it."

D'Hérelle went on to explain that in early 1915, while in Tunisia during World War I, he again had the opportunity to investigate an invasion of locusts. In attempting to spread the diarrheal disease so as to slow the invasion, he once again observed clear spots and began to suspect that the culprit might be a virus. He then claimed that he returned to France in August of 1915, a year earlier than suggested in his official obituary, and began to investigate another outbreak of dysentery, the case upon which he based his 1917 paper.

Now having access to a laboratory, he could look into the possibilities as he would like.

But why didn't d'Hérelle mention the incidents earlier? Why wait until years after his initial article, when he had already been accused of claiming a discovery that was by rights not really his? Least charitable explanations suggest that it was because the events never happened, that in the heat of the attacks against him d'Hérelle simply fabricated the experience to give added chronological credence to his contention that the discovery of bacteriophages was his and his only. If so, however, it is certainly a convincing fabrication, and vividly expressed. And such an interpretation runs counter to the contemporary view of d'Hérelle — that although abrasive and self-serving, he was not dishonest. His scientific work was considered reliable, his technique and capabilities superior.

A more likely explanation is that unable to study his strange observations during the locust diarrhea epidemic, and, like everyone else, unaware of Twort's work, d'Hérelle saw no need to include them in his early articles concerning bacteriophages. It wasn't until Twort had been rediscovered and his own reputation attacked that d'Hérelle felt the need for added evidence. But no matter his motives, the effect turned out to be counter to that which he might have hoped for. Fueled by suspicion concerning the truth of his new evidence, the attacks on him simply increased and continued through the 1920s and into the 1930s, finally, however, focusing on the origin and nature of these destructive agents, rather than on who discovered them first. D'Hérelle unswervingly continued to assert that they were viruses; his opponents, that they were enzymes or other sorts of compositions. Thus protracted, the debate sometimes became ludicrous, as when one researcher, attempting to explain the origin of the destructive agents, whatever they might be, actually invoked the long-discredited doctrine of spontaneous generation. Pasteur must have been spinning in his grave.

But despite the enormous energy invested in the arguments, resolving the problem was as yet impossible, for, as Twort had said way back in 1915, "in the first place, we do not know for certain the nature of an ultra-microscopic virus." It would take until 1940 and the advent of the electron microscope before d'Hérelle's theory of the nature of the bacteriophages could be vindicated. As far as the dispute over who first discovered the phenomenon, however, no such resolution has been possible, to this day. Suffice it to say that, as in the case of Ivanovsky and Beijerinck, if Twort was indeed the first to lay eyes on the phenomenon of bacteriophages, he simply didn't know what he had. It was d'Hérelle who discerned their true nature. And not only do we owe our understanding of the bacteriophage to d'Hérelle, in a way we owe our understanding of viruses in general. For in a world in which Beijerinck's conclusions were anything but proved, d'Hérelle offered compelling evidence that viruses did indeed exist. By shouting so loudly and long about the bacteriophage, he put all viruses on the map. By the time he was through, there were few inside or outside of the scientific world who had not heard of viruses.

Though d'Hérelle was right that bacteriophages are viruses, the jury is still out as to their therapeutic value. He investigated the possibilities so thoroughly that of his three books on bacteriophages, two of them, *The Bacteriophage: Its Role in Immunity* and *The Bacteriophage and Its Clinical Applications*, are devoted to the disease-fighting possibilities of the virus. But despite d'Hérelle's efforts and those of many other researchers, treatment with bacteriophages never caught on in the mainstream of medicine. Although d'Hérelle and others reported success in treating dysentery and cholera with bacteriophages, and while the viruses easily obliterated bacteria in the lab, they often failed to perform nearly as well in ill people themselves.

Just why they failed, however, no one quite understood. It may have been that when bacteriophages were injected into people in large doses, the body's antibody response simply overwhelmed the tiny viruses, preventing them from completing their expected bacteria-destroying mission. When people took their bacteriophage dose orally, perhaps the digestive tract's gastric juices killed the viruses; such is often the case with a wide variety of viral invaders. And bacteria themselves were capable of developing immunity against their tiny predators, or at least resisting them successfully.

With the advent of penicillin in 1928 and the addition of other antibiotics soon thereafter, the question of why bacteriophages didn't fight disease as well as expected became academic. These new medicines provided astonishing control of bacterial diseases. Slowly the idea of bacteriophage treatment was all but abandoned.

Yet the story is not quite complete. In the last few years, researchers have again taken up the cudgels of bacteriophage therapy, and with promising results. Working with farm animals, microbiologists H. Williams Smith and Michael Huggins of the Houghton Poultry Research Station in Huntingdon, England, have demonstrated that bacteriophages can indeed prevent and treat such severe maladies as meningitis and enteritis. In Poland, Stefan Slopek and his colleagues at the Academy of Sciences have tested bacteriophages in humans to find that they can help control bacteria in blood infections.

If these preliminary results hold true in further tests, they may pave the way for an entirely new era of bacteriophage-therapy research, with results that could be noteworthy for a number of reasons. First, although antibiotics have provided a previously unimaginable measure of control over bacterial disease, bacteria have the unfortunate capability (unfortunate unless you're a bacterium, that is) of mutating to become resistant to antibiotics, thus forcing the development of more and stronger antibiotics. Second, antibiotics kill more than simply their intended bacterial targets, thereby ridding the body of other, *useful* bacteria — for example, bacteria that aid in digestion. Moreover, to be effective, antibiotics must be taken in a number of doses, carefully controlled over time. Sometimes people don't have the patience to follow through with the recommended sequence, or they otherwise fail to administer the doses effectively. The result can be a renewed attack of the initial disease, this time borne by bacteria more resistant to the antibiotics involved.

Bacteriophages, on the other hand, attack no bacteria other than those they're supposed to. Their ability to reproduce and multiply within bacteria, thereby increasing their concentration rather than becoming diluted, means that they need not be taken more than once. And, in what must be considered a delightful irony in light of the most familiar role of viruses, by spreading from person to person along with their target bacteria, bacteriophages can actually function as a contagious cure. The potential is most obvious when it comes to animals. As bacteriophages are eliminated from the intestinal tract along with other bodily wastes, farm and domestic animals readily reingest them, allowing the cycle of attack, reproduction, and elimination to occur again.

The promise is great. Of course, that is what d'Hérelle and many others thought, and look what that optimism came to. But it may be that more refined modern methods of dealing with viruses and the microbial world, together with our greater understanding of the creatures involved, will make a difference (after all, at the least we know for sure that bacteriophages are viruses — an assurance unavailable to d'Hérelle).

Refined modern methods came much too late for Félix d'Hérelle, however. He did live to see proof provided by the electron microscope that bacteriophages were viruses; but he has not proved correct in protesting to the last that "therapy with bacteriophage provides the specific therapy par excellence and, it might be said, the only possible natural specific therapeusis, for it is the exact experimental reproduction of the natural process of recovery."

As for Twort, he lived one year longer than d'Hérelle, but his voice was stilled long before. He returned to the Brown Institution after World War I and continued to work quietly — not on bacteriophages, however, but on the problem of growing viruses in laboratory culture media. As World War I interrupted his early research in bacteriophages, so he experienced even more frustration with this next effort. In 1936, one of his major funding sources withdrew support; and then Twort found himself facing the disruption of yet another war.

"I knew that the technique which was being evolved was correct," he wrote in 1949. "Striking results of a positive nature were obtained, but, unfortunately, in 1944 the laboratories of the Institution were destroyed by a bomb, and the University deprived me of my post and all facilities for completing the research."

Thus stripped of his scientific life, Twort died in 1950, alone and virtually anonymous. His last published words, written a year before his death, reflect the poignancy of his life: "Yet my research . . . offering, I am convinced, such high prospects of success, has, in its last stages, been foiled."

On the other hand, d'Hérelle went out screaming and kicking. Because he was nominally a British citizen, of Canadian birth, the Nazis incarcerated him during their occupation of France. He was freed at the end of the war, but his health was broken. In 1949, however, at the age of seventy-six, nearly four decades removed from his first encounter with bacteriophages, and only a few months from death, he could still defiantly reply to the question of whether bacteriophages were living beings:

> [Existing] physiological proof not only means that bacteriophages are living, it proves that they are completely alive. Many biologists have seen in filterable viruses in general, and in the bacteriophages in particular, intermediates between chemical substances and truly living beings. . . . [My] conclusion is that life is one, that no intermediates exist: all things endowed with life are fully alive, and all living things are alive in the same way.

Life is one; all things endowed with life are fully alive; all living things are alive in the same way. If he only could have known how prophetic those words would turn out to be, and how central a role his bacteriophages would play in discovering the very essence of life. D'Hérelle lived long enough, though, to witness the first, groping beginnings toward an understanding of the motivation at the heart of life, the gene. And the entrée into that previously forbidden world was none other than the bacteriophage.

In 1930, in the full flush of his enthusiasm for bacteriophages, d'Hérelle had written, "I believe that I did not overstate matters in writing . . . that the discovery of the bacteriophage was of such a nature as to cause within the biological sciences a revolution comparable to that which the discovery of the electron caused in the physical sciences."

He was right. Little could he know how right. How proud he would be were he able to see the result of his prophecy — proud, that is, if not furious that he died too early to take full credit for it.

II

. . . .

The Science of Virology

. . . .

Six

. . . .

A Glimpse of the Eternal Miracle

. . . .

So now viruses were on the map — but so what? What did that really mean? No one had ever seen one (unless you call noticing a tiny, almost imperceptible dot, as Buist and others had done, "seeing"), and no one knew what they were. Despite their surprisingly modern ideas on the subject, what Beijerinck and d'Hérelle "knew" about viruses was based on their insight and intuition. For d'Hérelle, anyway, that might have been enough. When asked if he looked forward to the day when he would actually be able to see viruses, d'Hérelle replied that such verification "would be only a detail. I have never felt the slightest need for it to believe in the existence of my agent." But such bravado wasn't necessarily shared by his fellows, most of whom would settle for nothing less than the positive detection of this unculturable, filterable mystery.

In fact, the whole business of passing through filters only deepened the enigma of viruses. Rather than being able to find something left behind in their filters that they could describe, measure, and analyze, microbiologists were forced to acknowledge the presence of viruses by the fact that they *didn't* show up. They were conspicuous for their absence, viable because they weren't there. It was a terribly uncomfortable situation for researchers. Here was a paradox more congenial to poetry than science — thus the frequent unscientific references, from Beijerinck on, to viruses as being like fire. Now you see it, now you don't — but look out, you may get burned.

And what if viruses weren't alive after all? They multiplied, that was true, but they didn't seem to eat, as they refused to grow on nutrient lab culture. They were too small to assume the form of cells, then considered to be the identifying structure of all life. It was tempting to think of them as suspended between the realms of dust and mite, rock and animal, neither alive nor dead. Yet such speculation carried little weight with physicians, who, trained in the germ theory of disease, contended that agents that caused disease in living things must themselves be alive. The result of such confusion was often simply more confusion, as microbiologists (few were yet brave enough to call themselves "virologists") and physicians were obliged to deal with something about which they knew next to nothing.

Nevertheless, the results were occasionally stunning. As Pasteur did with rabies, Rockefeller Institute microbiologist Max Theiler in the mid-1930s developed two yellow-fever vaccines by taming a virus he knew relatively little about. But despite such isolated triumphs, the fledgling science of virology was stalled. Without their sometime, somehow being able to see, cultivate, and identify the object of their study as bacteriologists could with bacteria, the prospect of microbiologists understanding viruses was limited, indeed.

The time was ripe for a fresh, new approach to the problem, for someone who was not obliged to proceed solely from a biological point of view but could apply different criteria and different methods. The time was ripe for a nonbiologist, a nonphysician, a person who could see a virus as a substance, a composite of chemicals that follow certain laws of nature. Enter a chemist. Enter Wendell Stanley.

In a way, it's only appropriate that the next major advance in dealing with viruses should have come from a chemist, for Pasteur, who played such a large part in establishing the science of microbiology, was a chemist. And it's appropriate that the means should have involved crystals, for Pasteur began his career studying the ways in which molecules form into crystals. But for most people involved with viruses, it seemed inappropriate indeed, because it was common knowledge that crystals were associated only with nonliving matter. Ice, sugar, salt, snowflakes, rocks, jewels — these, not living, reproducing organisms, were the stuff of crystals. Crystals involved molecules attaching themselves to one another's surfaces to extend indefinitely in angular, geometric, repeating patterns determined by the structure of the molecule itself — Pasteur had shown that. Crystals do what they have to do, in other words. As chemicals, they form according to the physical forces inherent in their particular makeup. They "grow" differently than do living organisms, which, while themselves made up of chemicals that function according to inherent forces, are far more complex. Once you understand the pattern of the ice crystals latticing your windowpane, for example, you can predict the pattern of its further growth: it will be identical. No such predictions are likely with living cellular organisms, at least not nearly to the same degree.

Wendell Stanley, though, uninhibited by the traditional biological point of view, decided to try dealing with viruses as he knew best. If he could crystallize them, thereby fixing them in time as a crystal of quartz is fixed in time, he could then have a look at them. The premise was not as outlandish as it may have seemed, for Stanley had good reason to suspect that viruses were crystallizable. He thought it likely that they were made entirely of protein, and it had been demonstrated more than once that protein could be crystallized.

Who has not heard the statement that "proteins are the building blocks of life"? It's true: they are the molecular "bricks" that nature uses to construct life-forms. It has been estimated that there are over a hundred thousand different kinds of proteins in the human body. Hair, skin, bones, muscles, tendons, ligaments, nerves, tissues — all these are primarily protein. So are enzymes and hormones, antibodies and blood. A large part of each cell is made up of protein. Surely viruses, too small to be cellular, were comprised of protein. (D'Hérelle had intimated as much, noting that life is not bound to the cell "but derives from another phenomenon, which cannot reside else-where than in the physico-chemical constitution of a protein parti-cle.") Accordingly, in 1932, at the age of twenty-eight, Stanley began to try to crystallize a virus. The one he picked was an old friend, Beijerinck's tobacco mosaic virus, by this time commonly known in scientific circles as TMV.

It was the ideal choice. More was known about TMV than any other plant virus (more than any animal virus as well, with the pos-sible exception of vaccinia). Highly stable, almost indestructible, TMV is so infectious that just rubbing it onto a leaf of tobacco will produce the characteristic mottling within a day or two. You can dilute a TMV solution a million times, a *billion* times — it'll still infect. And the number of spots of infection on a leaf corresponds to the number of viruses in the solution. An investigator can count the spots and estimate the concentration of viruses.

Even if TMV itself weren't so accommodating, however, Stanley would've chosen a plant virus, for plant viruses are much easier to work with than viruses that infect animals. Remember that viruses, unlike bacteria, won't grow on laboratory culture media. For them, nothing will do but to infect living cells. In the case of animal vi-ruses, that means an animal, and working with animals presents many problems. Besides humane considerations, and the necessity of feeding and caring for it, what if you can't infect the animal — does that mean it is immune or that your virus solution is too weak? What if the animal gets sick from another disease along the way, ruining your experiment? What if it inadvertently dies partway through? What if it bites? Plants present fewer problems, and al-though some plants are immune to the diseases of another, those that aren't won't acquire resistance along the way. Once susceptible, always susceptible. Unlike animals, plants don't develop antibodies.

So Stanley, working at the Rockefeller Institute's plant pathology laboratory in Princeton, New Jersey, decided to go ahead with to-bacco mosaic virus. Cooperative plant virus or no, the work ahead of him was daunting. It would take him three years to reach his goal. He had to become as much farmer as chemist, raising literally a ton of Turkish tobacco on the institute's farm in the summer and in its greenhouse in the winter. Lucky for him that he was hardy and tough. Born in Indiana, in the midst of farm country, prematurely bald, thick-necked and solidly built, Stanley had played football in college and had intended to become a coach rather than a chemist. Now he was cultivating tobacco rather than future All-Americans, toward the goal of producing a TMV solution purer than had ever been accomplished before — so pure that it would form a crystal.

Stanley began by infecting his plants with TMV, using a piece of gauze to rub a solution containing the virus onto the leaves of the tobacco plants. Soon they became mottled and blistered, the telltale signs of infection. After three weeks, he harvested the now heavily infected plants and stored them in a deep freeze at about ten degrees Fahrenheit. The cold ruptured the cells of the plants, thereby beginning the process of recovering virus.

Stanley then ran the frozen plants through a power meat-grinder, catching the pulp in five-gallon dairy pails. He stirred in hot water, added a chemical to help liquefy the mash, and dumped the mess into large gauze bags, which he then hung up to drain. It was the "tobacco juice" (as Stanley called it) filtering through the gauze that he was after. He even squeezed the bags in a fruit press to extract more liquid.

Now, awash in tobacco juice, Stanley could begin the tedious process of ridding it of everything it contained except virus. He decanted the liquid from one container to another, filtered, it, precipitated it, evaporated it. He added chemicals to the brew to help separate out the enormous mix of proteins that constituted the juice; the only protein he wanted left was virus protein. He poured the juice through a succession of filters. It took twenty-four hours to filter one container of juice and that much time again to determine whether the virus had passed through or stayed behind. He rubbed the filtered solution onto new tobacco plants. If the leaves became mottled, the juice still contained virus; if they remained clear, the virus had been trapped in the filters. Soon the blistering showed up, proof that the virus was still in the filtered liquid. Time to filter again.

Gradually, Stanley removed the miscellaneous ingredients from the juice. The rich brown brew became yellow, then pale yellow, and finally lost color completely. Ultimately, when Stanley rubbed the clear solution on tobacco leaves, the leaves remained smooth. Using filters far finer than the early Chamberland candles, he had actually trapped the virus behind. There were virtually no proteins of any kind left in the juice. It was pure.

Now came the last, crucial step. Stanley washed the virus from the filters and returned it to the clear, purified liquid, and, in a highly sophisticated version of a high-school chemistry student's method of making crystals by adding salt or sugar to a solution and letting it precipitate out, he stirred in chemicals that would separate the virus from what was left of his tobacco juice. If he had done the preceding steps correctly, and if, as he suspected, the virus was actually a protein, the result would be TMV crystals.

Sure enough, the liquid soon sported a glistening sheen, denoting the presence of crystals of pure TMV. Under the microscope, they looked like needles, millions upon millions of viruses adhering together to form miniature stilettos about 3/25,000 of an inch long. To test if they were truly TMV, Stanley redissolved some of the crystals in water and rubbed the solution onto tobacco leaves. This time, tobacco mosaic spots appeared more quickly than ever. The crystals were indeed TMV, pure TMV, about a hundred times more infective than the tobacco juice he had started with. Stanley had done it. He had crystallized a virus.

Or so he thought. Other people weren't quite so sure. As with so many other important advances, much of the scientific reaction ranged from noncommittal to downright skeptical. Stanley's immediate superior and the head of the Rockefeller Institute's plant pathology laboratory at Princeton, Dr. Louis O. Kunkel, wrote the institute's board of trustees that Stanley's report "that the material is a protein possessing the properties of tobacco mosaic virus has aroused great interest, because the work indicates that this virus may be a chemical compound. . . . It is difficult to prove either that the . . . protein is the virus or that it is not the virus. Dr. Stanley is engaged in accumulating further evidence bearing on this important point."

Stanley might have expected as much, for he knew that Kunkel was, like most people, convinced that viruses were nothing more than especially tiny living organisms. But other associates were even more skeptical, to the extent of warning Stanley that he'd be entirely on his own if he published his study. Publish his study he did, however — only to encounter more of the same.

Some physicians, less than taken with plant virus work that at first glance seemed to have little to do with problems of flesh-and-blood human patients, belittled the significance of Stanley's efforts. Some contended that viruses that infected plants were a lower form of life and thereby different from those in higher animals. Perhaps Stanley had only succeeded in isolating some kind of chemical poison, suggested others. And others charged that his crystals may have been contaminated with an agent of the mosaic disease but were not the actual virus.

But if the scientific community was slow to acknowledge the significance of Stanley's work, it was different in the outside world. "Crystals Isolated at Princeton Believed Unseen Disease Virus," trumpeted the headline on the front page of the *New York Times*. "This . . . marks the first scent on the trail of one of the 'big game hunts' of science, and may mean that the road has at least been opened for the similar isolation of the deadly viruses that attack men." Indeed. In the eyes of the general public, Stanley had accomplished the impossible. Now, finally, viruses were real, visible, *there*.

But one mystery intensified: was the virus alive or dead? Live things, cellular organisms, couldn't be crystallized, yet here were crystals of viruses. That suggested that viruses weren't living. Still, as an infectious agent, a substance that could reproduce, a virus must have the capacity for life on some level.

"Enough is known about matter, organized and unorganized, to assure us that there may be things 'twixt heaven and earth which are not so alive as an eel or so dead as a rock," commented the *New York Times*.

In the light of Dr. Stanley's discovery the old distinction between death and life loses something of its validity. In the last analysis, all matter is composed of atoms. Somewhere between the simplest atom and the simplest cell life begins. Biologists have long suspected that the filtrable viruses may belong to this intermediate group — that in them we see life in the making. Perhaps Dr. Stanley has caught a glimpse of the eternal miracle. Note that his crystals are proteins — the stuff of which protoplasm, beefsteak and white of egg is made. Note, too, that living organisms are largely composed of proteins. Life is to one of Dr. Stanley's protein crystals like a licking flame. The crystal is ignited and itself bursts into life.

Again, it seemed that the only way to describe a thing so evocative as a virus was through the metaphor of fire, the language of poetry.

For Stanley himself, however, the question had been answered once and for all. Viruses were proteins, and, while living things were comprised of proteins, proteins themselves were not alive. "The evidence presented . . . , " he wrote, "indicates that the virus of tobacco mosaic, the first to be described in the literature, is not alive. . . . The fact that this virus seems to be a protein may necessitate a fundamental change in the conception of the nature of other viruses." And he continued to work to establish his case. He published further articles on the subject, made a lecture tour of the United States and Great Britain, and, with others, discovered that many other plant viruses could be similarly crystallized. It seemed more likely than ever that all viruses were indeed proteins and that all could be crystallized.

Then, in 1937, two years after Stanley announced his feat, came a discovery suggesting that some of the skepticism that greeted Stanley's work had been justified. Not that it wasn't the tobacco mosaic virus that he had crystallized — it was indeed. And not that the virus wasn't made of protein — it was. But not completely. Two English biochemists, Frederick C. Bawden and Norman W. Pirie, confirmed Stanley's accomplishment by themselves crystallizing TMV; but then, going on to analyze the crystals more carefully than he had, they found something that Stanley had not: TMV wasn't made up entirely of protein. There was a smidgeon of acid there, of the type known as ribonucleic acid, or RNA. As its name implied, it was one of the *nucleic* acids, a family especially abundant in the nucleus of cells.

Nucleic acids were nothing new. They had been discovered as long ago as 1869 in the cell debris on bandages that had covered wounds. Later on, they had been isolated in yeasts and other plant and animal cells. No one knew their role in the nucleus of cells — only that they consistently showed up there. It was known, however, that there were two types: ribonucleic acid, or RNA, and deoxyribonucleic acid, DNA. Structurally, they differed only in the kind of sugar molecule — ribose or deoxyribose — that helped to make up the whole.

The amount of RNA in the tobacco mosaic virus wasn't much, really — only about 6 percent of the entire virus (Stanley himself later confirmed that fact). Protein made up the remaining 94 percent. But it was clear that the RNA did belong — it wasn't the result of contamination from elsewhere. At the same time, another discovery, that bacteriophages contained a comparatively large portion of DNA, reinforced the view that nucleic acids were an integral part of the makeup of viruses. In TMV, with so little nucleic acid compared to the overwhelming amount of protein, Stanley had just plain missed it.

It didn't seem an important error. Stanley duly modified his view to include the presence of nucleic acid, renamed his TMV protein molecule a *nucleo*protein, and proceeded with his research. In 1946, Stanley was awarded the Nobel Prize in chemistry for "preparation of virus protein in a pure form." Two years later, he established the Laboratory for Viral Research at the University of California, Berkeley. When he died in 1971, he was as admired for his scientific leadership as he was for his pioneering effort in making the invisible visible and the unimaginable a fact of human understanding.

As it has turned out, however, not noticing the presence of RNA in tobacco mosaic virus was an important omission, indeed. For while the preponderance of protein may be the most obvious distinguishing characteristic of the virus, it is that tiny portion of nucleic acid that affords it its essential life-force. It wouldn't be too many years after Stanley crystallized the virus that the previously innocuous letters DNA and RNA would become almost as well known to the world as the term *virus* itself. It may be that the most famous scientific accomplishment of our century was the 1953 deciphering of the molecular structure of DNA by James Watson and Francis Crick.

The age of molecular biology gained enduring momentum with Stanley and his crystallization of TMV, as did the science of virology. Because of Stanley, viruses became for ever more chemical substances amenable to chemical investigation. No longer would it be possible to consider viruses ultrasmall bacteria that happened to pass through the finest filters. No longer would it be possible to refer to them as some other, as yet unknown kind of miniature life-form, distinguished primarily by their size, or lack of it. No, viruses were something entirely different from what had often been surmised. If they weren't exactly living molecules, as Beijerinck had suggested, they were molecules that could infect like living creatures but that crystallized like inert chemicals. Stanley threw open the doors of the viral world, inviting in chemists, crystallographers, and even physicists.

While Wendell Stanley was learning what viruses were made of, others were dealing with two other fundamental necessities: being able to work with viruses in the laboratory and actually seeing the mysterious creatures in their natural, uncrystallized form. The latter was on the verge of being accomplished, for in the early 1930s German physicists developed the electron microscope. It operated on the same principle as the ordinary light microscope, but with one major difference: instead of illuminating the subject to be magnified with light waves, the electron microscope used electrons, the small, negative units of electricity. The advantage was that the wavelengths of electrons were much smaller than those of light. Objects such as viruses that were tinier than the length of light waves didn't show up under ordinary microscopes, but they could not hide from the electron microscope. The electron wavelength was so small — about five thousandths of a millimicron — that it became theoretically possible to spy on atoms themselves (although technical problems made realization of that prospect impossible). Rather than focusing with lenses, electron microscopes used magnets, which readily directed a beam of electrons in the right direction. The beam bounced off the subject onto photographic film, thereby giving a record of the form it had just encountered.

The invention liberated the science of microscopy, and in 1939

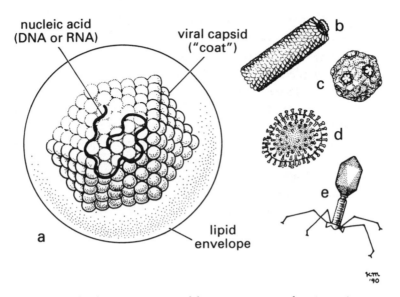

The standard components and basic structures of various viruses

Most viruses are quite simple in structure, consisting of DNA or RNA surrounded by a coat of protein. Many have an outer fatty envelope torn from the walls of their host cells as well (figure a). Such viruses are called "enveloped." Figures b through e show variations on the theme: b, tobacco mosaic virus, with a cylindrical coat; c, rhinovirus, an unenveloped type with icosahedral symmetry; d, flu virus, which has a fatty envelope studded with proteins; and e, T-4 bacteriophage, an unenveloped virus with a complex structure.

three German scientists, G. A. Kausche, E. Pfankuch, and Helmut Ruska, took an electron micrograph of — what else? — the tobacco mosaic virus. Finally, *finally,* scientists could see their old friend face to face. As it turned out, there wasn't much to see — no more than a shadowy little piece of straw — but it might as well have been a beautiful rose. And the next year, Ruska took a picture of a bacteriophage, which looked for all the world like a miniature tadpole, bunched with other bacteriophages as though cavorting in a shallow stream. ("*Mein Gott!* They've got tails" was the startled reaction of one German-born virologist.)

When in 1944 physicist Robley Williams and chemist Ralph

Wyckoff of the University of Michigan perfected the technique of plating objects with vaporized metal — gold, uranium, platinum, tungsten — the micrographs suddenly gained depth and clarity. Now TMV revealed itself to look almost precisely like a cigarette, real enough to smoke. And Jenner's cowpox virus, the one that started it all, appeared much like an ashed-over charcoal briquette, down to the mottled texture of the ash.

Though they were finally disclosing their form, however, viruses remained difficult to handle in the lab. While bacterial vaccines proliferated, microbiologists had been unable to develop many new virus vaccines because they simply couldn't figure out a manageable way to raise viruses. To develop a vaccine, not to mention simply study a virus, researchers must have access to huge amounts of the elusive creature. That being so, how can they know when they have enough? By the virus's effect on the environment in which it grows. With TMV, that meant mottled and blistered tobacco leaves. With the bacteriophage, it meant the rapid clearing of test tubes that initially were full of cloudy bacterial solutions. So far, so good; specific cause leads to specific effects in an easily manageable environment. But with an animal virus, the precise equation blurs. Here, it's necessary to gauge the effect of the virus on the animal in question, a task that sounds much easier than it is to accomplish.

Jenner could see evidence of cowpox virus in the sores on udders of cows and on the hands and arms of people also infected. But such gross ways of determining the presence of virus were simply unsatisfactory when it came to a more detailed study. And since the virus was only available when a cowpox epidemic raged through the countryside, Jenner had no consistent access to the microbe. Pasteur overcame that problem by growing his rabies virus in rabbits, an approach that, while ingenious, still necessitated keeping scores of the animals. And the only way he could be certain he had the virus was if it killed the rabbits.

Later researchers working with other viruses tried smaller, less troublesome animals — mice, for example. In the mid-1930s, Max Theiler developed his yellow-fever vaccine by injecting the virus into mouse brains. Influenza, and even polio, it was later found, would grow in mice. But while much easier to work with than larger animals, and much less expensive, mice still posed the familiar problems associated with animal research: the threat of random disease, the possibility of the animal acquiring immunity to the virus in question, and not entirely reliable results. Even if all these difficulties could be surmounted, you still faced the frustration that vaccines grown in animal tissue can cause dangerous allergic reactions when injected into humans. Clearly, growing viruses in animals simply wouldn't do.

As early as 1913 someone had tried to find a different approach to the problem. A Columbia University researcher, Edna Steinhardt, placed tiny bits of cornea from the eyes of guinea pigs into a drop of blood plasma on a glass slide, and in that solution she was able to keep vaccinia viruses alive for several weeks. In effect, Steinhardt substituted a few living animal cells for the whole beast. In the years following that early effort, researchers grew vaccinia viruses in such various environments as rabbit testicle tissue and minced chicken-kidney tissue. But in every case, the promise was thwarted; for sooner or later, the solutions of animal tissue would show evidence of another, unwelcome kind of growth: bacterial contamination. Ultimately, the slide or test tube of precariously persisting viruses would be overrun by invading colonies of bacteria. And that was the end of that. Without an environment that could remain sterile, prospects would be bleak, indeed.

The first to make substantial inroads into the problem was a Vanderbilt University School of Medicine researcher, Ernest Goodpasture (what a wonderful name for someone trying to find a congenial grazing ground for viruses!). With the help of his research fellow, Alice Woodruff, he succeeded in growing a number of different viruses in chicken eggs.

It was an approach that made sense, for an egg is a naturally sterile environment — nature's self-contained laboratory — and the chicken embryo growing within offered the virus a living organism in which to multiply. As any cook knows, eggs are easy to handle and convenient to store — all that in contrast to living animals. In 1933, Goodpasture grew in chick embryos a vaccine against smallpox that worked as well as that grown the traditional way, in calves. Although the vaccine never caught on, the point was clear: eggs could be used to make vaccines. And in the years to come, Good-pasture grew in eggs the viruses that cause rabies, mumps, meningitis, and shingles.

By the end of the 1930s, then, the world of virology had seen three immense hurdles at least partially overcome. Thanks to Good-pasture, certain viruses could now be grown in the laboratory; thanks to the electron microscope, what they looked like was a mystery no longer; and thanks to Stanley and the researchers that followed him, viruses had been shown to be made primarily of inert protein rather than organic cells. But there was much that virologists didn't yet know. Although they could now see viruses, and so had a good idea of their shape and size, they didn't know just how the various structural components were arranged in a virus, nor what glued them together. They suspected that although TMV and its relatively simple cousins were composed only of protein and nucleic acid, larger, more complicated viruses such as vaccinia, say, most likely contained other ingredients, but no one knew for sure. And although it was clear that viruses invaded living cells, reproduced there, and then somehow made their way back into the larger environment, no one knew just how. No one knew how viruses got into cells, what happened once they were inside, or how they made their way out.

And despite the growing ability of virologists to produce vaccines against viral diseases, they simply hadn't come up with any that prevented those afflictions that seemed to matter most to the general public. Theiler's yellow-fever vaccine was a notable exception, as were University of Michigan epidemiologist Thomas Francis's early influenza vaccines — both did valiant duty during World War II — but mumps, chickenpox, measles, certain strains of flu, and hepatitis were still around. Clearly, much work needed to be done.

Beyond that, most of these problems paled in the general imagination next to yet another viral disease, although it was a less widespread one than many others. By the middle of the century, it was perhaps the most terrifying viral disease of them all. It came to symbolize the helplessness many people felt in the face of the power of viruses, as well as the capriciousness of life and its dangers. And this disease was all the more poignant because it primarily struck children. It was, of course, polio.

Seven

· · · · · · · · · · · · · · ·

Catching Lightning in a
Test Tube

· · · · · · · · · · · · · · ·

On August 9, 1921, the dashing, thirty-nine-year-old former assistant secretary of the navy Franklin Delano Roosevelt fell into the water while sailing near his summer home on Campobello Island, off New Brunswick, Canada. Roosevelt became chilled. The next day, feeling tired, he took a swim to refresh himself. But the swimming didn't refresh. And by the next morning, Roosevelt recalled, "my left leg lagged. . . . Presently it refused to work, and then the other." Three days later, he couldn't move his legs at all, much less stand up. His back, arms, and hands were partially paralyzed. He couldn't even hold a pen to write.

The future president had contracted polio, almost certainly in the week or two before he fell into the water. While eventually he regained the use of his hands, arms, and back, he would never again move his legs. In the years to come, the world became familiar with the sight of Roosevelt, seated, conferring with dignitaries who, out of respect for the president, also remained in their chairs, or happily touring in an automobile or jeep, cigarette and holder thrust out at a jaunty angle, as though mocking the fate that had dealt him such a blow. He made a point of delivering speeches on his feet, supported by steel leg braces, and rarely made public mention of his infirmity.

In 1945, the last year of his life, thin, weak, suffering from a succession of illnesses, Roosevelt made an exception to that rule. While reporting to Congress on the Yalta meeting with Churchill and Stalin, he took his place behind the microphones sitting down. "I hope that you will pardon me for this unusual posture of sitting down . . . ," he apologized, "but . . . it makes it a lot easier for me not to have to carry about ten pounds of steel around at the bottom of my legs."

In a way, Roosevelt had been lucky — at least he didn't die from the disease, as did as many as one in every four people afflicted during those years. But he had been unlucky as well, for polio never has attacked a high proportion of the population, and those it does, overwhelmingly, are children.

The disease, at least in epidemic proportions, seems to be of fairly recent origin. The first written record did not occur until 1835, in the village of Worksop, England, which suffered "four remarkable cases of suddenly induced paralysis, occurring in children." But there's some evidence that polio was around as far back as antiquity, for an Egyptian monument of about 1500 BC depicts a priest with one leg withered as though from polio, and mummies have been found with one leg shorter than the other.

After the first recorded cases of polio in England, the nineteenth century witnessed other outbreaks, and by midcentury physicians had associated the paralysis caused by the disease with damage to the spinal cord. As children were stricken most frequently, the disease came to be called "spinal infantile paralysis." Another name, "poliomyelitis" — inflammation (*itis*) of the gray (*polios*) marrow of the spinal cord (*myelos*) — also came into use. The inflammation seemed to destroy nerve cells, thereby paralyzing the muscles they controlled.

The disease first presents itself with fever, headache, sore throat, vomiting. These symptoms may last only twenty-four hours or so and then disappear. In that case it may have been a mild form of polio — estimates are that for every recognized case, there are a hundred of these mild attacks — but as the signs resemble those of so many other diseases, no one knows for sure. When the symptoms don't abate, however, the nature of the illness soon becomes clear. The back and neck become stiff, the muscles weak, movement difficult. Soon the legs begin to hurt, especially when stretched or straightened, and muscles become sensitive to the touch. Within a few days, the result can be paralysis or even death. But in the early days, no one understood the cause of the affliction or its agent.

Around the turn of the twentieth century, the incidence of polio started to rise — for what reason, no one knew. Nor did anyone know much more about the cause of the disease than had been discerned years earlier. Physicians suspected that polio was an infectious disease, but the agent, whatever it was, couldn't be found under the microscope. Nevertheless, it multiplied in living cells and could be passed from victim to victim, infecting each along the way. And, as a final test, it did not show up as residue when infectious matter was passed through the finest Chamberland filters. Sound familiar? It was determined that it must be one of the so-called filterable viruses — whatever that meant.

Soon something else became clear: laboratory monkeys that survived polio — monkeys were the only animals other than humans that seemed susceptible to the disease — became immune thereafter. Moreover, the blood of these recovered monkeys contained a substance that, when mixed with the virus in a test tube, neutralized it, rendering it incapable of causing an infection. Like rabies, then, perhaps polio produced the means for its own prevention.

All such hopeful speculation seemed futile, however, when the incidence of the disease in the United States reached a peak in the summer of 1916: 27,363 cases nationally, 7,179 deaths, with the most concentrated outbreak in New York City. It was as though the medieval plague had returned. Thousands of New Yorkers fled the city, taking their infections with them. Others remained, but transformed their homes into bastions, locking doors, shutting windows, drawing blinds. But nothing seemed to work. Soon rural communities began aiming guns at the refugees from New York, redirecting them back to the city to spread their polio in their own neighborhoods. And when the city's health commissioner compelled recalcitrant hospitals to set aside beds for polio care, police sometimes had to use force to invade homes and carry sick children away from their frightened families.

Quack remedies abounded. Frog blood; radium water; wine of pepsin; a drink made of rum, brandy, champagne, and mustard plasters (which presumably either nauseated or inebriated patients to the extent that they didn't care if they caught the disease) — all these were offered as cures. A former state legislator claimed that cedar shavings scattered about a house would protect children and sold bags of them at a profit. He was fined $250 and sentenced to thirty days in jail. When fresh ox blood was touted as a cure, people showed up at East Side slaughterhouses with buckets ready. Nothing worked — nothing, that is, except the coming of fall. The disease seemed to be a seasonal affliction, for by October New York City was back to normal.

But not really. Neither New York nor any other American city would be normal for another forty years. The nightmare of polio had seeped deeply into the consciousness of the country. For American families, the heat of summer now brought with it the terrifying realization that your child might be the next to suffer the headaches, fever, and muscle pains that led to paralysis or death.

Such was the prevailing climate in 1948, when a thirty-four-year-old University of Pittsburgh virologist named Jonas Salk entered the campaign against polio. Over the years, millions of dollars, primarily raised by Roosevelt's National Foundation for Infantile Paralysis and its March of Dimes, had been poured into the endeavor. But despite the extravagant funding and the efforts of such other researchers as Russian-born Albert Sabin of Children's Hospital in Cincinnati, the prospect of winning the battle was still remote. There was one basic, seemingly insoluble problem: poliovirus simply refused to grow in the laboratory other than in nerve tissue, and nerve tissue was useless for producing a vaccine, because injecting into the human body a solution containing foreign nerve tissue could cause an allergic inflammation of the brain, a condition that can sometimes be fatal. Without a reliable, manageable, renewable supply of virus to work with, neither Salk nor Sabin nor anyone else would be able to come up with a vaccine.

The solution came from an unanticipated source. While respected in the scientific community, he was unknown to the general public and had had nothing to do with the foundation's highly publicized fight against polio. He was a man who might have been of another century, self-effacing to the point that, when asked how he came to make his great discovery, he answered, "I've wondered about that a good many times."

Despite the importance of his work — without which there would be no modern virology or molecular biology, not to mention a polio vaccine — and his status as a Nobel Prize laureate, he remains to this day virtually unknown to the general public. But he is a legend among virologists. His name is John Enders.

In 1948, Enders was fifty-one years old, an associate professor of bacteriology and immunology at Harvard Medical School. Through the years he had developed a style all his own. Large-faced, with a square jaw, prominent nose, and receding brown hair, Enders habitually clenched a pipe between his teeth; the corners of his mouth curled in an ironic grin. Bow ties, tweed suits, and a buttoned vest were his mode of dress, with a gold watch and heavy chain conspicuous across his stomach.

The effect was an appearance of calm and refinement, qualities that defined his approach to work as well as dress. Perhaps it came with his breeding and freedom from material concerns (he had been born into a family that included a grandfather who was president of the Aetna Life Insurance Company and a father who was president and board chairman of the Hartford National Bank), perhaps it was plain, old Yankee stubbornness, but Enders simply refused to be rushed into directions he had no inclination toward following. His very demeanor was a hedge against such unwelcome intrusions. Commented an associate:

> The superficial Enders guise was that of a round-shouldered, overburdened, sometimes meek, pedantic scientist — a caricature he deliberately fostered and exploited effectively as a shield against the

unwelcome intrusion of assignments to distracting committee or administrative chores. In fact, he was a strong, competitive, thoroughly contemporary, artful academician who conserved his energies for those challenges he judged worthy.

One of those challenges was to cultivate not only viruses but a feeling of warmth and mutual concern in his lab. Enders made a point of discussing work, politics, literature, and the arts with his younger lab associates, conversations that might continue for hours. His wit was wry and gentle, his patience was great, and his modesty, about his own accomplishments and his disinclination to suffer other people's boasting about theirs, was absolute. "Every day, Enders made the rounds of the benches to talk with each fellow and ask, 'What's new?'" recalled a coworker. "Those two words were a wonderful stimulus to productivity, because a fresh answer earned one extra time with the Chief."

It was thus the quality of the man that attracted people to him — certainly the lab itself, an old building in Boston's Back Bay, wasn't much. It consisted of only two rooms, plus a small office for Enders, and a storage area. Research animals were housed more than a block away. All the lab workers pitched in to wash glassware, prepare culture media, and check incubators on weekends. In 1948 two of those people were Thomas Weller and Frederick Robbins.

Weller had worked with Enders before. In 1940, as a strong-minded, twenty-five-year-old Harvard medical student, he had assisted Enders in one of his early investigations into growing viruses in tissue culture by helping to keep vaccinia viruses alive in chick tissue for months at a time. Weller's roommate was Robbins, a year younger, also a medical student, and not quite sure where he was heading. "Enders took me in because of my relationship with Tom," recalled Robbins. "I was brainwashed by Tom. So I was prepared to be impressed."

What he was not prepared for — nor were any of the trio — was the magnitude of the discovery they were about to make. By chance their research was supported by the National Foundation for Infantile Paralysis, which had awarded $200,000 to the Harvard Department of Bacteriology for a five-year study of viruses. The grant did not specify that the viruses in question be polioviruses, which was lucky, because neither Enders nor his two associates were particularly interested in working with polio. In fact, Robbins had come aboard stating that he'd be interested in working on any virus *except* polio. "It was the bandwagon," he explained, "and I wasn't much of a bandwagon type. I'm afraid none of us was." Robbins wanted to investigate infant epidemic diarrhea (a disease that occasionally raised havoc in hospital nurseries), whose cause he suspected to be a virus. To do so he began to work on techniques for maintaining tissue from the intestines, the seat of the disease, in laboratory culture. At the same time, Weller was investigating the virus of chickenpox and, with Enders, had for some time been looking into the possibility of growing mumps virus in tissue culture.

In early March of 1948, they succeeded. Weller grew mumps virus in a soup made of ox blood with bits of chick embryo membrane suspended in it. Rather than transplanting the viruses (aboard a small piece of infected membrane) from one flask of membrane-tissue soup to another every three or four days as most researchers would have done, Weller changed the nutrient ox-blood soup surrounding the tissue every few days but kept the original tissue growing for as long as forty days. He avoided an overgrowth of bacteria by exemplary technique and hygiene, while allowing the slow-growing mumps virus to flourish undisturbed. And flourish it did, suggesting to virologists that slow, steady incubation was the way to cultivate certain viruses in the lab.

With mumps virus growing well, and with Robbins busy cultivating intestinal tissue from mice in hopes of eventually growing the sought-after virus that caused infant diarrhea, Weller turned to chickenpox virus. To flasks of nutrient soup he added tissues of skin and muscle taken from babies stillborn or dead after being delivered prematurely at Boston Lying-In Hospital across the street; then he inoculated the brew with chickenpox throat washings taken from children. By chance, he found four flasks full of tissue culture left over. Instead of throwing them away, Weller decided to try something. For some time Enders had been storing a tube of poliovirus in his deep freeze. None of the researchers was particularly interested in polio research, but since the virus was at hand, and since the lab had received a National Foundation grant, and since they did have these four extra flasks, what had they to lose? Why not try it? Weller dabbed a little poliovirus into his leftover flasks containing human skin and muscle culture.

Learning of the experiment, Enders became intrigued. Turning to Robbins, he said, "I've been wondering about poliovirus growing in the intestine. Why don't you try it?"

From such offhand impulses come major breakthroughs. Remember that poliovirus in the lab would grow nowhere except in nerve tissue, and that to cultivate the virus in nerve tissue for purposes of a vaccine exposed the recipient to the possibility of an allergic reaction more dangerous than the virus itself. Polio research was locked deep in a quandary: how could you work with a virus you couldn't cultivate safely in the lab? The answer was, you couldn't.

So Enders, Weller, and Robbins, who weren't even engaged in polio research, on an impulse made the polio vaccines possible. Although the poliovirus did not do well on Robbins's mouse intestine, in the flasks containing Weller's human embryo tissue it did just fine. The chickenpox virus would have nothing to do with Weller's cultures, but in the excitement surrounding the unexpected success of the poliovirus experiments, chickenpox was for the time being forgotten. Poliovirus would grow in nonnervous tissue! The impossible had come true.

The occasion called for klieg lights, a drumroll, a chorus of angels, at least a bottle of champagne. But such wasn't Enders's style (although Robbins remembered writing in the lab book, in *red* ink, "This is the first time in the history of the world somebody has done this"). He refused to be rushed. First it was necessary to duplicate the experiment to make sure there was no mistake; then they had to see how long the poliovirus would continue to multiply. It turned out that there was no mistake — once again poliovirus grew in nonnervous tissue, and it continued to grow for as long as twenty days. They measured this duration by the common technique of injecting the culture fluid into the brains of mice to see if they would develop paralysis. If the answer was yes, the researchers knew they were working with live, virulent virus. Then they proceeded to dilute the infectious fluid and again inject it, to see if the virus was multiplying.

If it was, greater and greater dilutions would still contain enough virus to infect. In this case, Weller and Robbins were able to dilute their poliovirus cultures by a factor of 1,000,000,000,000,000,000,000 and still paralyze mice with polio. But still Enders wasn't convinced. "We had to prove it three times over before Dr. Enders would accept the fact," Robbins recalled.

Then the trio noticed something that made the dilution experiment they had just performed virtually obsolete. Normally researchers add a special red dye to a tissue culture to determine if it is growing. When tissues flourish they produce acid as a by-product, and the dye reacts to this presence of acid by becoming orange and then yellow. But these poliovirus-infected cultures weren't turning yellow as quickly as cultures containing nothing but tissue. Why? "It meant that the tissue was dead," explained Robbins, "and that the virus had killed it." By killing the tissue, the virus was inhibiting the production of acid, thereby preventing the change in color from red to yellow. Suddenly the team was face to face with a simple, accurate way to monitor the presence of virus without having to use laboratory animals: simply add a dye and check to see if it changes color. If the color remains unchanged, the virus is present and multiplying.

But that wasn't all. The researchers then went further and transplanted some of the poliovirus-infected tissue onto glass slides so as to observe it under the microscope. Although the virus itself was much too small to see, it was apparent that some of the tissue cells were dying. The culture was not spreading, not growing out around the edges of the slide as it would normally do. The reason: poliovirus was killing it off. "The key to the whole business was that we observed the polioviruses slowly destroying the cells in which they grew," said Weller. "Under direct microscopic examination you could see they were necrotic, going dead."

Thus, taking off from their unexpected success in growing the poliovirus in tissue culture, Enders, Weller, and Robbins had found two simple, reliable ways to detect the presence of viruses in living cells: monitor the color of the dye, and check under the microscope for reduced tissue growth. Enders called this effect "cytopathic" — *pathic,* "producing disease in"; *cyto,* "cells." In the face of such compelling evidence, even he had to conclude that indeed they had discovered something special.

Finally, the team decided to employ a relatively new development in medical science, antibiotics. Penicillin, the first of these "wonder drugs," had already been used with some success to sterilize tissue cultures against unwanted bacteria. To further reduce the chance of contamination, Enders's team added a healthy dose of a new antibiotic, streptomycin, to their polio cultures. It was a successful ploy. The cultures remained pure.

Still, however, Enders would not think of publishing his team's findings without a great deal more information. All but forgetting mumps, chickenpox, and infant diarrhea, they worked through the summer and fall refining their discovery. They kept their poliovirus (the so-called Lansing strain, as it had been taken from an eighteen-year-old boy who had died of the disease in Lansing, Michigan) alive and growing in tissue culture for 224 days. They tried the virus on other kinds of tissue samples from the Lying-In and Children's hospitals — brain tissue and muscle from the intestines of human embryos — and the virus grew there as well (pleasing Robbins especially, who had had no luck with his mouse intestine culture).

It wasn't surprising that poliovirus would grow in the nerve tissue of brain culture — Sabin and collaborator Peter Olitsky had shown that a decade earlier — but it was most definitely news that the virus would grow in intestines, news that had enormous implications. For if the virus did indeed multiply in the intestines, perhaps its route of infection was not directly through the nervous system as had been supposed but through the intestinal tract.

Then Weller went a step further by testing the virus in mature human tissue. Fetal tissue grows rapidly; if the virus were to grow in any human tissue, it would most likely grow there. But what about relatively stable, slow-growing mature tissue? To find out, Weller went after the most readily obtainable healthy tissue he could find, foreskin from circumcised boys four to eleven years old. Poliovirus grew here as well.

Now, having discovered how to grow the virus, Enders and his team tried to stop it. Since poliovirus produced immunity in those who recovered from it, what would an application of blood from immunized animals do to the tissue cultures? From Johns Hopkins University they obtained blood serum from monkeys who had survived the disease. Sure enough, when they added immunized serum to their thriving tissue cultures, the virus stopped multiplying.

But what about other strains of poliovirus? Perhaps the success of the experiments was characteristic of the Lansing strain alone. Not so. When they tried the two other main types of virus, they grew also.

Now Enders had to concede that they had waited long enough. On January 28, 1949, he, Weller, and Robbins published their findings in the journal *Science,* under the relatively innocuous title "Cultivation of Lansing Strain of Poliomyelitis Virus in Cultures of Various Human Embryonic Tissues."

It was, as the crusty virologist (and head of the Rockefeller Institute hospital) Tom Rivers said, "like hearing a cannon go off."

> Please bear in mind that, until 1949, most virologists believed that it was impossible to cultivate poliovirus in nonnervous tissue. . . . in 1936 Dr. Olitsky and Dr. Sabin proved that it couldn't be done. . . . That work was so meticulously done that I believed it was absolutely correct. Hell, it was correct, and every working virologist that I know believed it, with the possible exception of John Enders at Harvard.
>
> To this day, I don't know why John didn't believe that work. I suppose it's his nature. He is a great old skeptic who never believes anyone right off, and I expect he just didn't take this work as proved. . . . Dr. Enders' achievement in 1949 lay in the fact that he and his coworkers proved the exact opposite. . . . Now if that wasn't shooting off a cannon, I don't know what is. I'll tell you one thing, that report sure as hell captured everybody's attention.

And with good reason — for the discovery absolutely liberated polio research. Now it was possible to work with the virus in the lab, to grow it there, measure it there, and, perhaps, modify it there, *without having to inject it into a live animal.* Now test tubes and petri dishes could diminish the role of monkeys in polio research (a boon to both the virologist and monkey populations). "In essence," said Weller, "one tissue culture became the equivalent of one mon-

key. That was the main breakthrough." For the first time, researchers could look forward to wedding their dreams of a polio vaccine
to the reality of actually producing one.

In 1954, Enders, Weller, and Robbins received the Nobel Prize
"for their discovery of the ability of poliomyelitis viruses to grow in
cultures of various types of tissue." And in 1955, Jonas Salk presented to the world the first effective polio vaccine. He knew who
had made his feat possible. Some months earlier, while being interviewed by Edward R. Murrow on the television program "See It
Now," Salk remarked, "Enders threw a long forward pass, and we
happened to be at the place where the ball could be caught."

While the recognition accorded Enders and his colleagues focused
on the importance of their work to polio research, their discoveries
made an impact far beyond that. "These discoveries incited a restless
activity in the virus laboratories the world over," stated the representative of the Nobel committee in presenting them with their prize.
"The tissue culture technique was rapidly made one of the standard
methods of medical virus research."

Enders, Weller, and Robbins had made a huge conceptual leap.
Now it was clear that tissue culture was the way of the future in
virus research. It has been called "such an important tool in both
basic and applied research in medical science that many have felt
that its impact has been as important as the introduction of the
compound microscope." Enders and his colleagues rendered viruses
virtually as accessible as bacteria. Viruses — invisible, dangerous,
mysterious, almost unimaginable — were now *available*. Because of
Enders and his collaborators, microbiologists could now discover
where and how viruses lived, and *how to track them down*. Viruses
could no longer hit and run, no longer hide. They were now within
reach, open to investigation.

The immediate results of ensuing investigations were manyfold.
Once Enders and his team had demonstrated how to grow the poliovirus in tissue culture, everyone wanted to get in on the act, with
viruses of all kinds. Within just a few years of the 1949 *Science*
article began the so-called golden age of virology. As researchers
were now able to recognize signs of the presence of virus in tissue

culture — in Enders's term, the process of "cytopathology" — many new viruses came to light, a number of them found by Enders and his colleagues. Among those discovered during this explosion of research were the rhinovirus, which causes most cases of the common cold, and the echoviruses and coxsackieviruses, which cause aseptic meningitis. And virtually every virus discovery since then — including the discovery of the virus that causes hepatitis A; of HTLV, the human leukemia virus; and of HIV, the AIDS virus — has been accomplished using techniques based on those introduced by the Enders team.

Tissue culture techniques made it possible to produce vaccines efficiently against a number of these newly discovered viruses as well as others that had been known, and suffered, for some time. The polio vaccines are the most conspicuous example, but in 1954 Enders himself, with his associate Thomas Peebles, isolated the virus that caused the much more widespread disease of measles, and later, with Samuel Katz and others, he developed an attenuated virus vaccine against the disease.

Thomas Weller went on finally to pin down the elusive chicken-pox virus, and he discovered as well the cytomegalovirus, a member of the herpes family, and the rubella virus, which causes German measles. (He had a particular interest in tracking down this last culprit, as it was at the time making life miserable for his own family; the rubella virus he grew in tissue culture came from the body of his ten-year-old son.) The rubella virus discovery paved the way for a vaccine against German measles.

And while these virus hunters and vaccine makers were following Enders's lead with highly publicized results, others were much more quietly fomenting another, perhaps even more profound, revolution. Because now, no longer hindered by the inconsistencies and variations of chicken eggs or animals, researchers were able to initiate a viral infection at a precise point in time in precisely uniform cells, all of which became infected. It thus became possible to study precisely how a virus infects. Microbiologists began to use the electron microscope to find out when and where viruses were multiplying in tissue culture cells. They began to affix viruses with radioactive markers that allowed them to follow their route into infected cells and determine what processes went on there. Since viruses commandeer the cells' reproductive machinery, scientists were able to follow along and uncover how protein is produced in cells, how DNA and RNA are produced. They began to uncover, in other words, the most basic workings of life itself. In a nutshell, Enders's leap into tissue culture provided much of the impetus and technique for the fledgling fields of molecular biology and genetic engineering.

All this — some of it by now old news, much of it research that is still in its infancy — stems from those days in 1948 when John Enders, Tom Weller, and Fred Robbins decided to take the seemingly foolhardy step of trying to grow poliovirus in nonnervous human tissue.

POSTSCRIPT: In 1954, Thomas Weller became full professor and head of the Department of Tropical Public Health at the Harvard University School of Public Health (outranking his old boss, Enders, who was still an associate professor), where he continued extensive research into viruses and bacteria, taught, and cared for patients until his retirement in 1986. Frederick Robbins in 1966 became dean of the school of medicine at Case Western Reserve University in Cleveland. He is now professor emeritus.

Enders continued to work in his lab into his eighties. After his work with the measles virus, he went on to investigate the virus-fighting properties of interferon, a substance that cells under attack by viruses produce in self-defense. He was the first to describe the ability of some viruses to transform normally growing cells into cells that go berserk, dividing and multiplying so quickly that, if unchecked, they will soon take over the body — in other words, cancer cells. And in his last days, finally retired, he became particularly interested in the strange phenomenon of AIDS, especially what happens to the body during the disease's long period of incubation. In 1985, at the age of eighty-eight, long since a full professor of microbiology at Harvard and recipient of numerous honors (as diverse as the presidential Medal of Freedom and Commander of the Republic of the Upper Volta), John Enders died. He was buried, where he began, in West Hartford, Connecticut.

For all his industry, all his love for science and openness to directions unsought and unanticipated, one path that Enders resolutely did *not* take was that of developing his own polio vaccine. After he and Weller and Robbins made their great discoveries, they knew very well that they had the means in their grasp to make a vaccine and, because they alone already understood the tissue culture techniques that would take some time to be assimilated by others, had a head start on the leading lights in the field, Salk and Sabin. Enders knew how to make vaccines. He had already developed a feline distemper vaccine, and he was later to develop his vaccine against measles. Why, then, didn't he grasp for the brass ring and set out to produce what might have been known as the Enders vaccine, forever sealing his fame? "Dr. Enders felt that wasn't our cup of tea," says Robbins. "That required a fairly large-scale operation — it was more the kind of thing a drug firm could do. He just didn't feel that was our role in the world."

There was a story that Enders liked to tell about himself. At the end of the day, he often took a taxi home, and he came to know most of the drivers well. One particular evening, long after Enders had won his Nobel Prize, a cabbie, undoubtedly sympathetic to this aging, quiet, dignified man, offered a bit of encouragement.

"Keep trying," he said. "Perhaps someday, like Dr. Salk, you might discover something wonderful."

Eight

. . . .

The Children's Disease

. . . .

On April 12, 1955, the tenth anniversary of Franklin Delano Roosevelt's death, the world was presented a vaccine against polio. In the history of medical science, there had never been a day like it. The setting was an elegant auditorium on the campus of the University of Michigan at Ann Arbor, where Jonas Salk had worked during the 1940s. The specific purpose of the presentation was to announce the results of a field trial of the vaccine Salk had developed. The scope of the testing had been unprecedented. The vaccine had been given to over 420,000 children; over 200,000 others had been injected with a placebo solution; and almost 1,200,000, acting as controls, had received nothing at all. In total, 1,829,916 children took part in the study. To administer such an enormous undertaking required the efforts of some 20,000 doctors and health authorities, 40,000 nurses, 50,000 teachers, 14,000 school principals, as well as 200,000 other volunteers. And the task of compiling and analyzing the results had fallen to one of the most respected scientists around, Michigan's Thomas Francis, who some years earlier, with Salk's assistance, had developed an influenza vaccine.

An invited audience of some 500 influential scientists and physicians crowded the hall to hear Francis announce his results. But that was just the live audience. A roundtable discussion was later carried on closed-circuit television to 54,000 doctors in sixty-one cities, and in the pressroom three stories above the auditorium more than 150 newspaper, radio, and television reporters strained to get the news out. Francis's presentation was to begin at 10:20 in the morning.

Thus ensued what the *New York Times* described as "Medical History . . . Written in Hollywood Atmosphere." Another witness described the performance as "the rockets' red glare and flashbulbs bursting in air." At 9:15 the elevators outside the pressroom opened to reveal messengers bearing advance notices of the trial results. It was as though they had stepped into a feeding frenzy of sharks. Reporters surged toward the elevator, shoving and elbowing each other aside in their rush to be first to get the news. Frightened, the messengers panicked and began flinging their information packets into the advancing hordes, who then fought one another for the precious bundles. "In a word," said virologist Tom Rivers, "the newspaper people and photographers created a madhouse. I don't know when I have seen such wild people."

By common agreement, there were to be no news leaks. The reporters could announce the findings to the world the moment Francis started to speak. But at 9:20, Dave Garroway of NBC decided that agreement or no, the news was too good to delay — he announced on the air that the Salk vaccine worked. So by the time Francis stepped to the podium at 10:20 as scheduled, the scientists in the hall were virtually the only ones in the land who hadn't already heard that, as Francis put it, "the vaccination was 80–90 percent effective against paralytic poliomyelitis." Francis finished to polite applause.

Outside the hall, however, pandemonium ruled. Church bells pealed across the country; sirens wailed. Courtrooms, schools, offices, and factories asked for moments of silence to give thanks to Jonas Salk. Banners sprung up — *"Thank you, Dr. Salk"* — and synagogues and churches called special prayer meetings. Salk himself didn't yet know it, but suddenly he was no longer a promising young virologist; he had been transformed into a national hero.

"You can't imagine how acutely uncomfortable I was," Salk recalls. "This whole thing was a great opportunity for the University of Michigan, but I didn't know until an hour and a half before the meeting began what the essence of it was going to be. I walked in, saw what was happening — My God! I didn't want any of it." Want it or not, however, Salk's life had changed. And it may be that the next few moments, during which he rose to a great ovation from the scientists present in the hall, provided the sweetest recognition Salk was to experience in his long polio odyssey; for the moment he started to speak, things began to go sour. In trying to deflect the spotlight from himself to the many researchers through the years who had contributed to the realization of an effective polio vaccine, Salk neglected to thank his own staff, thereby alienating those who should have been most gratified. And in attempting to explain that his vaccine might have been 100 percent effective had his directions for its use been followed more closely during the field trial, he offended a number of the scientists present, who took his remarks as dubious to say the least — no vaccine ever had attained 100-percent effectiveness — and as an implied criticism of the highly respected Francis as well.

But nothing could possibly darken the brightness of the day as far as the nation was concerned. In just hours the Department of Health, Education and Welfare licensed the vaccine for distribution, and in just two weeks *ten million* doses were prepared for injection. Salk, who very quickly began to long for the days when he was unknown, was inundated with honors, gifts, propositions. He turned down Hollywood offers to film his life story (rumors were that Marlon Brando would play the lead role), gifts of cash, automobiles, even farm tillers. He turned down a ticker-tape parade in New York City; he turned down a million-dollar contract with a public-relations firm. One day, a 208-foot-long telegram signed with seven thousand names came to his door from Winnipeg, Canada; ten thousand letters and telegrams flooded into his office in Pittsburgh during the first week alone. Streets were named in his honor, scholarships created, professorships established. President Eisenhower presented him with a special citation during a ceremony at the White House. Edward R. Murrow, who interviewed Salk on television, summed up what was happening to him: "Young man," he told Salk, "a great tragedy has befallen you. You've just lost your anonymity."

But the kind of recognition that Salk might have welcomed, that of the scientific community, was as niggardly as praise from the world at large was overwhelming. A Nobel Prize? Well, Enders, Weller, and Robbins had won their Nobel just the year before, and Max Theiler had won the prize in 1951 for developing a vaccine against yellow fever; virologists had been too much recognized too recently for there to be a Nobel forthcoming for Salk. Membership in the National Academy of Sciences? That, too, was denied him. The explanation was that he had done no "original" work; but the germ of the real reason may well have been spread when he took the podium in Ann Arbor. Salk the public hero, the savior of children, suddenly the most recognized scientist in the country if not the world, had unleashed deep-rooted jealousy and anger — resentment exacerbated by the fact that his lifesaving vaccine flew in the face of scientific convention. For by making a killed — or "inactivated" — virus vaccine, he had taken an approach not supported by widespread precedent. Pasteur had attenuated rather than inactivated the virus in his rabies vaccine. Jenner's smallpox vaccine, while containing no smallpox virus, in effect operated under the same principle — a less virulent form of a live virus provided protection against the real thing. And Theiler's yellow-fever vaccine incorporated that approach also: he had attenuated rather than inactivated.

That Salk would lean in this unconventional direction was no surprise, however, for the flu vaccine on which Salk had worked in Thomas Francis's lab was a killed-virus preparation. Francis and Salk had used formaldehyde to kill flu virus, thereby challenging the prevailing wisdom that inactivated vaccines, in destroying a virus's ability to infect, ran the risk of destroying its ability to bring about immunity as well. Their approach worked; the vaccine they developed became the prototype of the many flu vaccines that followed. And Salk's inactivated polio vaccine worked as well.

"Jealousy and anger," says Salk, "they're innate. I didn't have to do anything — I just had to exist. But who the hell cared to be lionized. Why does anybody do anything? For its own sake, not for the rewards." He would be content with the satisfaction of knowing that no matter what anyone said or thought, his vaccine would improve the lot of thousands of children who otherwise would be facing the horrors of paralysis and death.

Soon enough, however, even that boon was called into question. It began innocuously enough on April 25, just thirteen days after the presentation and licensing of the Salk vaccine: a Chicago child became paralyzed after inoculation. It was sad news but not necessarily alarming, for the child may well have already been infected before being vaccinated. Besides, Salk's vaccine, like any other, was not 100-percent effective; rare cases of polio in vaccinated children were to be expected. But the next day brought news of five more cases of paralysis, these in California children. And here the news was alarming indeed, for although polio usually struck the legs, every one of these children had become paralyzed in the left arm — the very arm that had received the injected vaccine. Finally, one more startling fact surfaced to complete the picture: every one of the paralyzed children had received vaccine prepared by the same manufacturer, Cutter Laboratories of California.

Faced with such information, the surgeon general, Leonard Scheele, within hours requested that Cutter withdraw its vaccine from the market. But it was too little, too late. The same day, Pocatello, Idaho, reported two more cases of polio, one fatal. Sure enough, Cutter vaccine had been distributed in Idaho as well.

Before the Cutter epidemic was done, over 20 Idaho residents contracted polio and in turn infected more than 50 family members or other contacts. Five of these people died. In the country as a whole, of the slightly more than 400,000 people who were inoculated with the Cutter vaccine, 79 contracted polio, and 125 others became infected through contact with those vaccinated. Three-quarters of the cases involved paralysis — and 11 were fatal.

As soon as he had a chance to examine the Cutter records, Salk concluded that the reason for the vaccine's virulence was obvious: "they did not follow the prescribed procedure." As a killed-virus vaccine, the crux in its manufacture was to make sure that every last virus was dead from exposure to formaldehyde before the solution went out the door. But since every dose of vaccine contained at least ten million polioviruses, absolute assurance was impossible. To deal with the problem, Salk devised an "inactivation graph" on which he plotted a curve that reflected how long it took to kill those viruses that could be measured and approximated the time it would take to kill the rest. Then, at the end of the time curve, he added a margin of safety. Thus extended, the curve reached zero — Salk was sure of it. In other words, even though there was no definitive way to assess the condition of each virus, all viruses simply would have to have been killed after undergoing such an extended procedure.

"You couldn't possibly determine from any one sample whether all the viruses in the rest of the batch were dead," explains Salk. "Therefore you had to establish that the *process* was correct. The product would be right if the process was right — that's the important thing."

In the batches of vaccine that had been made up for the field trial, such was certainly the case, as there were no problems whatsoever. But once commercial manufacturing of the vaccine began in earnest, the trouble began: in the Cutter labs, Salk's inactivation process was shortened. At the same time, bottles containing poliovirus-infected tissue culture were stored over weeks and months before being processed. During that time bits of tissue debris settled to the bottom, clumping together to engulf virus particles. With some of the virus thus protected, the shortened process couldn't inactivate the entire solution; virus remained alive and virulent.

"It was criminal negligence, gross negligence," says Salk, still angry at the memory thirty-five years later. "All of the rules that were set up would have precluded any problems. They were simply not able to make the vaccine. They should never have been given a license." He sighs. "That was a bad time."

So, the difficulty was in Cutter's taking shortcuts in the manufacturing process, not in the vaccine itself. The other four companies making the vaccine experienced no such problems, and within the first month over five million Americans were vaccinated. Of these, the grand total of polio cases was 113, about one in fifty thousand. It was an admirable record.

No matter. The Cutter incidents shook public — and scientific — confidence in the Salk vaccine. The *New England Journal of Medicine,* one of the most respected medical publications in the country, called the situation a "polio panic." Some people began referring to the vaccine funding organization as the "Infantile Foundation for National Paralysis." Congress demanded a hearing on the matter before a panel of scientific experts, with the result that, after great and emotional deliberation, it was finally decided that the Cutter incident was an aberration and the vaccination program should be continued. The vote was eight to three. One of the three naysayers was Albert Sabin.

Years earlier, Sabin had begun working on a vaccine of his own — an attenuated live-virus vaccine. By taking such an approach, of course, he had the great majority of virologists on his side; in fact, Lederle Laboratories scientists Hilary Koprowski and Harold Cox were working on live-virus vaccines also. Sabin and Salk had never gotten on well — a conflict that intensified when the National Foundation, while continuing to support Sabin, placed most of its eggs in Salk's basket. Sabin had spoken out against Salk's vaccine from the beginning. Now, with Salk's luster dimmed and his own research well on its way, Sabin stepped up the attack.

"I am not against the vaccine program," Sabin announced at the end of June. "I am against continuation of it with the present Salk vaccine. The milder viruses I have developed would be substituted for the dangerous ones." And Sabin wasn't alone. Other virologists, including John Enders, who had never been happy with Salk's killed-virus approach, offered support for Sabin's attenuated-virus approach. But support alone wouldn't do the trick. What Sabin needed was a field trial of his own vaccine, and that required a large number of people who might be expected to have very little natural resistance to the disease and who would be willing to go along with testing an unproved product. In the United States those were the very people the Salk-vaccine trials and subsequent distribution had already immunized. Sabin would have to look elsewhere. As it turned out, he didn't have to look long, although he looked far. While Salk's vaccine continued to be the only one licensed for use in the United States, Sabin turned to the land of his birth, Russia.

It was a brilliant move. The USSR had only recently been afflicted with polio — in 1957 the disease struck ninety-four of every million Soviet citizens — and was hungry for an effective vaccine. The Russians had tried the Salk vaccine, with success, but had found the manufacturing process as difficult as did American companies who had the benefit of much more experience. And the logistics of providing three doses of vaccine, the initial shot and two boosters, to 200 million people were forbidding, indeed. The Sabin vaccine, on the other hand, was much simpler to produce and, consisting as it did of vaccine-drenched sugar drops to be swallowed, was a comparative breeze to distribute. What could be easier?

For Sabin's purposes, the match was equally attractive. With the Soviet government behind the vaccine, there would be no problems with protesters, adverse publicity, or scrutiny by the news media. If problems showed up, he could deal with them in relative obscurity. And, with such a huge population to work with, it would be a trial more impressive than any yet attempted. It would be ideal, like conducting a field trial on an immense, isolated island. Distribution of his vaccine began in Russia and other Iron Curtain countries in the spring of 1957.

Not until 1959 did any results come back, but they were worth waiting for. At the International Conference on Live-Virus Vaccines in Washington, DC, it was announced that Sabin's vaccine had been given to 4.5 million people in the USSR, Poland, and Czechoslovakia (and to others in Singapore and Mexico, as well) — a figure that dwarfed Salk's once-imposing field trial of 1.8 million people. The results were overwhelming. Although there was a slight tendency of the attenuated virus to increase in virulence over time, the vaccine had achieved almost 100-percent effectiveness. And given that kind of spectacular performance, it seemed not to matter that these were results that could hardly be verified. Put up against the sheer numbers involved in Sabin's program, the similarly good performances of Koprowski's and Cox's live-virus vaccines were mostly overlooked. By 1960, the Russians and others had fed Sabin's vaccine to more than 100 million people, with equally impressive results.

"It was like sitting in on the plans for one's own assassination," Salk was quoted as saying, in describing the many polio conferences during those years.

> The atmosphere of intrigue and hostility was even more intense than in 1953 and 1954. In those years mine was the only vaccine in the picture and the battle lines almost drew themselves. But now you

had Sabin and his vaccine, Koprowski and *his* vaccine, and Harold Cox and his Lederle vaccine, and each of them had [his] coterie. The vying for position was absolutely brutal. There were actual plots hatched to keep one or another vaccine report off one or another scientific program. And, long before a particular report was read, you heard a dozen whispered allegations about the lies contained in it and the number of deaths unmentioned by it. It was like Lisbon during the war — everybody hawking secrets to everybody else.

But now the tide was definitely turning in Sabin's direction — all the more so because in 1958, after dropping steadily ever since the Salk vaccine had appeared in 1955 (at which time there were almost 40,000 new US polio cases yearly), the incidence of polio rose slightly in the United States before absolutely jumping in 1959. In 1958 there were 5,787 new polio cases, an increase of 300 over the previous year; in 1959 the number shot up all the way to 8,425. Salk attributed the rise to a failure to immunize the people who most needed it, the urban poor, and the incidence figures for the succeeding years seemed to bear him out, as with more widespread vaccinations new cases dropped to 3,190 in 1960 and only 1,312 in 1961. And in 1962 the figures plummeted sharply again, as only 910 people came down with the disease. But by that time, it was too late. For while Salk saw his vaccine virtually rid the country of what only a few years before had been perhaps the most feared disease of them all, he also saw the American Medical Association perform an act unprecedented in its 114-year history. In June of 1961, it voted its approval to a product that had not yet been licensed for public use: the Sabin polio vaccine. In just months it supplanted the Salk preparation as the vaccine of choice in this country and in most places in the world.

There are a number of likely reasons why the Sabin vaccine displaced that of Salk. The most obvious one, however — the Cutter vaccine incident — had little impact in the long run. In view of its excellent immunizing record before, during, and after the Cutter episode, the vaccine more than atoned for a situation that really had nothing to do with its inherent safety or effectiveness. There had to be something else.

That something was for the most part the nature of the two vaccines themselves. The live-virus vaccine was simply much easier to produce, distribute, and administer — a prime consideration in rural areas and developing countries — than the killed-virus variety. Who would not rather suck a sugar cube than have a needle stuck into an arm, especially when lasting immunization required three separate shots? Salk certainly had a point when he argued that other successful vaccines had necessitated the dreaded needle, but in light of Sabin's spectacular field trial, Salk's vaccine no longer seemed to warrant the discomfort and inconvenience. And there were other, less obvious considerations. Most researchers agreed that while the killed-virus vaccine did an excellent job of immunizing against polio, the protection it afforded could take months to develop, particularly in the very young, and it didn't last very long — thus the need for repeated annual boosters. In contrast, the attenuated-virus vaccine took only days to bring about immunity, and so might be more effective in the face of an impending epidemic. And it provided resistance in the intestinal tract as well as the bloodstream. In short, being ingested, it mimicked natural infection, thereby providing a long-lasting immunity akin to that experienced after an actual infection. Or so it seemed. These supposed benefits were — and still are — disputed, but they were instrumental in turning the tide against Salk's vaccine.

There was, however, anxiety that the oral vaccine might not be entirely safe. The attenuated virus seemed to display a tendency to revert to its original wild and virulent state. Yet in the face of the pristine results of the vaccine's Russian field trial, not to mention the sheer persistence and power of Sabin's personality, those fears accounted for little. Older than Salk by eight years, more established than he, and the champion of the mainstream approach to vaccination, Sabin had long since gathered the support of the bulk of the scientific and medical world. He was tireless, brilliant, articulate. Supporting his oral vaccine would be a popular move. Salk himself has acknowledged as much. "The idea of avoiding injections, taking it by mouth, had great appeal. And, of course," he says, dryly, "Sabin's personality was not inconsequential."

And perhaps there was yet another reason why Sabin's vaccine took precedence, a reason that had little to do with type, ease of administration, or effectiveness. It would have had a great deal to do, however, with that familiar bottom line, politics. For in those days of intense cold-war rivalry between the capitalist and communist worlds, adopting the Sabin vaccine was seen as a way to put an end to what was being referred to as the "polio gap." "They shamed the US," says Hilary Koprowski. "Here was a thing developed in the United States, and the *Russkies* were doing it. So there was a change in the bandwagon."

In any case, in the winter of 1962 the US Public Health Service licensed the Sabin vaccine for distribution to the American public. Although Salk's vaccine continued to be licensed as well, almost immediately the Sabin vaccine took over. Jonas Salk had won the initial battle and reaped the honors, but it looked as though Albert Sabin would win the war.

Today polio has virtually disappeared from the Western world. The Salk vaccine and then Sabin's have done what they were developed to do. In 1988 only 350 cases of polio were reported in the Western Hemisphere. In the United States, there are no more than 10 cases annually. But in the rest of the world, primarily in Africa and Asia, the disease still strikes as often as ever. Estimates vary, but it's thought that between 250,000 and 2 million people each year experience the pain and paralysis of polio. Another 25 million may experience minor or inapparent polio infections. Worldwide the disease is emphatically *not* dead.

But why? Both vaccines have been proved effective. What could be causing the continued danger?

There are a number of answers to the question, but they boil down to essentially three. In the first place, although the Sabin vaccine, the predominant vaccine worldwide, is effective and available, it's difficult to administer under certain conditions. For example, it must be kept cold, as close to freezing as possible, to retain its potency. Where refrigeration is unavailable, the vaccine is ineffective. Such a limitation sharply curtails the possibility of distributing it in many remote areas and developing countries, especially those near the equator. People therefore go unvaccinated.

Second, sometimes the vaccine simply fails to "take." No one knows exactly why, but it seems as though existing intestinal-tract infections can interfere with the attenuated virus. And in the world's tropics, an enormous number of infectious microorganisms inhabit the gut all the time. The oral vaccine has not been particularly effective in these areas.

Finally, as suspected at the beginning, the Sabin vaccine is less than completely benign. In rare instances, the attenuated virus does indeed revert to virulence, and the vaccine actually causes the disease it is intended to prevent. It's likely that *all* the handful of annual polio cases in the United States are the fault of the vaccine rather than natural infection, and elsewhere the situation is far more dramatic. For example, a 1986 Brazilian epidemic involving more than three hundred paralytic cases of the disease was caused by the oral vaccine. In such cases the attenuated virus replicates and regains strength in the intestines, then passes on by way of mouth or feces into the outside world. The newly virulent virus may then infect another person and pass on once more. The result: an outbreak of polio.

Admittedly, the risk of contracting polio from the oral vaccine is very small — only about one case in every 560,000 first doses. In areas that have never enjoyed any protection from polio, such odds may seem well worth taking. But health officials don't see it that way; they want something better. And in the United States, a country accustomed for over three decades to living without the fear of polio, many people consider even one case unacceptable — especially those people who become the "one" case. The last few years have seen a number of lawsuits filed against the manufacturers and distributors of the vaccine, including the US government, and awards of as much as $10 million granted to angry polio victims.

The solution to the problem, in the United States at least, may be to use both vaccines. As both are "merely" excellent products, not foolproof ones, perhaps in combination they might provide more effective protection. Such thinking is behind a recommendation of the National Academy of Science's Institute of Medicine. A report released in May of 1988 suggests that within five years the Food and Drug Administration will approve a new vaccine that combines an enhanced version of the Salk killed-virus vaccine, called E-IPV, with the familiar diphtheria-pertussis-tetanus (DPT) vaccine that most children already get. E-IPV could be added to at least the first two of the three DPT shots, which are typically given at two months, four months, and eighteen months of age, and then the child could take oral doses of live-virus vaccine thereafter. In this way, all bases would be covered: the killed-virus vaccine would provide immunity against oral-vaccine-induced disease. The report concludes that the scheme "would reduce or even eliminate cases of vaccine-associated paralysis."

Predictably, it's an approach that does not satisfy everybody — notably Salk and Sabin. "I would say that there is no need for a change," Sabin has written, "and that the changes proposed . . . could very well increase the probability of a return of outbreaks of paralytic poliomyelitis caused by wild polioviruses in the United States." Salk scoffs at the notion. "Is it that IPV [inactivated polio vaccine] is needed to make OPV [oral poliovirus vaccine] safe? There is no evidence that OPV is needed to make IPV effective," he has written. "The preference to continue to use OPV at any cost . . . is based on unverified beliefs rather than demonstrable evidence."

So the debate continues. The irony, in a story replete with ironies, is that if the program is implemented, Salk and Sabin, antagonists for so long, may inadvertently have been teamed up to do the job that each by himself has not yet been quite able to accomplish.

The larger goal, of course, is to rid the entire world of polio, thus relegating the disease to the status of smallpox, an unmourned relic of the past. The World Health Organization, so successful in purging smallpox in 1977, has called for the eradication of polio by the end of the century.

It won't be easy. For a disease to be entirely eliminated, not simply controlled, the natural transmission of the causative agent must cease entirely. And the result must be verified over several years. Then, and only then, can people stop being immunized, for then the agent no longer survives, either in humans or the environment. In the case of smallpox, the only virus still existing is housed in high-security laboratories at the Centers for Disease Control in Atlanta and in the Soviet Union.

Before a virus-caused disease can even be targeted for eradication, however, it must meet a number of conditions. First, it must naturally infect only humans, and it must not recur after the initial infection. Both smallpox and polio receive high marks there. Beyond those shared characteristics, though, the diseases are quite different from each other. With smallpox, people infected almost always became sick. There were almost no invisible cases of the disease, as there are with polio. In fact, by far the greatest number of polio infections are inapparent; for every instance of paralysis, there may be hundreds of undetected cases. It would be next to impossible to identify every infection. So while physicians could track down virtually every case of smallpox and vaccinate everyone surrounding the sick person, nothing of the kind can be done with polio. Because polioviruses are easily transmitted through food and water as well as through close person-to-person contact, by the time investigators are able to recognize any particular case of the disease, infection has already spread widely.

Polio also has a tendency to infect especially young children. Twenty percent of all cases in the Americas occur in children before their first birthday, and an additional 60 percent show up in children between the ages of two and four. Given that early onslaught, and the difficulty in recognizing infections in the first place, it would be hard indeed to vaccinate enough people soon enough.

That is, assuming there were a vaccine equal to the task. It comes back to that: the inadequacies of present polio vaccines. While suitable for some of the world, they cannot meet the requirements presented by much of the rest. In contrast, the smallpox vaccine, virtually identical to Jenner's original cowpox-virus solution, was a marvel. It offered long-lasting protection from just one dose and retained its potency even when stored at body temperature for as long as thirty days. It traveled well, demanded little in the way of special conditions, immunized widely and lastingly. All else being equal, a new polio vaccine may be necessary if the World Health Organization is to see its summons answered.

And a new vaccine is precisely what a number of people are working to develop. Assuming that the oral vaccine remains the vaccine of choice, it may be possible to improve its resistance to heat simply by changing the proportions of the vaccine's strains of polio-virus or by varying the amount of stabilizing additives in the mix. "That would require very minimal alteration because you're not changing the virus, you're just changing the soup it comes in," says polio researcher James Hogle of the Research Institute of Scripps Clinic in La Jolla, California. "Beyond that there's the possibility of manipulating the virus's genes to improve its stability and resistance to heat."

Molecular biologists have done just that by making so-called chimeras (after the mythological monster that was part lion, part goat, and part serpent). These are hybrid polioviruses that employ genetically altered genes to express proteins from the less stable strains of poliovirus onto the surface of the most stable strain. Incorporated into a vaccine, these brand-new viruses may prevent polio effectively while resisting the tendency to mutate to a virulent form.

Meanwhile, researchers are also looking for other ways to prevent polioviruses from gaining a foothold in the body. Vincent Racaniello of Columbia University has identified the cellular receptor where the poliovirus begins its attack. This feat may lead to new drugs that block the receptor, thereby preventing the virus from invading the cell. But, as promising as such achievements are, they're just a beginning.

"We may begin to see genetically engineered improvements by the year 2000," says Hogle, "but until then there's no reason why improved versions of the existing vaccines can't knock out much of the disease. The recombinant vaccines can then mop up what's left." But all that will take money, and lots of it — an estimated $155 million over the next decade. The World Health Organization has asked several international agencies to fund a "Manhattan Project" to develop a better polio vaccine. Meanwhile, the end of the century is fast approaching, and right now, somewhere in the developing world, someone is coming down with polio.

So the story of the "children's disease" is far from over. Even if, like smallpox, the polio story had an ending, however, it would remain significant for the changes it wrought in the world of viruses. In large part because of the attention given polio, and the vast resources committed to controlling it, viral research and understanding grew tremendously. Here was a viral disease that caused consternation and concern to a degree never seen before, at least in terms of a formal, organized response. Here was a viral disease that commanded a research effort unheard of to this point. Here, for better or for worse, was established the expectation that medical science, through vaccines, can produce miracles to prevent the ills that assail us. Here, in the middle of the twentieth century, the study of viruses became a scientific profession as widespread, respected, and important as any. No longer were virologists the neglected stepchildren of medical science — now they were at its forefront.

And, obliquely, the fight against polio spawned the discoveries of John Enders and his colleagues, discoveries that raised virology to a new level and helped to spawn the new world of molecular biology.

The "children's disease" has indeed engendered much. Because of it and the new techniques it generated, exploration into the world of viruses — for so long a matter largely of hunches and intuition followed by the first gropings toward reliable and effective procedures — is now an art that combines abundant scientific method with the insight of the investigator. In the next section, we'll examine the state of that art.

III

The State of the Art

Nine

· · · · · · · · · · · · · · · ·

Star Wars within Us

· · · · · · · · · · · · · · · ·

The physician and essayist Lewis Thomas described Oswald Avery as a "small, vanishingly thin man with a constantly startled expression and a very large and agile brain." As if caressing that brain, Avery was wont to stroke his bald head when thinking deeply, which is what he did much of the time, and often about the bacterium that causes pneumonia. Avery spent the bulk of his career, from 1913 to 1948, investigating the workings of this one organism, the pneumococcus, a preoccupation he called "digging a deep hole in one place." The hole he dug was so deep that it descended beyond pneumonia to encompass the mystery of how life perpetuates itself.

How life perpetuates itself — it may seem odd to encounter the phrase in the context of viruses, which all too often had been considered not properly alive at all. Yet by the middle of the century it had become clear that, alive or no, viruses had in common with cells a number of the basic processes necessary to perpetuate life. Indeed, viruses even usurped some of those cell processes for their own purposes. The secrets of cells were in part the secrets of viruses; they shared a common inspiriting spark. To learn about one was to discover the other.

It was Avery who provided a major insight toward the realization of all this and more, but such scope was hardly his intent. He might have been content to labor away in his small, spare laboratory at New York's Rockefeller Institute, forever investigating the intricacies of pneumonia bacteria, had it not been for a puzzling experiment conducted in England in 1928 by a physician named Frederick Griffith. Griffith had learned that the pneumococcus that caused lobar pneumonia, the most lethal type of the disease, came in two forms, smooth and rough (S and R, for short). The S-form was the deadly type. Its glossy, gelatinlike covering protected it from the body's immune system. Not so the R-form. The result of a mutation, the R-form was unable to produce a smooth, protective surface and thus was easily dealt with by the body's immune defenses. It was harmless.

In order to learn more about these dissimilar bacterial siblings, Griffith injected mice with a dose of the harmless rough form and a dose of the lethal smooth form that he had killed beforehand by exposing it to heat, thereby rendering the bacteria harmless. He expected, therefore, that the mice would continue to scurry around unharmed, as before.

Griffith was in for a shock. Not only did many of the mice stop scurrying, they died — of pneumonia. And when Griffith examined the dead mice, he found their blood swarming with living, virulent S-form pneumococci.

Where could the live, lethal bacteria have come from? As far as Griffith was concerned, there was only one answer: he had botched the experiment by inadvertently injecting live S-form bacteria into the mice. That had to be the explanation, for virulent microorganisms couldn't appear from nowhere. But when he tested his rationale by injecting only the heated S-form into a fresh batch of mice, without first injecting the R-form, nothing happened. The heated bacteria didn't affect the mice; the pneumococci must have been dead after all.

Puzzled, Griffith performed his original experiment once more, inoculating another bunch of mice with the live rough bacteria and the heated smooth form — which he knew to be dead and harmless — and waited to see what happened. He didn't have to wait long. Soon mice started dropping as before. Something was going on within these mice to allow dead bacteria to transform live, harmless bacteria into a virulent form. And not only were these particular bacteria transformed, but the change was permanent and inheritable from generation to generation. Those newly created S-form bacteria gave rise to more S-form bacteria, which themselves produced yet more S-form bacteria. Whatever had happened within those mice had altered the bacteria forever.

When Griffith published his results, however, people were incredulous. Life simply wasn't like that. Things didn't turn into other things for no apparent reason — things didn't turn into other things, period. And no one was more skeptical than Oswald Avery. Such apparent craziness had no place in Avery's meticulous, thoughtful scientific life. But when a year later the experiment was repeated by others, and when two years after that Avery's own associates found that they could obtain the same results without mice — that is, just combining live R-form and heat-killed S-form pneumococci in a glass dish was enough to produce brand-new live and virulent S-form bacteria — Avery could no longer ignore the matter.

Then, several months later, another Rockefeller microbiologist performed an experiment that engaged Avery for good (in fact, he was working in Avery's own lab). J. L. Alloway grew a culture of S-form bacteria and crushed it, setting free the contents of the bacteria. He then passed the mix through filters fine enough to remove the broken bacterial shells and any other cellular material. Finally, he added the resulting filtrate, which contained no whole bacterial cells at all, to a culture of the R-form. The result: transformation once again. A new generation of virulent S-form began to grow. Not only were mice not necessary to effect the change, but whole bacteria themselves were not necessary . All that was needed was the extract from the bacterial innards. When Alloway added alcohol to this extract, he produced a "thick, syrupy precipitate." People in the lab began calling it the "transforming principle."

Now Avery was engaged indeed. Alloway's thick syrup worked, yes, but it contained all sorts of impurities, so many that it was impossible to know just what in the viscous liquid actually caused the transformation. Avery decided to find out. Usually given to erudite monologues, his characteristic cry now became, "What is the substance responsible?" It took him over ten years to find out.

Enlisting the help of colleagues Colin MacLeod and Maclyn McCarty, he grew the virulent S-form bacteria in twenty-gallon vats of nutrient solution, and then, in a series of laborious steps similar to those used by Wendell Stanley to purify his tobacco-mosaic-virus juice, painstakingly removed everything that was not the transforming principle. He used a centrifuge to whip away extraneous material, filters to sift them away, various chemicals to eat them away. The last step involved gently stirring alcohol into the purified liquid that remained. Soon, before the eyes of the astonished technicians, cobwebs of whitish, translucent threads appeared and began to cling to the stirring rod. From the twenty gallons they had started with, Avery and his colleagues had produced just under a hundredth of an ounce of this stringy material. It was DNA.

DNA. When Avery published his findings in 1944 (at the age of sixty-seven, "the ever rarer instance of an old man making a great scientific discovery," as a colleague put it), DNA, or deoxyribonucleic acid, was know as a substance that inhabited the nuclei of cells and viruses alike, but what it was doing there wasn't clear. In fact, many considered DNA to have no function whatsoever other than that of providing a protective coating for the real stuff of genes, protein. DNA was a small molecule made up of just four subunits. It was boring, monotonous stuff, as interesting as a layer of mortar between bricks. Protein, on the other hand — complex, versatile protein — excited people. It was protein that most likely made possible the flow of genetic inheritance from one generation to another, that probably constituted the transforming principle.

But now, here was prudent, careful Oswald Avery suggesting that no, it wasn't protein at all. DNA wasn't simply a structural component — it might actually be the transforming principle itself. He was too cautious to say so in his published paper, but in a letter to his brother he revealed his excitement.

If we are right, & of course that's not yet proven, then it means that nucleic acids are *not* merely structurally important but functionally active substances in determining the biochemical activities and specific characteristics of cells — & that by means of a known chemical substance it is possible to induce *predictable* and *hereditary* changes in cells. This is something that has long been the dream of geneticists. . . . Sounds like a virus — may be a gene.

May be a gene, indeed. Today it's common knowledge that it is indeed DNA that contains the genetic instructions in the heart of every cell, and it is DNA, or its fellow nucleic acid, RNA, that occupies the heart of every virus as well. Avery, however, never did receive appropriate recognition for what he had accomplished. When he died in 1955, late enough to have witnessed Watson and Crick's elucidation of the molecular structure of DNA, he was still primarily known, if at all, as the shy and quiet "Professor" who studied pneumonia. The plaudits and honors, the celebrity and notoriety, devolved elsewhere, on those who followed and expanded upon his lead. "He was a quiet man," observed one of his colleagues, "and it would have honored the world more, had it honored him more."

But if Avery's accomplishment has been overlooked, what it led to, molecular biology, was no less than a revolution. In 1979 Horace Freeland Judson wrote a book chronicling the beginnings of molecular biology, a work he entitled *The Eighth Day of Creation*. The phrase was not an exaggeration. Now scientists could deal with the basic processes of life — creation, reproduction, death — on a *physical, chemical* basis. You could touch these phenomena, examine them, even manipulate them. Now the invisible world, from cells to viruses, became a fantasylike realm of amazing "creatures" doing amazing things — but these creatures were *real*. For example, thanks to Watson and Crick it became known that a molecule of DNA resembles two long strands of twine spiraling about each other — the famous double helix. Each of these strands is composed of four distinct molecular units called nucleotides, strung along repeatedly one after the other, in various combinations, like links in a chain. And each nucleotide, or link, lines up across from a complementary nucleotide, or link, in the other strand of the curled structure. It is these combinations of nucleotides, like the combinations of dots and dashes of a telegraph message, that carry the genetic information of DNA, the so-called genetic code.

That in itself was a revelation; the way this helical structure works was yet another. Aided by special protein molecules called enzymes, whose function it is to facilitate various processes — to perform as catalysts, in other words — the curled strands of DNA begin the process of reproduction by uncurling and disengaging, as though being unzipped. Then, free and separated from its partner, each strand begins to attract nucleotide links complementary to its own from amongst free-floating molecules in a cell, eventually forming two new strands of nucleotide links. Then each original strand curls about its new complementary strand to form two new coiled pairs — two new double helices — each identical to the original. What was one is now two. When the cell divides into two cells, there is now enough DNA for both.

But that process, while it explained reproduction of genes, didn't account for growth and development. How does DNA enable its particular cell to become uniquely what it is — that is, a hair cell, or a blood cell, or a bone cell, or a muscle cell? DNA can produce new copies of itself, that much was now clear, but how does it communicate its genetic information to cells so that a bacterium and a whale, a daisy and a virus, you and I each develop at our own pace and in our own way to become uniquely us, different from anyone and anything else? The answer to that question turned out to be no less amazing and involved the other nucleic acid, RNA (for ribonucleic acid), which is identical to DNA except for three characteristics: it is a single strand, not a double helix; of its four repeating nucleotide links, three are the same as those of DNA, while one is different, unique to RNA; and its sugar component differs from that of DNA.

Growth and development enter the picture by virtue of the fact that sometimes DNA may not produce a new strand of itself but instead — in similar fashion, by uncurling and attracting complementary links — may produce a strand of RNA, thereby transferring to the RNA its genetic code. It's then that the fireworks start. RNA is much more mobile and adventurous than DNA, which, like the queen bee in a hive, is confined to the nucleus, spitting out genetic instructions. RNA, on the other hand, is free to explore all around the cell. And so it does. Aided by a variety of enzymes and other proteins, it slips out of the nucleus and travels into the watery world of the cell (called the "cytoplasm"), where it pulls up at one of the cell's protein-making factories, a "ribosome." The strand of RNA, having left the nucleus, becomes messenger RNA (mRNA, for short) and feeds into the ribosome factories like tape into a tape recorder, allowing the ribosome to "read" the genetic code. The ribosome's function is to make proteins, and this it now does, translating each of the "words" in the code into a section of a protein molecule.

It turns out that proteins, far from being the "solid," massive structures that their popular description, "the building blocks of life," would suggest, are really much like DNA and RNA in that they, too, are long, sinuous chains made up of smaller units arranged as links, one after the other. In this case, however, the units are called "amino acids," and instead of four different types repeated in various combinations throughout the length of the chain, amino acids come in twenty different varieties. So for each "word" in the RNA code, the ribosome assembles an amino acid, which links up with already produced amino acids to form a long chain, the protein.

There are a number of these ribosome factories scattered throughout the cell, and as new strands of messenger RNA continue making their way out from the nucleus, the ribosomes continue making chains of protein, each custom-built according to the coded instructions carried by the mRNA. Those chains then begin their own dance. Because certain of their constituent amino acids chemically attract or repel others on the chain, the protein folds in on itself, much like a length of twine that you've twisted and folded over and over and then crumpled in your hand. What you've come up with is a mass of twine rather than a strip of it. Similarly, what was once a lengthy protein chain becomes more nearly a solid structure with an inside and outside of sorts, an exterior and interior. The newly folded protein then migrates to its appropriate position in the cell and takes up its duties. Some proteins are enzymes, others hormones. Some form muscle fibers, others hemoglobin, which carries oxygen from the lungs through the bloodstream to body tissues. Some form receptors on the surface of cells to receive that oxygen, others receptors that receive different callers, such as viruses, which themselves are mostly protein. Proteins make up a large part of almost every cell in the human body.

Such is the process of life, a process that occurs, again and again, in every cell of every living thing and has, in similar if not identical fashion, since life began. The great continuum from Adam — or from the first one-celled organism broken free from the undifferentiated primordial slime — has been a function of these chains coiling and uncoiling, these molecular codes replicating and translating into the stuff of life. We are eternally connected to our forebears by this unceasing dance; through it we carry them always with us. It is the discovery of all this, no less, to which Avery provided the inspirational spark.

But if *attaining* such humbling knowledge might have been difficult to anticipate, who could have predicted the extent to which we have learned to interfere in the process — that is, *manipulate* it? Not only are we now able to observe this molecular dance, we are able to direct it. We are able to introduce our own hoofers, which carry out the steps we have choreographed for them, changing genetic instructions to reflect the outcome we desire. We are able to create, almost from scratch, the proteins and nucleic acids involved by manipulating in the laboratory the amino acids and nucleotides of which they are made. We are able to take strands of DNA from their accustomed place in a living cell and ensconce them in another kind of cell for our own purposes. For example, by transferring a certain protein-making gene from our body into a simple bacterium, we can force the bacterium to make that same protein. And, as bacteria reproduce and grow many times more quickly than humans, they in effect will mass-produce the protein, supplying us with huge amounts of it. The process is called "cloning" and is being used, for instance, to produce large quantities of a cellular receptor that, attacked by the AIDS virus, allows the virus entry into our cells. If we can divert and obstruct the virus by introducing huge numbers of these cloned receptors into the bloodstream, perhaps we can prevent the disease.

Such procedures fall under the rubric of genetic engineering, and that brings us back to viruses. For viruses have become a primary focus of genetic engineering, not only to try to prevent the diseases they cause but to better understand the creatures themselves. For decades we have known that viruses are essentially bundles of nucleic acid enclosed in a shell of protein. As we'll see, there are many varieties on the theme, but that's the basic idea. As such, viruses lack the machinery for reproduction. In contrast to cells, they have neither the enzymes to catalyze the process (although some viruses do carry some enzymes), nor the ribosome factories to carry it out. In order to survive, then, they must hijack the cell's reproductive capability. In so doing, a virus is a model of molecular simplicity. Its protein coat serves to protect its vital nucleic acid until the virus finds itself in proximity to a cell, at which time the coat binds to the cellular surface so as to facilitate the virus's entry into the unfortunate cell. Then the protective protein falls away, exposing the viral nucleic acid, which seats itself in the command post formerly occupied by the cell's own genes, and proceeds to direct the cell to start making new viruses.

Molecular biologists not only know all this, they are able to break the process down into its constituent phases, stop and start it along the way, discern the function of virtually every protein and nucleic-acid molecule, and actually reconstitute the makeup of viruses so as to produce varieties never known in nature. Viruses have actually facilitated the rise of genetic engineering. As they are little more than a collection of genes inside a protein shell, they allow scientists an unmatched access to DNA and RNA and to the processes they initiate.

What Pasteur had intimated, Beijerinck had attempted to define, and Stanley had made apparent was now obvious: the invisible world contained the seeds of the visible. The processes of the submicroscopic world shaped all life. Nowhere was that truth more apparent than in the realm of viruses. And with the new capabilities of growing viruses in the lab, suddenly there were more viruses to work with than anyone had ever anticipated. If the late nineteenth century, following the discoveries of Pasteur and Koch, was the golden age of bacteriology, the middle twentieth century, following the discoveries of Enders and the emergence of molecular biology, was the golden age of virology. It seemed as though new viruses were popping up wherever you looked. Rhinoviruses, hepatitis viruses, a seemingly endless stream of new influenza and herpesviruses, and then, somewhat later, the human retroviruses — viruses were busting out all over.

Most of these viruses were not new, of course — it was our emerging knowledge that was new. But our bodies had known of them all along. Our immune systems had been confronting the subminiature geodesic domes and soccer balls and spaceship-like invaders of the viral world all along, but in the past we knew very little of such encounters. Pasteur saw that his rabies vaccine prevented the disease, but he didn't know how. The preventive mechanisms his vaccine prompted in the body were a mystery. Theiler and the other vaccine makers who followed Pasteur were aware of much more — they knew of the existence of antibodies, for example — but it would take many more years before an awareness of the multifaceted nature of the immune system permeated the everyday thinking of medical science.

Now we realize that the armamentarium of the immune system is as fantastic as the creatures they protect us against. Antibodies, once thought to provide most of the immune system's defense against disease, are now understood to be only an advance line of protection. They are Y-shaped proteins, produced early in the immune response, that seek out and stick to foreign intruders such as viruses and bacteria, disabling them until other immune-system defenses can properly dispose of the unwelcome visitors. More early protection is provided by macrophages, large microbe-eating cells (the term means "big eater") that can not only devour the invader by themselves but also initiate further immune defenses. Hard on the heels of antibodies and macrophages come helper T cells, killer T cells, suppressor T cells, and other bodyguards, all coordinated as precisely as any modern task force could be, communicating through contact with specific mechanisms on their respective surfaces called receptors.

Receptors, made of protein, are the gateways to our cells — they "receive" visitors. It is by binding to receptors on other cells that cells communicate, if only, as in the case of the immune system, to say, "I'm holding fast to a foreign invader. Help!" But cellular receptors may also invite unwelcome visitors. As noted, viruses attach themselves to receptors when they infect cells. It follows, then, that one of the major avenues currently being pursued to stop viral infection is to block these receptors.

By now it's clear that a comparison to science fiction may not do justice to the disease-fighting skirmishes that go on constantly inside our bodies. These are Star Wars with a vengeance, Star Wars on a daily, hourly, second-by-second basis. These are Star Wars whose stakes are easily as great as those in a science fiction plot — Star Wars in which the stakes are survival itself.

And now, as we've seen, we ourselves have consciously joined the fray on a scale never before possible. We have learned to combine forces with the naturally occurring processes inside our bodies and to add to their effectiveness our own implements and designs. Medical science now makes the transition from the outside to the interior world almost as effectively as the invisible invaders, the viruses themselves.

Consequently, as recently as a decade ago, people considered it possible that we had at long last conquered infectious disease. Antibiotics had rendered much of bacterial disease controllable, and the many modern vaccines had greatly reduced the incidence of viral diseases. In the wake of such triumphs, the golden age of virology found itself in premature eclipse, a victim of its own success. Virologists working in research facilities across the United States began to despair at the paucity of PhDs and post-docs available to work in their labs. The study of viruses seemed to be heading toward a quiet semiretirement.

Today, of course, we know that the dream of eliminating infectious disease, seemingly so near to fruition, was only that — a dream. At the beginning of the 1980s AIDS appeared. At the same time, scientists were making other disquieting discoveries that quickly and, it seems, permanently put to rest any notions that there was no longer much of a need to study viruses. Researchers found what seemed to be brand-new viruses — at least brand-new in their propensity to invade human beings. And they found that cancer — long one of the most dreaded of afflictions, but always assumed to be the result of inherited or environmental determinations rather than an infectious agent — was in some cases caused by viruses.

With the surfacing of these newly discovered consequences of viruses, the question became, what else might be the handiwork of these fantastic organisms? It's a question that continues to be asked, and answered, with no end in sight. Influenced by the fact that there are more discovered viruses than there are known diseases with which to connect them, virologists have begun to suspect that the role viruses play in our lives is greater, more profound, and certainly more complicated than merely causing discomfort.

"Viruses are in us all the time and may be the arbiters of our health and futures," says Paul Cheney. Cheney, a physician at the Nalle Clinic in Charlotte, North Carolina, is best known as one of the people who first recognized chronic fatigue syndrome, a disease that has afflicted many people in recent years. It is a disease whose cause, first thought to be a virus, is now debated. If a virus is involved, it certainly is working in concert with some kind of immune-system abnormalities — a situation that at once points out how far we've come in our ability to deal with the submicroscopic world and how far we have to go. No wonder Cheney says, "We're like aborigines dealing with fire."

The metaphor again is apt. Viruses may "burn" us — and they do, more often than we'd like. At the same time, they offer the possibility of light and understanding. The chapters to follow explore the power of viruses in the context of major viral concerns, and major viral discoveries, of our day.

Ten

. . . .

The Yuppie Plague, the Kissing Disease, and a Link to Cancer

. . . .

"I first noticed it after I had a wisdom tooth extracted five years ago. The first couple of days afterward I thought that I must be taking a long time to get over the anesthesia. I even called the dentist and said, 'Could it still be lasting?' and he said, 'No, it should be out of your system by now. Maybe you have the flu.' "

Karen Rian knew that it wasn't the flu. She had no fever, her throat wasn't sore, her head didn't ache — not yet, anyway. She simply felt tired. No, not just tired — exhausted. So exhausted, it was as though she were still partially anesthetized. And the exhaustion dragged on for weeks, then months. It was so total that she couldn't even muster the energy to make an appointment with a doctor.

In time she developed a minor sore throat, and every once in a while her head ached a bit, but mostly it was the exhaustion and the depression that came with it. "I was feeling horrible about being alive, about everything," she says. "Finally I made the energy to see a doctor. I had to because I couldn't keep living like this."

The doctor, a general practitioner who had been recommended by friends, examined Rian, did routine blood tests, and announced finally that except for her high cholesterol level, Rian was perfectly fine. It was probably just a bout of flu. Not to worry, it would go away.

"He didn't go so far as to say, 'It's all in your head,' " explains Rian "but that was the implication. According to him I was okay physically, so there was only one other alternative. But it was unspoken, at least. I appreciated that."

There she was, thirty-nine years old at the time, the deputy director of a large social-services agency in Santa Cruz, California. She was intelligent, keenly analytical, and articulate, the holder of a PhD in history of consciousness from the University of California at Santa Cruz and a BA in philosophy from Cornell University. And here was some doctor implying that the months and months of sudden, debilitating exhaustion and depression were solely the result of a psychological problem.

"So I said, 'You don't understand. I can't do *anything*. I can't move. This isn't normal.' And he said, 'But I can't find anything wrong with you.' "

And that was it. Rian was left to her life — what remained of it, that is. She would drag herself to the nearby mom-and-pop market for a frozen dinner or some fast food ("I could never quite get it together to make a shopping list and actually go to the grocery store"), drag herself home to her two cats, turn on the TV, and collapse into bed. Friendships and social life simply disappeared. The exhaustion was all-consuming.

But Rian continued to work. "It was only by tremendous exertion of will," she says. "I literally had to force myself out of bed in the morning. Once at work, I wasn't really there. I remember being in meetings and thinking, 'I'm not paying any attention, so why am I here?' I would write a letter and be completely detached from it — like a robot writing a letter." Amazingly, her performance didn't suffer, at least not enough to cost her her job. But the exhaustion continued.

"I started forgetting names a lot," Rian says. "Names of things, people, places. There were times when I'd feel so depressed that I wanted to volunteer for vivisection. Maybe if they cut my brain open and saw what neurotransmitters were doing what, they might learn something about it. But mostly I adopted an intense intellectual curiosity about it. It was a way of saving myself. Otherwise I might've just given in."

It was now the fall of 1985; Rian had been ill for a year. On the lookout for any information about her condition, she began to notice a number of interesting articles in newspapers and magazines. They sported such titles as "Like a Flu That Won't Subside" and "Woman Battles Incurable Disease." The articles described people who were exhausted all the time and suffered sore throats, muscle and joint pain, weight loss, memory loss, and had difficulty concentrating — precisely the same symptoms Rian was experiencing. These people had run the gamut of doctors who either weren't able to diagnose their problem or who tended to doubt that they had an actual illness. Some of the physicians had bluntly recommended that their patients see a psychiatrist. Just to be taken seriously was a victory. "I was so happy," one woman exclaimed. "I didn't care that there was no cure. Here, finally, was a doctor who believed me — and he had a name for my illness."

Rian felt an immediate empathy. She understood the predicament only too well. And she felt as well a strong sense of relief. There were people out there who were going through what she was going through, and there were doctors who were taking people like her seriously. She wasn't crazy after all — there was indeed a real disease at the root of her problems. And the ailment even had a name. It was a cumbersome, tentative name, applied to a tentative diagnosis, but it was a name: chronic Epstein-Barr viral infection — CEBV, for short. An informal label, on the other hand, caught the media's fancy and reflected the people the disease seemed to strike as well as the skepticism surrounding it: yuppie plague. (It was also given another, equally flip designation: affluenza.) For evidence — albeit more subjective than confirmed — was accumulating that for some unknown reason this mysterious disease, if that was indeed what it was, primarily struck intelligent, ambitious, and energetic professionals between the ages of eighteen and forty-two, most of whom were women. Rian fit the profile well enough. The culprit was suspected to be a virus — the Epstein-Barr virus, a relatively obscure member of the herpes family of viruses known to physicians primarily because it causes infectious mononucleosis.

Bolstered by the information, Rian decided to see another doctor (it was now the fall of 1986). This time the result was a new diagnosis: Rian indeed was suffering from CEBV, as well as from candida, a rare disease caused by a yeastlike fungus. Though some of the symptoms of candida are similar to those Rian was experiencing, she doubted the candida diagnosis from the beginning — the lab results seemed to her critical eye inconsistent. But any diagnosis was better than none, and she began taking the candida medication the doctor prescribed. It was a mistake. She began to feel worse than ever and soon stopped using the prescription. For her, it was back to square one.

Meanwhile, her most likely ailment was obscure no longer. *Newsweek* magazine dubbed chronic Epstein-Barr viral infection "the malaise of the 80's." Major features about the fatigue syndrome proliferated through the entire range of media, from *Science* magazine, to the *New York Times,* to *Rolling Stone,* to ABC's TV newsmagazine "20/20." Estimates varied as to the number of people at risk. One researcher suggested that twelve million people would come down with the disease in the next few years in the United States alone; others dismissed the number as extravagant but still envisioned the syndrome afflicting millions; other declined to predict.

None of the predictions was wholly preposterous, however, because sooner or later in the course of our lives virtually all of us come into contact with the Epstein-Barr virus and will then carry it until the day we die. It's a ubiquitous virus: it eventually shows up virtually everywhere — in virtually everyone — in the entire world.

Ubiquity is a new concept in our chronicle of viruses. The viruses we've looked at so far — poliovirus, smallpox virus — occasionally strike and then, whether repulsed by the immune system or passed on after having caused disease, depart. They are travelers to whom we'd rather not extend hospitality, against whom we've developed defenses in the form of vaccines to ensure that their sporadic visits will be brief and without incident. But here is a virus that visits everyone and comes to stay. The fact is unsettling. The idea that we must, eventually and subsequently forever, suffer the intrusion of such submicroscopic hitchhikers does not sit well. When we're sick, yes — it's easy enough to swallow the notion that something foreign within us is causing our discomfort. But always, and without our suspecting it? If so, why aren't all of us in Karen Rian's shoes? There should be so many cases of CEBV — and mononucleosis, for that matter — that the estimates of millions of cases of the disease would seem like small change.

The answer is not easy to come by. It involves the nature of viruses and the nature of our bodies. It has to do with our ability to defend ourselves against such invisible invaders. The answer shapes unfamiliar notions of the nature of disease and of the immune system and shakes our confidence in modern medicine's ability to bail us out of such tight spots. It is an answer that is at once hopeful, disquieting, and humbling.

It is also supremely interesting. In an irony appropriate in the often surprising world of viruses, researchers have experienced a twisting adventure that now actually casts doubt on the Epstein-Barr virus's role in causing this debilitating disease of exhaustion, which, among other designations, has been newly christened chronic fatigue syndrome, or CFS, to reflect that doubt. We will jump back into the story in its relatively recent stages and in a most unlikely place: a tiny resort community on the shores of Lake Tahoe.

Straddling the border between California and Nevada, Lake Tahoe has it all: swimming, boating, waterskiing, fishing, hiking, camping, snow skiing, and, on the Nevada side, a nightlife of casinos and entertainment. Incline Village, a town of between four thousand and five thousand people on the north shore of the lake, is one of a number of flourishing Tahoe communities that attract well-heeled vacationers and second-home owners to the area. These communities depend on publicity, but the publicity that was appearing in 1985 was not the kind the gambling casinos and chambers of commerce had in mind.

"National disease experts are investigating reports of an unusually high occurrence of a mysterious, mononucleosis-style illness striking residents on the north shore of Lake Tahoe." So began a typical article, this one in the *Sacramento Bee*. It went on to announce that at least eighty residents of the village were experiencing headaches, sore throats, and, worst of all, overwhelming fatigue. The exhaustion was so bad, said one woman, that she felt like Raggedy Ann without the stuffing. Thus was coined yet another name, the Raggedy Ann syndrome, for the disease that had struck Karen Rian and thousands of others across the country. And some of these people had other, even more disturbing complaints — weakness in the limbs, blackouts, disorientation. One person driving in the Incline Village business district, which has only three intersections, couldn't even remember how to get home.

The unafflicted locals were aghast at the unfavorable national exposure. And when ABC's "20/20" came to town, the local reaction intensified. Chris Guthire, a young fatigue syndrome sufferer who was featured in the TV report, was warned while standing in line at the checkout counter of the local supermarket that her TV appearance would ruin business around Lake Tahoe. "I just felt devastated," she later told a national magazine. "I was afraid I would be ostracized or run out of town with a scarlet EBV on my chest."

But there was no stopping the publicity, or the disease. By October of 1985, Paul Cheney and Dan Peterson, the physicians who had first diagnosed the ailment in town over a year before and who, as the only doctors who admitted to the problem, were seeing virtually all of the area's fatigue syndrome patients, were now diagnosing as many as fifteen new cases every week. One-third of the faculty of a high school came down with the disease. So did an entire girls' basketball team and members of the Hyatt Lake Tahoe Hotel and Casino staff. "We were horrified," Cheney said to a reporter. "We felt like we were in a nightmare that wouldn't end."

So they called for help. It arrived, such as it was, under the auspices of the Centers for Disease Control.

When it comes to public-health enemies, the Centers for Disease Control in Atlanta is the medical community's FBI. An agency of the federal Public Health Service, the CDC, as one of its duties, investigates and tracks the course of major diseases in an effort to separate real dangers from imagined ones.

Now Paul Cheney was relieved. He hoped a number of questions would be answered. Where did the problem come from? How many people were involved? What was its cause? Why was it rearing its head now? In other words, just what in the hell was going on? It was simply too much for a couple of small-town doctors to handle by themselves.

But instead of answers, what they got was more confusion. Once the publicity-sensitive townspeople got wind that a couple of federal investigators were nosing around trying to track down a mysterious disease, Cheney and Peterson, whose reputations were already suspect, became even more unpopular. Their cause wasn't helped when misleading snatches of the news drifted all the way to San Francisco, causing Bay Area newspapers to begin inquiring about the supposed outbreak of AIDS at Lake Tahoe. And when a reporter polled the other doctors in town and learned that they hadn't seen any cases of this so-called illness, the surreal atmosphere intensified.

CDC epidemiologists Gary Holmes and Jonathan Kaplan stayed in Incline Village for a week, interviewing patients, taking blood tests, and conducting a case-control study. Their conclusions, published eight months later, did nothing to clear things up. In fact, they pushed the whole business back to where it had been when Cheney and Peterson and other physicians around the country had first begun to see cases of chronic fatigue.

"Our study indicates that EBV serology was of little value in diagnosing individual patients thought to have the fatigue syndrome," the report stated. Besides that, Holmes and Kaplan found that far from being exhausted and confused, some of the people who were infected with the Epstein-Barr virus were perfectly healthy. The report went on to suggest that there wasn't enough evidence to consider the problem a single disease and therefore proposed a more generalized term — "chronic mononucleosis-like syndrome" — instead of CEBV. It ended by urging physicians to search for "more definable and often treatable conditions" before diagnosing the problem as chronic mononucleosis-like syndrome — conditions that included "anxiety and depression." In other words, a tip of the cap to the old insinuation that it was all in the afflicteds' heads.

To say the least, Cheney and Peterson were disappointed. They suspected that the CDC, by this time up to its neck in reports of the mysterious ailment, simply wanted to put a stop once and for all to what it considered a case of overkill. As far as they were concerned, Holmes and Kaplan's methods were not even close to being adequate. "Why anyone would think that a week invested here in a simple case control of Epstein-Barr-virus serologies would solve this puzzle is remarkable," complained Cheney. "This is far too complex."

But the CDC's Holmes stuck to his guns. "I was less than convinced that there was an epidemic. Some of these people appeared to be quite ill, but it's hard to get overly excited about this thing when you can't even be certain that they're telling you the straight truth. You can't be totally convinced that they're not just telling you a string of stories to make you think they're sick. And I think that's probably going on in a viable percentage of these people — they *want* to be diagnosed as having it."

But why? According to Holmes, the answer has to do with the kind of people who complain of the problem. "It tends to show up in young to middle-aged people who tend to be kind of high-stressed, high-strung individuals who may be overdoing themselves. They may be looking for a way to be able to cut back on their activity." In other words, if you can't quite make it in the professional world but can't bear to admit it, a debilitating but not life-threatening illness is the perfect way out.

If Holmes and the CDC's stance made sense to some people, it did little to dissuade Cheney or the growing numbers complaining of the mysterious ailment. For his part, Cheney had never stopped investigating the problem. He virtually submerged himself in the fatigue syndrome, running up phone bills of two thousand dollars a month talking to interested physicians around the country, and devoting two or three unpaid days every week to the problem. Soon his judgment that something real and important was going on seemed to be vindicated. In December, just three months after the CDC visit, the farming town of Yerington, Nevada, fifty miles to the southwest of Incline Village, reported its own outbreak of the disease — 105 patients out of 2,800 residents. And two months later, in February of 1986, Cheney's efforts were given a boost by another player: Harvard Medical School professor Anthony Komaroff agreed to come to Incline Village.

Komaroff, the director of general medicine at the Brigham and Women's Hospital in Boston, had been seeing CEBV cases of his own since 1983. In fact, he had suspected the existence of a chronic fatigue syndrome as early as 1977 but hadn't been able to find research collaborators who considered it a plausible illness. Like Cheney, he was impressed by the suddenness of the syndrome's onset and how it completely turns around the lives of the people it strikes.

"It's generally an adult," Cheney explains, "who has been in excellent health, until a moment in time. They can often give you the date, and sometimes the day and time that all that changed. It usually changes with the onset of a viral syndrome, usually upper respiratory or flulike, following which there's the development of a fatigue syndrome, waxing and waning most of the time, with some other characteristic complaints. That's the classic case."

"One day they were in the prime of life, the next day they became ill, and they were never the same again," Komaroff adds. "When I hear that story, I find it hard to believe I am dealing with a primary psychoneurosis. The patients say, 'It all started with that virus.' I think they're right."

Komaroff and Cheney had struck up a correspondence in 1985. Now Komaroff accepted Cheney's invitation to visit. He and his team conducted a thorough epidemiological study at Incline Village and also at Yerington. Their conclusion reinforced what Cheney knew all too well: something extraordinary was happening. Komaroff concluded that there were at least five convincing reasons to consider the fatigue syndrome a nonpsychiatric disease. They ranged from the fact that a number of the symptoms of the syndrome are similar to those of other viral diseases, to the evidence offered by Incline Village and Yerington that it can strike clusters of people at the same time and therefore may be the result of common exposure to some external infectious agent, to the remarkably similar stories fatigue syndrome sufferers tell about the onset of the disease. It just didn't act primarily like a psychiatric problem.

Cheney and Komaroff decided to join forces in investigating the disease. They made an unlikely team — the lanky, fair, small-town doctor and the stocky, dark, big-city medical-school professor — but they shared an intense determination to get to the bottom of the problem. With Komaroff's help, Cheney made contact with several other interested physicians, among them James Jones of the National Jewish Center for Immunology and Respiratory Medicine in Denver, who had been seeing cases of the disease for years, and Stephen Straus, head of the medical-virology section of the National Institute of Allergy and Infectious Diseases in Washington. Jones and Straus had written the two influential early papers on the syndrome. With others at such places as Duke University and the University of Massachusetts, the researchers formed a collaborative network aimed at discovering as much as possible about the disease.

In time they found out a great deal. "It'll often start out with one set of symptoms and then evolve to another set of symptoms as the first set begins to wane," explains Cheney. The first phase involves aching muscles, low-grade fever, sore throat, and, of course, intense fatigue. No wonder the disease was labeled chronic mononucleosis at the beginning. The later period is at once less debilitating and more worrisome. "When they're in the mono-like phase, they're often sicker and more dysfunctional," Cheney says. "When they enter the later phase, their functional capacity often improves. So you have the impression, my gosh, they're getting better. But actually they're in the part of the disease that's more prolonged and whose evolution, whether improving or worsening, is less obvious." This is the period of memory loss, difficulty in concentrating, problems with balance and coordination and, waxing and waning through it all, the ever-present fatigue.

While Cheney and his collaborators were discovering more about the disease, sufferers began forming their own support groups. The largest of them, the National Chronic Epstein-Barr Virus Syndrome Association in Portland, Oregon, enrolled more than eleven thousand members in its first year. Since they were receiving little help elsewhere, these people decided to take matters into their own hands: they would challenge the entrenched medical profession by themselves.

If they found it hard to muster enough energy to make their concerns known in person, they had little difficulty putting their grievances down on paper. The National CEBV Association in Portland distributed some sixty thousand brochures in its first two years. Letter-writing campaigns were so effective that one physician jokingly suggested that polygraphia — voluminous letter writing — be added to the list of fatigue syndrome symptoms. CEBV was no longer a little-known ailment. Even the CDC's Gary Holmes, while scoffing at such figures as twelve million sufferers, admitted that "untold thousands are being diagnosed. How many have the disease is incalculable." And the disease remains difficult and evasive. "It drives me crazy," Holmes says. "I wonder every day about my continued work in this area. I can't get anything firm to work with. The problem is, who do you define as a case? It's extremely murky."

Has there ever been such a well-publicized, thoroughly debilitating, and potentially epidemic-sized illness that is such a morass of confusion? "Virtually all the important questions about chronic viral fatigue syndrome remain unanswered," states Komaroff.

Then what can we be sure that we do know? In March of 1988 Holmes and fifteen other physicians attempted at least partly to answer that question by coming up with the CDC's official working case definition of the disease. The major criteria were that a patient must have "debilitating fatigue" lasting more than six months and reducing activity by 50 percent, and that no other disease that might bring on similar symptoms be present, diseases that include cancer, autoimmune ailments, and AIDS. In addition, the patient must show other common symptoms of the syndrome, some of which are fever, sore throat, headaches, unexplained muscle weakness, confusion, and forgetfulness.

But while it was a welcome step in the direction of recognizing the disease as a real entity and in providing physicians with specific criteria for diagnosis, the case definition did nothing to help decipher the cause — except to suggest that, because its origin was so cloudy, the disease be given a new name: chronic fatigue syndrome, or CFS. That nomenclature reflected, as noted, the growing suspicion that the Epstein-Barr virus — EBV, as it came to be known — may not be the sole cause.

Still, many researchers feel that somehow, to one extent or another, EBV may be involved, if not in all patients, at least in some. And like all viruses, others of which may play a role in the syndrome, it works in insidious ways. The Epstein-Barr virus attacks our immune system itself, infecting certain white blood cells called B lymphocytes. These are cells that are formed primarily in the lymph nodes and bone marrow and that circulate through the blood, manufacturing antibodies against disease. Once infected with the Epstein-Barr virus, however, B lymphocytes become transformed into mutant cells that reproduce wildly.

This unchecked proliferation alerts other inhabitants of the immune system, the killer T cells. These cells are equipped to target the virus-infected B lymphocytes and destroy them. The upshot is that the body literally goes to war with itself, and it's this internal battle — the body actually killing off parts of itself — that makes the patient sick. T cell after T cell attacks infected B lymphocyte after B lymphocyte until the Epstein-Barr invasion is finally under control. But while the battle hangs in the balance, the patient feels terrible.

At least that's the scenario in infectious mononucleosis; no one knows what happens in CFS. While the battle with mononucleosis usually lasts no longer than a few weeks, the CFS battle may go on for months, or even years. Something as yet unknown tips the scales.

A final twist, in the case of mono, is that no matter how thoroughly you might recover, your immune system never completely wins the battle. A few transformed B cells containing the Epstein-Barr virus — about one in a million — remain, forever threatening to run wild again, an event that the ever-vigilant T cells do their best to keep from happening. It's a kind of armed truce.

Before CFS, the effect of this standoff was a lifelong immunity to similar ravages of the virus — if you couldn't completely rid yourself of it, at least it could no longer gain the upper hand. It may be that that assurance exists no longer. For people with CFS are almost always people who have already fought the battle with the Epstein-Barr virus, having suffered through mononucleosis or a mild, flulike, EBV-caused childhood ailment. Now they're getting sick again. Something seems to be upsetting the balance. Whether the virus itself is reactivating somehow, or another microorganism is entering the fray, disrupting and overwhelming the body's immune system, no one is quite certain. Thus, as suggested, CFS seems to be an illness that may or may not involve the Epstein-Barr virus and something — or some *things* — else.

But while digesting that unwholesome notion, recall that in mononucleosis, EBV infects white blood cells, causing the cells to mutate so as to reproduce unchecked. Well, an illness of transformed cells multiplying out of control resembles nothing so much as perhaps the most dread disease of all: cancer. In fact, it's almost an exact description of leukemia, cancer of the blood. Leukemia too is caused by uncontrolled reproduction of blood cells. While primarily affecting different white blood cells, called leukocytes — thus the name, leukemia — which normally help guard the body against infection, the disease can also strike red cells, which carry oxygen, and platelets, the tiny disk-shaped structures that help coagulate blood. The result can be the destruction of the body's ability to protect itself against infection, as well as anemia and blood-clotting disorders. In all cases, the aftermath can be death.

Sometimes leukemia at first expresses itself much like mononucleosis or, for that matter, CFS, causing fatigue, lethargy, and weakness. Some people lose their appetite and therefore lose weight. The difference is that in mononucleosis, anyway, while the body may not completely win the battle, neither does it lose. After a period of weeks or months, mono relaxes its grip. In contrast to leukemia, mononucleosis is not a life-threatening disease. Accordingly, it has been called "a self-limited leukemia," a description that underscores its resemblance to cancer.

But while mono only *resembles* cancer, other kinds of EBV infections may actually *cause* cancer. In fact, the Epstein-Barr virus *is one of only four viruses ever shown without doubt to be involved with cancer*. The fight against the Epstein-Barr virus, then, has important ramifications beyond mononucleosis and CFS. Chronic exhaustion is one thing — death is another. Understanding how the Epstein-Barr virus works may help us understand — and prevent — cancer.

It's an effort that began some thirty years ago in tropical Africa, when the world had never heard of a virus called Epstein-Barr. And like so many great scientific beginnings, it might never have happened without the intervention of sheer chance.

The year was 1957. Denis Burkitt, a Scottish surgeon stationed at Mulago Hospital in Kampala, Uganda, was examining a frightened five-year-old boy named, appropriately, Africa. The boy was in bad shape. He had grotesque swellings on both sides of his upper and lower jaw. Burkitt had seen tumors of the jaw in children before (he usually removed them surgically — a harsh, disfiguring surgery that only delayed the inevitable, death) but never before had he seen four symmetrical swellings on both sides of the face.

A few weeks later, while Burkitt was visiting a different hospital about fifty miles away, he happened to glance out a window of one of the wards. There, on the grass, sat another child with a swollen, lumpy face. Burkitt hurried outside and examined the little girl. "It was just the same as the mouth of the little boy I had seen in Mulago Hospital," he said later. "I took photographs again, and I made notes. And I began thinking about these things seriously." He arranged that the sick little girl and her mother be driven the fifty miles to Kampala, so that he could examine and try to treat her. It was to no avail. Soon, helplessly, he watched her die.

Burkitt was, and is, an especially tenacious and determined fellow. He had prevailed over the loss of an eye in an accident and the hardships of living in a developing country far removed from his native Scotland. Moreover, he was a family man himself. He was married, with three young daughters. And he was deeply religious — he had come to Africa to serve God. What better way to answer his calling than by searching for any possible clues as to what had afflicted the two children? And there was a clue, tenuous to be sure, but better than nothing: the little girl not only had malignant tumors on her face, she had them in other parts of her body as well. That was a surprise. Was it just a coincidence, or could there be a connection between the jaw tumors and the others? Might they even have the same cause?

Burkitt began to search through past clinical records and autopsy reports at Mulago Hospital. Sure enough, the documents showed that where there were jaw tumors, there were invariably growths in other parts of the body as well. As the prevailing wisdom maintained that tumors that developed in specific organs had their own specific causes, no one had ever before drawn a connection between the various tumors.

The next step was to find out the age distribution of the tumors. Burkitt knew that different forms of cancer had different age patterns. If there were some consistency in the ages of the children with multiple tumors, it might be a sign that the cause of all of them was the same. Again the facts supported his suspicions: the tumors afflicted children between the ages of two and fourteen, with the peak incidence occurring at five years old. The age distribution indicated that this was different from any other known malignancy in children. This was a unique form of cancer.

Burkitt knew he was onto something important, but as a working surgeon he had neither the time nor resources to track it down by himself. So he went for help, first to a nearby medical school. Doctors there examined a collection of 106 slides of children's cancers similar to the jaw tumors. They found that under the microscope Burkitt's jaw tumor cancers looked exactly the same as those in other parts of the body. Burkitt had been right: no matter where in the body they showed up, the tumors were all the same type of cancer. It was a lymphoma, or cancer of the lymph system. Moreover, it was a lymphoma that had never been classified before. And beyond that, it was responsible for nearly half of the children's tumors recorded in the Kampala Cancer Registry. Extrapolating from that surprising fact, Burkitt suspected that this kind of cancer might account for half of all childhood cancers in tropical Africa. If so, it would be the commonest of all children's cancers.

Burkitt went back to the records, this time to try to find historical evidence of the tumors. Sure enough, there were pictures and descriptions of cancerous growths dating from the turn of the century, with specific reference to jaw tumors. It was clear that Burkitt was not dealing with a new disease, but rather one that had never been acknowledged or classified. This cancer had been afflicting African children for over half a century, at least (and venerable wooden carvings depicting grotesque swellings around the jaw suggested that it had been around for a great deal longer than that). And nothing had been done to stop it. The stakes were getting very high very quickly.

Soon Burkitt entertained a visitor, one George Oettlé, the director of the cancer research unit of the South African Institute for Medical Research in Johannesburg. Oettlé examined Burkitt's photographs of tumor patients and heard his growing concern, and then announced, with total certainty, that no such tumors occurred in South Africa.

Burkitt was stunned. The tumor occurred, and had occurred for at least half a century, in Uganda, but it didn't show up at all in South Africa. Why? And if there were no tumors in South Africa, some two thousand miles to the south, where did they cease occurring? There had to be a line — he called it an "edge" — that divided tumor territory from nontumor territory. If he could find that edge, it might provide a great deal of information. What kind of people lived on one side of the edge, what kind on the other? What were the environmental conditions on one side as opposed to the other? What about diet on one side compared to the other? Answers to these questions might provide some hints as to the tumor's cause.

So Burkitt applied for government grants — they came through, for the princely sum of twenty-five pounds (about sixty-seven dollars) — and used the money to print twelve hundred leaflets illustrated with his photographs of tumor-infected children. "We obtained addresses from everybody we could," Burkitt recalled. "Then we sent the leaflets to government and mission doctors throughout Africa, with a little questionnaire asking, 'How long have you worked in your hospital, and have you seen this condition?' "

It took three years to get the answers. "It was a part-time hobby, because I was employed full time as a government surgeon. But, at a guess, I might have had three to four hundred replies from different parts of Africa. Whenever I spoke at a conference I would get addresses from people. I would meet Dr. Brown, who knew Dr. Smith, who worked at such and such a place, and I would write to him. It wasn't all done at once, and the answers came in dribs and drabs."

But they came, and on maps hung behind his desk, using thumbtacks (home-colored to save money), Burkitt plotted the results. What emerged was a band across tropical Africa that descended west to east from about fifteen degrees north of the equator to about fifteen degrees south, with a tail that ran down the east coast of the continent to the Mozambique/South Africa border. It was the distribution of the tumor. It had discrete boundaries, an observable shape. This cancer wasn't simply a random occurrence.

Burkitt had found his edge — on the map, at least. But there were areas within the plotted territory where the tumor didn't seem to show up. In fact, despite the fact that it was a densely populated region, the southwest portion of his own Uganda reported little evidence of the tumors. Why? Perhaps the answer lay in the fact that the southwest was a mountainous area, and a closer look at the map revealed that a number of the other sparse tumor regions were high country, about five thousand feet. Could this cancer somehow be related to altitude, and therefore climate?

These days environmental links to cancer are taken for granted, but in the early 1960s it was a startling idea. As Burkitt studied his maps, he was ever more convinced that he was onto something. He was embarking on an entirely new approach to cancer research, what he called a "geographical biopsy." But studying maps was one thing — firsthand knowledge was another. It became clear to Burkitt that the only way he'd ever be able to find out why the cancer struck in one place rather than another was to take a look in person.

Easier said than done. The tumor territory — what came to be known as the "lymphoma belt" — covered five million square miles. He could never hope to navigate all that. And the social and geographical realities of the area further narrowed his options. Much of the northern part of the belt merged into the sparsely populated Sahara of Ethiopia and the Sudan. Travel there was pretty much out of the question. Due west of Uganda was the Congo, now Zaire, where Simba revolutionaries were on the rampage, and south of there was Angola, itself in the midst of revolution. They were no place for a European.

That left the eastern portion of the belt, a ten-thousand-mile expanse from Kenya south to the South Africa border. It was a densely populated area, with many hospitals. English was spoken over most of it, the roads were relatively good, and it was an area of vast contrasts in altitude and climate. In any case, it would have to do.

So, on October 7, 1961, accompanied by two other doctors from nearby areas, fortified by a paid leave from his surgeon duties and four hundred pounds (a little over a thousand dollars) in grant money, in a 1953 Ford station wagon purchased from a retiring missionary who had just escaped from the war-ravaged Congo, Burkitt set forth on a trek through eastern, central, and southern Africa — an adventure that soon became known as the Long Safari. He had no way of knowing that the big game he hunted was the tiniest and most elusive of prey, a prey more dangerous than any number of rhino and lions, a prey that finally was about to give up some of its secrets (although it would be almost another decade in the doing): the Epstein-Barr virus.

It was, Burkitt wrote in his diary, "the safari of a lifetime." In ten weeks, exactly as planned, the three men covered the ten thousand miles — equal to three times across the United States — and visited twelve countries. The terrain they covered ranged from rain forest to dry bush, from perilous mountain passes to seemingly endless flatland. They navigated rough roads by day and invisible tracks at night, guided only by a compass. They passed through thickly populated villages and towns and interminable, empty country. In the process they crossed the longest bridge in the world, observed the great Victoria Falls, and, in a mud-walled Arab town called Ujiji on the eastern shores of Lake Tanganyika, visited the very spot where, ninety years earlier, the *New York Herald* reporter Henry Stanley had greeted the most famous missionary of his day with the words, "Doctor Livingstone, I presume." And all the while, Burkitt kept a record in his diary.

Wednesday, 11 October (8:20 P.M.)

We left Kigoma at 8:45 this morning. I have never traveled through such endless, monotonous African bush as we did today. Mile after mile, 10 mile after 10 mile, and hundred mile after hundred mile, for 250 miles on dusty earth roads. Everywhere was dry as dust, brown and burnt. Just occasionally we went down an escarpment or through a pass bordered by great rocks. On the first 60 miles to a Roman Catholic mission we passed only one car. For the next 170 miles we didn't pass a single moving vehicle — just three stationary ones. We saw hardly a soul.

Wednesday, 8 November

(Pretorius Kop camp, at lower end of Kruger National Park) We have just returned from seven hours in the park, and had a wonderful time. We must have seen a thousand impala in all. We also saw lots of wart hogs, water buck, gnu, kudu, one elephant, one sable antelope (a very large animal). We had some remarkable experiences. I was about to photograph some vervet monkeys when one with a baby on her breast jumped up and actually sat with her baby on the ledge of the car window at my shoulder. It was too close to photograph!

At the end of such adventures were hospitals, missions, and clinics. At each one, Burkitt showed the staff his collection of tumor photos, toured the wards, and checked registers for evidence of the malignant lymphoma. He also gave lectures and treated patients of all kinds, occasionally even performing surgery. And slowly a pattern emerged that seemed to reinforce what Burkitt had noticed when studying his wall map. "We suddenly realized," Burkitt recalled, "that we weren't finding an *edge* to the belt. We were finding an *altitude barrier*."

Indeed, the tumor might commonly occur in any given area, but, depending on the locality, it seemed to disappear at a certain altitude. For example, near the equator no cases of the lymphoma showed up over about 5,000 feet. As they journeyed south, the altitude barrier changed. In Nyasaland, now Malawi, farther removed from the equator, no cases occurred above 3,000 feet. In Swaziland, even farther to the south, the altitude barrier was only 1,000 feet. And, of course, the tumors didn't occur at all in South Africa.

The findings were very strange, another oddity in the unfolding series of oddities connected with this bizarre cancer. By the time the men reached Johannesburg, the southernmost tip of the safari, they had amassed enough information to keep them busy for some time deciphering the results — which was fortunate, because soon thereafter the rains came.

Saturday, 16 December

Utterly and completely stuck on a detour off the impassable road between Kisumu and Uganda. Before me is a bus across the road with one side buried deep in mud so that the body is sitting on the ground. Beyond this is another bus in a ditch and beyond that a third right across the road from ditch to ditch. This has now been pulled into one ditch by a grader. Three lorries are also stuck. There is a row of cars at each end of this confusion optimistically hoping to get across sometime. It is difficult even to stand and I have fallen flat once. . . . Of course all the local populace are lining the road for the day's entertainment.

On Sunday, December 17, Burkitt and his compatriots arrived home. All in all, they had visited some fifty-seven hospitals and discovered over two hundred cases of the tumors. The cost of the ten-thousand-mile safari, including gains realized by reselling the station wagon, was 678 pounds, a little over $1,800 — absolutely cut-rate cancer research. And the news they brought back caused a sensation. No one had ever heard of a cancer dependent on altitude, not to mention a variety of altitudes. As the information got out, the surprised reaction was shared by other than Burkitt's African colleagues. Interested parties from all over the world, including *Time*-magazine journalists and American TV crews, began calling on Burkitt.

Burkitt's wall maps were now dense with colored pins, some four hundred of them. However, it wasn't until he showed his findings to a friend, Professor Alexander Haddow, the director of the East African Virus Research Institute in the Ugandan town of Entebbe, that the significance of the varying pattern became clear. Haddow checked the various altitude barriers against the patterns of climate in the regions involved and came to the conclusion that the real barrier wasn't altitude at all — it was *temperature*. The tumor simply didn't occur in areas where in the temperature fell below about sixty degrees Fahrenheit.

Of course. That explained why the altitude barrier was higher nearer to the equator and lower farther away: temperatures below sixty degrees occurred at lower and lower altitudes the farther the distance from the equator. To test the theory, Burkitt undertook two more journeys — short ones, this time: first to West Africa, then to Rwanda and Burundi, the two small East African countries the rains had forced him to bypass during the Long Safari. West Africa provided flat country in which the temperature consistently remains above sixty degrees; Rwanda and Burundi provided a useful contrast — high country where the temperature falls below sixty and relatively lower country with a hot and humid climate, both near the equator. If the pattern of tumor incidence with respect to temperature held true in these regions, it would clinch the argument.

To a point, it did. As Burkitt anticipated, tumors showed up in low-lying, warm Burundi but not in high, cool Rwanda. But West Africa didn't follow the rules. Whereas the temperatures were comparable in all of the countries there that Burkitt visited, the lymphoma didn't conform to the expected pattern. It occurred in southern Nigeria and Ghana but not in the northern areas of the two countries. And, more puzzling still, it was absent in the region around Accra, along the southern coast of Ghana.

Upon Burkitt's return, he and Alexander Haddow consulted government maps and statistical tables and found that in southern Nigeria and Ghana the rainfall ranged from two hundred to four hundred inches a year. But in the north, close to the Sahara, where the temperature was comparable, it fell to hardly twenty inches a year. And as for Accra, on the southern coast, it lay in what was called a rain-shadow — the rainfall in the area was negligible. The tumor occurred in the rainy country but disappeared where it was dry. Another dimension had been added to the mix: rainfall. This was one strange cancer, indeed.

The pattern of colored pins on Burkitt's wall map had become complicated. The edge he had sought was no edge at all but rather an amalgam of altitude, temperature, and rainfall factors that seemed to conform to no other discernible pattern. What map of Africa would show such a pattern? Haddow, who was an entomologist, came up with the answer: an insect distribution map.

An insect map. Could the cause of Burkitt's lymphoma have something to do with insects? No insect had ever been shown to cause cancer, yet a map showing the distribution of tsetse flies cor-

The "Lymphoma Belt"

Route of the
long safari (1961)

● Areas where there are several cases.

△ Areas where the tumour is known to be common but where there is no
specific documentation.

responded to Burkitt's tumor map. And charts of insect-carried diseases such as sleeping sickness, o'nyong nyong fever (an ailment that causes patients to feel as though their bones are broken), and yellow fever also exhibited a pattern similar to that of Burkitt's lymphoma.

There was as yet no conclusive evidence for the association, but for the time being it was enough. Here, for the first time, was a cancer that seemed to be caused by an infection, an infection mediated by an insect. "Once this thing got going," Burkitt recalled, "it went like a bombshell."

Although there are other diseases that kill, and, indeed, kill more frequently than does cancer, perhaps no other disease causes so much dread. Environmental conditions, heredity, diet, and lifestyle notwithstanding, cancer seems all too often to strike indiscriminately, a shadow that hovers near us in the brightest of days. The news that was coming out of Uganda, then, offered hope of a kind that had rarely been seen in cancer research. It was known that viruses can cause cancer in animals and that insects can transmit viruses, but no virus had ever been shown to cause cancer in humans. If an insect-borne virus actually caused this one cancer, it might be the breakthrough that the world desperately sought.

Burkitt had done all that he could. What was needed now was extensive laboratory research — something he was not prepared to undertake. He lacked the training and facilities for such difficult work. Besides, as usual his time was more than filled by his surgical duties and his continuing efforts to keep track of the occurrences of Burkitt's lymphoma, as the disease came to be called. So laboratories around the world inundated Burkitt with requests for tumor samples. One researcher, however, had an inside track: M. Anthony Epstein of Middlesex Hospital in London. In March of 1961, a good six months before Burkitt set off on the Long Safari, Epstein had heard him speak in London about his early suspicions concerning the tumors. Convinced that Burkitt was onto something important, Epstein visited Kampala and set up a system whereby tumor samples might be sent overnight by air to his laboratory. A year later, by the time the full impact of Burkitt's investigations had been sprung upon the world, Epstein was already hard at work, looking for the virus.

It wasn't easy to find. The first step was to grow the tumor in the lab. Only then would Epstein and his team have a consistent supply of tumor cells to study. The idea was to follow the lead John Enders had supplied a decade earlier and cultivate the cells in tissue culture — that is, in test tubes containing nutrients and maintained at the precise temperature at which the cells might flourish. The only problem was, while other kinds of cells from other sources had been successfully grown in tissue culture before, they did not include human lymphoma cells. Epstein had to figure out how to do something that had never been done before.

"For two years we tried to get these things to grow," he later recalled. "I tried twenty-four different samples, without success. Obviously, we were doing the wrong things in all our efforts."

Eventually, by trial and error, Epstein hit upon the right requirements of food and temperature. In 1963, having been joined by researcher Yvonne Barr, he finally succeeded in growing the tumor cells in his lab. The researchers gave the cells the code designation EB (*E* for Epstein, *B* for Barr). Now, with a steady supply of tumor cells, they could get on to the point of it all: finding out if the cells contained a virus. Their method involved looking at the cells under the intense magnification provided by the electron microscope.

Again it took time, but in 1964 a virus finally appeared under the microscope. Epstein was so excited, and so apprehensive that the heat of the microscope might alter the virus or in some way cause it to disappear, that he switched off the microscope and went for a walk to calm himself. When he came back and again turned on the machine, the virus was still there. Only then did he call in his team to share in the find.

So, there *was* a virus associated with Burkitt's lymphoma. A great question had been answered. But a bigger one remained: what did the virus have to do with the cancer? The fact that the virus showed up in tumor cells wasn't enough to prove that it was the cause of the tumors. It was the kind of circumstantial evidence that could raise hopes but simply wasn't enough to determine anything one way or another. The virus's presence might be the result of the cancer rather than its cause. It might be a hitchhiker, a by-product of the disease rather than its instigator. And, besides, another little problem remained: it was a virus no one had ever seen before.

"We were able to identify the *kind* of virus," Epstein remarked to a reporter, "but not exactly what it was. That is, we knew the family: it was one of the herpes group. But it wasn't any *known member* of the herpes group."

Ah, yes, the notorious herpesviruses. These complicated viruses are perhaps most widely known as the culprits behind genital herpes, painful sores that, before the onset of AIDS, were trumpeted by the media as the just deserts of the sexual revolution. But herpesviruses also cause problems as diverse as chickenpox, the common child-hood disease; shingles, the painful affliction of older age; and irritating and recurrent cold sores. In fact, herpes diseases are famous for coming back. Genital herpes and cold sores wax and wane. Shingles is a disease caused by a recurrence of the same virus that initially brings on chickenpox: the virus quietly hangs out in the body, buried deeply in the nervous system as though hibernating, during the long interim between eruptions. The name *herpes* itself comes from the Greek for "to creep" — an apt description for the viruses that cause these slowly erupting and recurring afflictions.

But this was an unknown variety of herpesvirus. Epstein was stumped — all the more so because herpesviruses just weren't associated with cancer, even in animals. He needed help. Epstein sent his latest batch of tumor cells, containing what was by then known as the Epstein-Barr virus, to Werner and Gertrude Henle ("*Hen*-ley") at Children's Hospital in Philadelphia.

A tall, elegant husband-and-wife virologist team from Germany, the Henles specialized in the study of children's viruses. They were already interested in Burkitt's lymphoma, having been alerted to its extraordinary possibilities by a colleague who had visited East Africa (a colleague by the name of C. Everett Koop, who, some twenty-five years later, as surgeon general of the United States, would find himself deeply involved in another viral disease, AIDS). They immediately confirmed Epstein's assessment that the Epstein-Barr virus was indeed a new virus by trying — unsuccessfully — to transmit it to cell cultures susceptible to known herpesviruses. Their next step was to investigate the link between EBV and Burkitt's lymphoma. Did the virus cause the tumors or didn't it?

To answer the question, the Henles decided to track down the antibodies in the blood of people with the disease. As we've seen, antibodies are special proteins that the body creates to help the immune system fight off intruders such as viruses. Once we're exposed to a virus, we manufacture specific antibodies to it that circulate in our bloodstream forever. So, the Henles reasoned, if EBV does play a role in Burkitt's lymphoma, anyone suffering the disease should have a healthy allotment of antibodies to the virus. Therefore, their first goal was to discover whether all patients with Burkitt's lymphoma have antibodies to the Epstein-Barr virus.

To find out, the Henles turned to a technique called immunofluorescence, in which they coated with fluorescent dye special antibodies that seek out all human antibodies, causing the latter to shine under ultraviolet radiation (just as fluorescent rocks glow in a museum display). When exposed to cells of Burkitt's lymphoma tumors, these fluorescent antibodies made a beeline for the antibodies in the cells and held tight. By illuminating the cells in ultraviolet light under a microscope, the Henles were then readily able to identify the EBV antibodies. When they tested the blood of a Nigerian child with Burkitt's lymphoma who had been flown to the United States for treatment, the antibodies glowed brightly in the microscope. Soon they found that every child with Burkitt's lymphoma carried such antibodies — EBV was present in all who suffered the disease. The finding was pretty much what they suspected: EBV was strongly associated with Burkitt's lymphoma.

But simply finding antibodies wasn't enough for the Henles. There were two other major questions they intended to answer. First, were the antibodies to the virus limited to patients with Burkitt's lymphoma, or were they also found in healthy individuals or people with other diseases? And second, if the antibodies weren't limited to people suffering Burkitt's lymphoma, did Burkitt's patients show a significantly higher concentration of antibodies to the Epstein-Barr virus than did other people?

Here the Henles encountered a surprise. As they later wrote in *Scientific American,*

> To our surprise antibodies to the Epstein-Barr virus were found not only in children with Burkitt's lymphoma but also in the serum of nearly all healthy African children tested. Even more unexpected was the finding that antibodies to the virus were present in children from

many parts of the world, which implied that the Epstein-Barr virus has a worldwide distribution and that almost no one escapes infection.

Almost no one escapes infection. Up to 95 percent of all of us — Africans, Europeans, Asians, and Americans alike — carried within us the Epstein-Barr virus. It was a ubiquitous virus.

After this shock, the answer to the second question came as an anticlimax. Yes, children with Burkitt's lymphoma had concentrations of EBV antibodies that were eight to ten times higher than those of healthy children. Their bodies were trying hard to fight off the intruding virus.

So the Henles had found that, yes, EBV shows up in everyone afflicted with Burkitt's lymphoma, but the virus is not limited to people suffering the disease — virtually everyone carries EBV; and yes, Burkitt's lymphoma patients have a much higher concentration of EBV antibodies than do the rest of us.

Then, another surprise: people with certain other diseases also showed high levels of antibodies to EBV. Foremost among these were sufferers of nasopharyngeal carcinoma, a cancer of the area behind the nose and above the roof of the mouth. In contrast to Burkitt's lymphoma, this cancer afflicts mostly adults, primarily in Southeast Asia, southern China, and northern Africa. Alaskan Eskimos also suffer from the disease. In these areas, there are at least fifty thousand cases of nasopharyngeal carcinoma reported each year — in southern China it accounts for fully 25 percent of all cancers — but in most of the world it's rare, and in the United States, virtually unknown.

The Henles were faced with a puzzle. If almost everyone in the world carries the virus, why don't more people come down with some EBV-caused ailment? How could it be possible that a ubiquitous virus such as EBV was associated only with such uncommon diseases? It simply wasn't logical. There must be some common disease that involves the Epstein-Barr virus. In the Henles' words, "We started to look for a disease to fit the virus."

Once again, chance favored the prepared mind. In the Henles' case, chance acted in the person of Elaine Hutkin, a nineteen-year-old lab technician. In the normal course of their investigations, the Henles had tested Hutkin's blood along with that of everyone else in the lab. In contrast to everyone else, however, Hutkin had shown no antibodies to the Epstein-Barr virus — that is, she had so far escaped exposure to the virus. She contributed her EBV-absent blood a number of times for the Henles' experiments.

One day late in 1967, Elaine Hutkin felt too sick to go to work. "I had chills, a sore throat, swollen glands in my neck," she recalls. "I felt awful — I was tired all the time." After six days, she went back to work, but she still wasn't feeling good. After a couple of days, she developed a rash. The Henles offered to do another blood test, hoping to find out what might be ailing Hutkin. In the course of the test, they decided to look once again for EBV antibodies. Soon the result was in: Hutkin was now positive. In other words, she had been infected by the EBV at roughly the same time that she had become sick. The surprise was complete when Hutkin's doctor diagnosed her ailment as infectious mononucleosis.

Mononucleosis, the "kissing disease," the scourge of college and university students. Although mononucleosis had been around for some time — it had first been described in 1885 — and although a blood test for diagnosing the illness had been in common use since the 1930s, the cause was unknown. And now Hutkin had come down with mono and at the same time had suddenly developed antibodies to EBV. Could the Epstein-Barr virus be the long-sought cause of the disease? Was mononucleosis the common EBV-caused disease the Henles were looking for?

If so, how interesting, and appropriate. For, as noted earlier, mononucleosis is a disease in which blood cells proliferate out of control. Here was a virus, EBV, that had first been detected in cancer tumors, and now seemed to be intimately involved in mononucleosis, a common cancerlike disease.

No matter how provocative their suspicions were, the Henles couldn't base their case on Elaine Hutkin alone. If they were to show for sure that EBV was involved with mono, they needed widespread evidence. Luckily for them, that evidence was waiting just a few hundred miles away in New Haven, Connecticut, where James Niederman of the Yale Medical School was engaged in a study of mononucleosis in the population that suffers it most — college students.

Niederman had collected blood samples from hundreds of entering freshmen at Yale since 1958. He had samples from students before they came down with mono, during the course of the disease, and after they were again well. It was a gold mine of EBV information — and, for the Henles, a godsend.

Niederman readily sent his blood samples to the Henles for testing, and soon the results were in. "We found," they wrote, "that antibodies to the Epstein-Barr virus were always absent before the illness but were present during and after the illness. These findings led us to conclude unequivocally that the Epstein-Barr virus is responsible for at least one disease: infectious mononucleosis."

There it was. The Henles had found their widespread disease for their ubiquitous virus. But although mononucleosis is certainly common enough — in the United States it afflicts about 50 in every 100,000 people (in colleges and universities the number may be as high as 1,500 per 100,000 people) — it certainly doesn't show up in numbers so high as to explain easily the fact that the Epstein-Barr virus is present in virtually *everyone*. Why?

Ironically enough, the explanation has to do with our relatively high standard of hygiene. It became evident to the Henles that the age at which people were exposed to the virus depended on their environment. Where hygiene was poor and living conditions cramped, as is common in developing countries, people encountered the virus early in life and manufactured antibodies to it by the age of three or four. In the cleaner and less crowded countries like the United States, a person might not be exposed to the virus until adolescence or even adulthood; Elaine Hutkin was graphic evidence of that fact.

But it wasn't usually *apparent* that people living in unhygienic environments had been exposed early in life. For some reason, whether because of the immaturity of their B lymphocytes (the cells the virus invades) or the immaturity of their immune system as a whole, children infected with EBV rarely come down with any kind of obvious illness — certainly nothing as serious and long-lasting as mononucleosis (with the obvious exception of Burkitt's lymphoma). A sore throat and cough, a few days of flulike symptoms, or nothing at all — usually that's the extent of EBV's effect on children.

But when the virus invades later, the result is usually more severe:

a case of mononucleosis. In causing a more serious illness in older people, EBV acts much like other viruses, hepatitis and poliovirus among them. The reason may be that in older individuals the immune system responds inappropriately to infection. In any case, as far as EBV is concerned, at least half of the people belatedly infected with EBV experience significant illness; again, Elaine Hutkin was a perfect example. Mononucleosis, therefore, is essentially a disease of developed countries. So as far as EBV-caused illness is concerned, mononucleosis is only the tip of the iceberg — it's the disease that shows. But in one way or another, apparent or not, the virus affects virtually all of us.

In the course of their investigations, then, the Henles developed a grudging respect for the Epstein-Barr virus. If the point of all life is to live and reproduce, and if a virus can only live and reproduce within living cells, then it behooves the virus to keep its host well and functioning for as long as possible. "In the course of its evolution the Epstein-Barr virus has established a nearly perfect symbiosis with its human host," the Henles wrote, "ensuring the survival of both." After the initial exposure, the virus persists in the body, infecting a relatively few cells that continue to produce new viruses. From time to time these young, infectious viruses find their way into the saliva, through which they may be passed on to people who may not be immune. And so it goes: the virus continues to live and reproduce, most of us who are infected with it continue to live and reproduce, and very few of us complain very much — a pretty slick arrangement.

But that state of respectful equanimity is shattered when we consider that, in the Henles' words, "the Epstein-Barr virus is the foremost candidate for being the first known human cancer virus." Certainly those afflicted with Burkitt's lymphoma (which by this time had also been discovered in New Guinea and, very rarely, in other parts of the world as well) do not admire the Epstein-Barr virus. Nor do those suffering from nasopharyngeal carcinoma. Just how is EBV involved with cancer, anyway?

The Henles' findings had exploded Burkitt's hard-earned hypothesis that his particular lymphoma was caused by an insect bite. Mononucleosis (and therefore the virus) wasn't transmitted by insects. Accordingly, Burkitt cheerfully retracted his hypothesis. "I have on many occasions publicly stated my conviction that no progress is made in medical research — or in happy marriage — unless mistakes are openly admitted." Soon he came up with another, more likely explanation.

Since the virus is already present, he suggested, it must act in concert with something else to cause Burkitt's lymphoma. Just what that something might be, he didn't know for sure, but perhaps it was transmitted by an insect — for example, a mosquito transmitting malaria, as malaria is common in the lymphoma belt.

It was a point of view shared by others. To discuss the possibility, some forty scientists met in Nairobi at the end of 1968. The group included the Henles and the French physician and epidemiologist Guy de Thé, of the International Agency for Research on Cancer, one of the subunits of the World Health Organization. It also included Denis Burkitt, who, two years earlier, had left Uganda for England after twenty years of African service. Burkitt, along with the other scientists, recommended periodically testing the blood of all children living in the West Nile region of Uganda, where the tumors were common. When, as was likely, some of them came down with Burkitt's lymphoma, their blood would be compared to earlier samples. It would then be possible to know if the Epstein-Barr virus was present in the child's blood before the onset of the cancer. It was, in effect, to be an Elaine Hutkin–type experiment, this time planned and on a large scale.

Guy de Thé was chosen to direct the project, and it was he who, in 1978, ten years later, wrote an account of the results. The researchers tested approximately 42,000 children up to eight years old. Immediately they found that, yes, EBV was as common in Africa as anywhere else. Virtually all the children carried the virus. Over the next five years, 14 of the children developed Burkitt's lymphoma (BL). Most of these children not only were already infected with EBV, but they had shown especially high levels of antibodies to the virus long before tumors developed; this suggested that they had been exposed to an unusually heavy dose of the virus. The information led the researchers to conclude that "children with long and heavy exposure to EBV are at increased risk of developing BL."

But, that being true, why didn't more of the children come down with the disease? Fourteen in 42,000 translates into a rate of only 1 in 3,000 children. Why so few?

The answer again seemed to be that something else was required to trigger the cancer — "other cofactors are required," as the researchers put it. And that something else might well be a weakened immune system. "It has been suggested," de Thé wrote, "that BL arises as a result of immunological disorders in children exposed since early infancy to heavy malarial infection." It couldn't be proved, but it seemed as likely a notion as any, and helped to explain the fact of high levels of antibody to the virus in these children. If their immune systems had been weakened, EBV might have had an easier time replicating, thus raising antibody levels.

Six years later, in 1984, another chance event an ocean away provided stronger evidence that a weakened immune system may be the cofactor that allows the Epstein-Barr virus to cause serious infections: David, "the bubble boy," died.

David was a twelve-year-old Texan who had been born without a functioning immune system — in medical parlance, severe combined immunodeficiency (SCID) — a disorder suffered by about two hundred children born each year. His options at birth were two: death within months with no intervention or life within a sterile plastic "bubble," awaiting the time when medical technology might be able to offer him the possibility of a fuller life. That time seemed to come late in David's eleventh year, when he underwent bone-marrow-transplant surgery.

It is in the bone marrow that a number of kinds of blood cells are produced, including those that David lacked — the B and T lymphocytes, which comprise an essential part of the immune system. Without these disease-fighting cells, David couldn't fend off the mildest infection; something as insignificant as the common cold could kill him. So the transplant surgery was done, in the hopes that new, healthy bone marrow donated by his own sister would establish itself in David's body and begin producing the very cells he lacked.

For the first couple of months after the surgery, life within David's bubble went on as usual. There was no indication that the surgery had been successful; nor was there an indication that it had failed. Then, on New Year's Day of 1984, seventy-three days after the surgery, David came down with a slight fever. Within a few days he was ill, with a temperature as high as 105 degrees Fahrenheit, and vomiting and diarrhea. It was the beginning of the end.

Soon his doctors brought David out of the bubble so as to be able to treat him more easily. It was virtually the first time he had ever felt the touch of another human being, the first time his family had been able to hug and comfort him without the intervention of the bubble's plastic wall. Still his condition worsened, and 124 days after the surgery, he died. At twelve years old, he had been the oldest known survivor of his condition.

The cause of death was deemed to be heart failure, probably caused by fluid that accumulated around David's heart and lungs in his last few days — or so it was immediately assumed. Later, an autopsy uncovered the real culprit: cancer. The B cells that David had obtained through the bone marrow transplant had run amok. He had died of cancer of the B lymphocytes, with tumors similar to Burkitt's lymphoma. All of the cancer cells contained the Epstein-Barr virus.

But how did David, who had lived all his life in a sterile, infection-free environment, become exposed to the Epstein-Barr virus? The answer is part of the strongest evidence yet that EBV infection can lead to cancer when there is no functioning immune system to hold it in check. David received the virus in the transplant itself. His sister had at some point been exposed without harm to EBV; she passed on this otherwise harmless dose to David through her bone marrow.

So David's life, and death, inadvertently provided an unprecedented glimpse into the unchecked workings of the Epstein-Barr virus. Had it been a controlled experiment, it hardly could've been clearer. Before the surgery, David had not encountered EBV, that was absolutely sure. Four months later, after this accidental infection, without an immune system to protect him, he was dead of cancer — a cancer in which EBV was intimately involved, and which resembled Burkitt's lymphoma.

Stanford University pathologist Jeffrey Sklar examined David's tumors. "The thing that was important was that we knew exactly when he acquired the Epstein-Barr virus because he was in a germ-free environment up to that point," he says. "And it probably demonstrates that the Epstein-Barr virus causes a very broad stimulation of B-cell growth. Out of that emerges a tumor."

But not always. Again, it seems a weakened immune system is the primary, perhaps necessary cofactor in EBV disease. "The strong hypothesis is that EBV works in tandem with some kind of other agent that compromises the immune system," says Sklar. In the case of Burkitt's lymphoma, that agent almost certainly is malaria.

"We know that very early viral infection can lead to Burkitt's lymphoma," says Guy de Thé. "It's a situation exactly like [that of] cigarette smoking and lung cancer. You don't fully understand the mechanism, but you can measure the risk. Very heavy and early exposure to EBV is as though you were smoking all your life, two packs a day. Then malaria enters at the second level, by promoting further proliferation of the B cells infected by EBV.

"We're all infected by EBV," de Thé notes, "but nothing happens to most of us because our immune system controls the infected B cells. Malaria specifically depresses the part of the immune system whose job it is to control the B cells. And after that, something, possibly a chance event, induced by nobody knows what, causes a change in chromosomes that transforms the cell into a tumor cell."

So the reality of EBV-caused cancer seems to be that if the immune system remains strong, the virus cannot do its damage. But if any chink in the immune armor appears, for whatever reason, the risk of cancer intensifies. That being so, what can be done to stop the disease? Given the fact that the virus is ubiquitous, the most effective strategy would seem to be, stay strong. But it's not a realistic strategy, for obvious reasons.

Yet de Thé can imagine Burkitt's lymphoma becoming a thing of the past. "It's very simple: social and economic development of Africa will see the disappearance of Burkitt's lymphoma. Simple hygienic rules, malaria control, putting netting on the baby's crib, which is now being done in the towns — you can cut the disease by ninety percent."

But is that happening widely enough to make a difference? "No," says de Thé. "It's an example of how difficult it is to change one's life habits."

And what if it were possible to do something about the virus itself? Ubiquitous or not, if there were a way to neutralize the Epstein-Barr virus, then it might be possible to eliminate EBV-caused disease.

Such is the hope of Anthony Epstein himself, who, among others, is developing a vaccine against EBV infection. He's already tested the vaccine on a species of monkey that ordinarily develops cancer within a few weeks of being infected with EBV, with the result that the vaccinated animals remained virtually free of tumors. His experiments prompt Epstein to conclude "that the vaccine is capable of protecting the animals against massive doses of tumor-inducing virus."

The next step is to try out the vaccine on humans. First Epstein will vaccinate volunteers who have not yet been infected by EBV to see if they develop protective antibodies. If so, he can then go on to see if the vaccine might prevent mononucleosis. Epstein envisions an experiment similar to James Neiderman's Yale study, which proved so important to the Henles' discovery that EBV causes mononucleosis. He would give the vaccine to one group of entering students who had no evidence of EBV infection, withhold the vaccine from another group without EBV antibodies, and compare the number of each group that eventually came down with mononucleosis. If the vaccine works, the likelihood is that few, if any, of the vaccinated students will fall ill, while at least half of the nonvaccinated students will get sick. And if he's successful up to that point, it's on to Africa and the Burkitt's lymphoma belt and to Southeast Asia and China, where nasopharyngeal carcinoma is such a concern.

If the vaccine works — still a large if — and if people can be vaccinated early enough (and, of course, that's the rub when it comes to EBV, which in many parts of the world infects children no more than a year or two old), EBV-caused cancers and mononucleosis, as well, may become diseases of the past.

All of this may or may not be good news for Karen Rian and the "untold thousands" trying to cope with chronic fatigue syndrome.

The problem here is that although the disease is most certainly real, the cause is still not known. EBV may be involved, yes — but to what extent and how, no one knows. So an EBV vaccine might have little impact on the disease. The questions about CFS still outnumber the answers. There are, however, some strong suspicions.

"My bias is that it's a syndrome that's probably not generated by a single agent," says CFS researcher James Jones, "but is the result of an already existing problem in the body which can be triggered by a number of viruses or other infectious agent." Jones's sentiments are shared, to a greater or lesser extent, by most of the people in the field. CFS seems to be a disease caused by an agent or agents that trigger — that's the key word — self-destructive processes in the body, in effect exacerbating tendencies that were there before the provocation. Disagreement comes in terms of the identity of the agent or agents and the nature of the physical processes they spark.

As far as Paul Cheney is concerned, there well may be a single infectious agent. He is now working in Charlotte, North Carolina, no longer looking over his shoulder at the casinos and chamber of commerce at Lake Tahoe. ("You don't do research on a virus in a resort, especially a resort controlled by casinos," he says with a laugh."You don't monkey with casino money. The lake is very deep. Not to imply that there were any threats or any great worries on my part, but over the long haul it wasn't the place to do research.") Cheney suspects that the culprit may be a newly discovered virus called HHV-6 (for human herpes virus number six), the sixth member of the by now infamous herpes family. The virus was discovered in the fall of 1986 in the laboratory of Robert Gallo, the National Cancer Institute virologist who also had a hand (along with French scientists) in discovering the virus that causes AIDS. Like its cousin EBV, HHV-6 attacks the white blood cells of the immune system. Also like EBV, it's a ubiquitous virus. But it attacks a wider range of disease-fighting cells than does EBV and seems to attack nervous-system cells as well. So it may be that in the already established presence of EBV, and working on an already malfunctioning immune system, HHV-6 is the added weight that tips the scales. Or it may be that the HHV-6 infection itself in some circumstances weakens the immune system to the point that the result is chronic fatigue.

If HHV-6 is at the heart of chronic fatigue syndrome, that is. "Maybe HHV-6 is involved," says James Jones, "but we don't have enough data on it to know for sure. There are individuals with the syndrome who have antibodies to the virus; there are individuals without it who have antibodies. There just isn't enough information to show cause and effect." Stephen Straus is even less likely to point a finger in the direction of the virus. "I think that it's possible that the virus could have something to do with the problem," he says, "but I don't think that HHV-6 is a major player here."

Now forty-two years old, Straus, like Jones, has been studying the disease for ten years. He raises possibilities that have made him persona non grata to a number of CFS sufferers. In the first place, Straus rejects the notion that in general CFS may be a long-term, eventually debilitating disease. "I think that this is an extremely common problem that occurs for a very short period of time," he says, "and it's uncommon for it to occur for a long period of time. Of those that do have it, some get completely better, no one gets much worse, and no one dies from this. And I don't care what someone wants to cite — show me!"

Why would such a comparatively optimistic point of view disturb those who suffer CFS? Because to some people, especially those hardest hit by the disease, it seems as though Straus is taking the syndrome lightly — an attitude that, as we've seen, was all too widespread in the early days of the problem. But if that outlook offends some people, Straus has gotten into even hotter water with the rest of his hypothesis.

"There are two types of symptoms that are caused by virus infections," he begins, building his argument carefully. "There's the type that the virus causes by destroying tissues. If you look at cold sores, or a blister, there's a virus causing that. If you look at polio — that's virus-induced destruction. But in most virus infections, some or all of the symptoms are due to the body's response to the virus and not direct destruction by the virus."

As an example, consider influenza. The flu virus attacks the respiratory tract only. It doesn't infect the muscles, the intestines, the liver. It doesn't circulate in the blood. Yet people who come down with flu hurt all over, with fever, headache, muscle aches. They're exhausted. Why? The answer lies in the body's *response* to the respiratory infection — a response that, in fighting the localized problem, releases chemicals throughout the body and makes the flu patient feel bad all over. The difference between infections that incite symptoms and those that don't may not necessarily depend on how much of the infectious agent is present, but on how aggressively the body is trying to throw it off.

So, as Jones also suggests, the miseries of CFS may more involve our body's response to an irritant than precisely what that irritant is. The irritant may be no more than a triggering device and may take many forms.

"If you listen to patients," Straus says, "you find that they begin the syndrome in a variety of different ways. Some begin with a respiratory-type infection, something like influenza or bronchitis. Some begin with a gastrointestinal-type problem — nausea, fever, diarrhea for a few days. Some begin with a mono-like illness. I've seen two cases that began with hepatitis and have been referred to others that began with chickenpox. There's a whole range of things. What all that's telling me is that a number of infectious agents can trigger this."

Rather than being the result of a particularly worrisome viral invader, then, CFS may involve the body's prolonged response to a variety of provocations. If so, the question becomes, why does the body respond so dramatically, and for so long a time? Jones offers one provocative suggestion. "What's peculiar in the majority of the patients we see is that they have a history of allergies — about seventy-five to eighty percent of them. And allergies are an immune problem. That's one possible predisposing set of circumstances."

The immune system again. It's a refrain that won't go away, and it makes good sense when you think about it. Allergies aren't the result of a weakened immune system, but rather a hypersensitive immune system, which overreacts to certain otherwise harmless substances that invade the body. These substances, called allergens, include everything from dust, to fur and feathers, to mold, to certain chemicals and certain foods — and, of course, pollen, an all too familiar subject to hay fever sufferers. Perhaps, Jones is suggesting, a hypersensitive immune system may be a cofactor in CFS just as a weakened immune system may be a cofactor in EBV-caused cancer. At heart, both cases involve an abnormally functioning immune system. Straus agrees: "If you argued that this is an individual who has the ability to overreact to some kinds of irritants, susceptibility to allergies could be an indication."

Yet another indication might be a person's personality and psychic makeup. "If you look at the people who suffer this syndrome," says Jones, "the majority of them do fall in a very specific age group and have a specific type of socioeconomic background, educational background, and so on. It raises the question of whether there is a certain personality type that predisposes to the disease."

What? Are we back to the "yuppie plague" business again? Is there really and truly a CFS personality? And if so, what in the world would explain a phenomenon like that? Straus, who advances the idea of a psychic component in CFS more strongly than most of his colleagues, explains by pointing to the lessons of recent history.

In the 1930s and 1940s, a disease called chronic brucellosis began to attract considerable attention. It was an ailment caused by bacteria called brucella that infected farm animals primarily but that also attacked humans. The disease was marked by fatigue, headache, loss of appetite, weakness, and fever — flulike symptoms, one might immediately say, but CFS-like symptoms as well. "The characteristics of that syndrome are *exactly* the same as the characteristics of this syndrome," Straus says. "Same age, same sex distribution, same type of onset, same complaints, same symptoms, same everything."

By the late forties, antibiotics were available to treat the disease, and many of the cases of chronic brucellosis cleared up. But some didn't. It turned out that the people who failed to recover showed no evidence of the brucella bacteria in their blood. That explained why the antibiotics didn't help, but it engendered a mystery: lacking the bacteria that caused the disease, how could these people have the sickness in the first place?

In the mid-1950s, a professor of medicine at Johns Hopkins named Leighton Cluff determined to find out and gave these patients a battery of psychological tests. What he found was that these people showed higher rates of neuroses than the people whose illness could be verified by the presence of bacteria.

By itself, the result proved nothing. But it was evocative. Could state of mind prolong a disease, even when there were no infectious agents around? Again, Cluff and his colleagues decided to find out. And it just so happened that the fates conspired to provide them a way to do it, for in 1956 and 1957 the Asian flu pandemic was crossing the Pacific to the United States. Cluff knew it was coming, so before the flu made its way to the East Coast, he arranged to give a battery of psychological tests to over four hundred men who worked for the army at Fort Detrick in Maryland. Soon after, the flu pandemic hit. Then Cluff and crew went back to check on the men he had evaluated to see if there were any differences in psychological profile between those who had suffered severe cases of the flu and those who had not. "They found that the likelihood of getting infected was no different regardless of one's psychiatric profile," explains Straus. "But what did seem to vary was how long they remained sick." Those with neurotic characteristics tended to stay sick much longer.

Straus falls quiet for a moment, considering how to phrase his next remark. For he knows that when you start to hint that there may be a psychological component to disease, and especially when the disease is chronic fatigue syndrome, you're heading for trouble. "There are many diseases and many physical processes that are affected by mood and emotion and psychic makeup," he says. "Everybody is willing to acknowledge that there are certain personality characteristics of persons who get ulcers, who get heart attacks. And there are plenty of data that none of us can explain well that suggest that people with cancer who live alone die sooner than people with a spouse or a pet. Survival after a heart attack in a coronary-care unit correlates with mood. Those are accepted observations. The question is, are there biological processes that are dependent on a person's psychic makeup? I have to believe that there are such processes."

So for Straus, CFS may involve a normal immune reaction to an original provocation, perhaps a viral infection, that for one reason or another simply doesn't shut down. And the reason the immune response doesn't shut down? It may involve an inherent tendency to overrespond, perhaps the result of a hypersensitivity hinted at by the presence of allergies, or having something to do with the psychic character of the person involved. Depression, anxiety, phobias — people with CFS often have had bouts with these problems before becoming fatigued, in some cases even years before.

What proof can Straus offer in support of the hypothesis? None, really. The state of CFS research does not yet include proof. It is still in the stage of suggestions and tendencies, of tentative conclusions drawn from circumstantial evidence. One of those bits of evidence comes from a study done at the Dana-Farber Cancer Institute in Boston, in which the blood of CFS patients was analyzed. What the researchers found was that natural killer T cells — a part of the immune system's armada whose function isn't well understood but that seem to attack and destroy infected cells — don't function as well in CFS victims. And it turns out that there are certain other diseases characterized by reduced effectiveness of these killer cells — and that one of these is depression.

But what does the correlation really mean? Does it mean that CFS sufferers have a tendency toward depression and so are especially vulnerable to the fatigue syndrome, or does it suggest that people who come down with CFS become, quite understandably, dejected and so exhibit the immunological signs of depression? No one knows. The correspondence is simply another evocative piece of the puzzle.

Yet another intriguing fact is that CFS patients tend to respond strongly to placebo treatments. Straus has found that "a very high percentage of the people feel better for a period of time on placebo. And a placebo is a very powerful thing. When we have an instance of a patient who feels so much better that her family comes up with her to tell us that they are so eternally grateful, this is the best their daughter has been in three years, and later we find out that she was on placebo — it's very powerful." But there are many instances of diseases retreating before the power of a placebo. By itself, the tendency isn't enough to show that CFS involves a healthy dose of mind.

Finally, there's the fact that by far the greater number of people plagued by CFS are women. Could the simple fact of being female predispose someone to catch this disease? It's possible, says Straus, if for no other reason than that "there are significant immunological differences between men and women. Most of the people who get connective-tissue disorders, rheumatoid arthritis, multiple sclerosis, and several other things are women. Ninety percent of all people with lupus are women. And all of these are immunologically mediated diseases." Again, the evidence is circumstantial, guilt by association, but the association is compelling.

So, anecdotes and full-blown experiments, fragments and thorough investigations — all of these combine to suggest what was suspected in the very beginning: that the mind may play a strong role in chronic fatigue syndrome. But now this hypothesis is couched in a much fuller context, one that casts a malfunctioning immune system in a central role, and one that readily acknowledges the probability of viral triggering agents, even if their identity remains unknown. And the net result? A terribly complicated disease, for which there may be no one solution, just as there may be no one cause.

Be all that as it may, what can we do about it? What help can Karen Rian and her army of fellow CFS sufferers count on? As of now, not much. An increasing public awareness of the problem, almost certainly. More interested and knowledgeable physicians? Perhaps — the medical profession is intractable and moves slowly. An effective treatment or cure? Not now, anyway. For it's important to realize that all hypotheses concerning the cause of the disease are just that, hypotheses.

Nevertheless, Paul Cheney sees reason for optimism. "I'm sure that as we uncover the basic biology of this problem, there will be ingenious ways to interfere with the process. I just don't know what time frame we're talking about. I don't know how fast this thing is moving."

And when — or if — the secret to CFS is found, along with it will almost certainly come knowledge concerning the interdependent nature of our physical processes. Not the least of those are the connections between disease and immune response, mind and body, spirit and health.

"We are a very complex, highly integrated organism," says Straus. "To presume that our systems are not integrated is naive. If you take a car apart, and you don't know much about an automobile, you pick one part, the carburetor, say, and you write whole textbooks about it. Then someone else writes whole textbooks about the axle. But eventually you come to realize that they're dependent on each other. Everything fits together."

So it is with our multifaceted bodies, and so it may be with one of the real mystery diseases of our day, chronic fatigue syndrome.

Eleven

· · · · · · · · · · · · · · · ·

Nothing to Sneeze At

· · · · · · · · · · · · · · · ·

William Osler, a nineteenth-century Johns Hopkins physician, said it best: "There is just one way to treat the common cold — with contempt." For colds are the cause of more sickness than all other illnesses combined. Americans spend about five billion dollars a year on colds; colds cause us to miss thirty million work- and schooldays a year. Chances are that one of every twenty of you reading this book is sniffling and sneezing right now. Chances are that you've suffered between one and six colds this past year. And chances are that you don't have the foggiest idea of how you got it, what causes it, or what to do about it — except wait it out.

If so, be comforted — you're in good company. The ancient Greeks considered colds the result of excess waste matter in the brain — it certainly feels that way sometimes — and thought that a runny nose released that excess, thereby restoring harmony. Hippocrates himself coined the archaic term for the common cold, *catarrh*, from the Greek for "flowing down," and, although he didn't recommend it, he noted that a common "cure" for the catarrh was bloodletting.

With the advent of Rome, the approach to the common cold

changed little. The Roman era did produce, however, perhaps the most contemptible treatment for a contemptible disease: "kissing the hairy muzzle of a mouse." It might well be argued that anyone able to do that could recover from anything.

And so it went. In the twelfth century, the famous rabbi-physician Maimonides recommended "soup from a fat hen," history's first mention of the benefits of chicken soup. Throughout the Middle Ages colds unleashed torrents of superstition. Many Europeans believed that colds could lead to possession by demons. Sneezes were particularly dangerous in this regard, as demons could rush in while bits of the soul were being forced out. The only ways around this unfortunate calamity were to cover your mouth while sneezing, so as to prevent the soul from escaping, and to have a bystander invoke the name of the Lord — "God bless you" — a sure deterrent to demons everywhere. Both customs, of course, survive today.

In the nineteenth century the aforementioned William Osler provided yet another sage recommendation: "Go to bed. Hang your hat on a bedpost. Drink whiskey until you see two hats. In the morning you'll be much better." In addition to being a practical measure — when you're that numb *nothing* can bother you — the prescription gave a nod to the fact that alcohol, acknowledged or not, has been a major ingredient in cold remedies for years. Nineteenth-century patent medicines claiming to cure everything from colds to tuberculosis included alcohol (as well as narcotics), and today a number of over-the-counter cold remedies continue that tradition.

Don't get your feet wet. Stay out of drafts. Eat onions. Wear garlic buds. Try honey in your tea. Breathe steam. Load up on vitamin C. Light a candle to your favorite saint. Colorful preventions and remedies for the common cold still abound. And until recently, effective or not, they were about all we had, for physicians and cold researchers alike knew relatively little about the affliction. The term itself mirrors that ignorance, for there is no such thing as *the* common cold. It's really a morass of infections caused by 150 to 200 different viruses, none of which is precisely like any other. The culprits go by such names as parainfluenza virus, coronavirus, respiratory syncytial ("sin-*sish*-al") virus, enterovirus, and adenovirus, yet most colds, perhaps up to half of them, are caused by one family of viruses, the rhinoviruses. (The name comes from the Greek for "nose" — think of *rhino*ceros — the site of the infection.) Precisely 100 rhinoviruses have been identified so far, each of which can give you a snootful.

Well, then, if colds are caused by viruses, what about a vaccine to prevent infection? Unfortunately, that's a long shot. It may well be possible to devise a vaccine against any particular cold virus, but so what? You may be protected against one version — look out, here come hundreds more. If we are to make any headway against this disease, it will most likely be through other means.

Happy to say, today there are some possibilities. Much has happened in the last few years to cause cold sufferers to cheer. The good news, and what led up to it, follows.

A cold virus enters your body through the twin portals of the nose. Swept along a thick carpet of mucus by hairlike cilia, the virus is borne to the back of the throat, a journey that takes no longer than ten or twelve minutes. Almost always this is the end of the line. Soon, along with dust, bacteria, and other foreign matter, the virus flows into the garbage pit of the adenoids, there to be destroyed. Or it's washed down the esophagus to the gut, to be totally obliterated at the first contact with digestive acids.

But occasionally a virus escapes. Before it can be safely transported away, it may lodge against a cell lining the nasal passage and, through a precise docking mechanism, bind tightly to a receptor on the surface. Almost immediately, an astonishing violation takes place. The virus inserts itself into the body of the cell, disassembles, and from its depths looses its genes, which usurp the cell's own genetic material. Thus provided with new instructions, the cell stops making new cells and starts making new viruses. The infection has begun.

At this point you're unaware of the invasion, but in about three days it'll be all too clear. Sneezing, coughing, runny nose, headache — the common cold has got you again. But what causes these symptoms? Is it the virus itself that somehow makes your nose run and your head ache? Well, yes . . . and no. For it's really your own body fighting off the infection that makes you miserable. The process works this way:

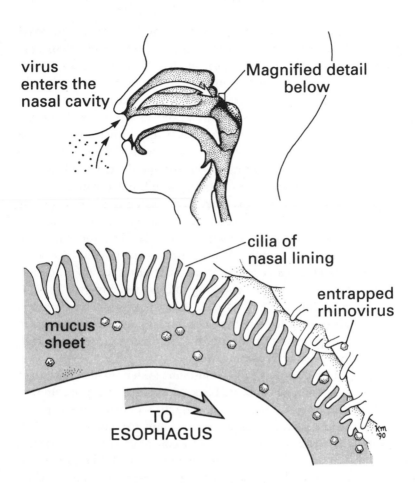

virus
enters the
nasal cavity

Magnified detail
below

cilia of
nasal lining

entrapped
rhinovirus

mucus
sheet

TO
ESOPHAGUS

How we catch a cold

Once cold virus particles enter the nose, they encounter the sheet of mucus covering the nasal lining. Beating cilia move the viruses along. In the vicinity of the adenoids, the lumpy tissue in the back of the throat, the cilia and the mucus sheet become thinner, and it's easier for rhinoviruses to gain access to the cells they infect. This explains why your first cold symptom is typically that sore feeling in the back of your throat.

Once the virus has commandeered some of the cells in your nasal passage and forced them to produce new viruses rather than new cells, the invasion increases exponentially. In a matter of hours, hundreds of new viruses may come bursting out of any single cell, looking for other cells to infect. Soon enough, however, the body begins to figure out what's happening and unleashes the immune system to turn the tide against the advancing viruses. But, at the same time, we accordingly begin to suffer the miseries we know so well. With the rush of healing blood, the nasal area reddens and warms. The mucus membranes begin to swell, narrowing the nasal passages, making it hard to breathe. At the same time, these mucus cells step up their production, creating so much fluid that there's no room for it within the nasal passage. The only place for it to go is out, which it does by dripping from your nose or flowing into your throat. But the body can't tolerate that unaccustomed flood, either. Soon you start coughing and sneezing to clear the passage.

Oh, no, you think, I've got a cold. But what you've really got is a battle against an infection that began days ago, a battle you're actually in the process of winning. By the time you feel miserable, you're on your way to becoming well.

So you can tip your hat to your immune system, which, although it makes you feel wretched, beats back the onslaught of the common cold. A much preferred alternative, though, would be to escape infection in the first place. Sounds good, but how can you escape when everyone in the office seems to have a cold? Or your child comes home from school with a cold? And how can your child escape when half the class is coughing and sneezing? It seems inevitable: when there's a cold around, it spreads.

How does it spread? While we know how a cold virus infects and how it causes discomfort, how colds spread is still the subject of great debate. Yet if researchers have had trouble agreeing, it hasn't been for lack of trying.

The first systematic experiments into the hows of cold transmission took place soon after World War II in Salisbury, England, at a sprawling institution called the Common Cold Unit of Harvard Hospital. There, under the direction of Sir Christopher Andrewes, a physician who in 1933 had first discovered a virus that causes respiratory disease — in this case, an influenza virus — volunteers were subjected to all sorts of flamboyant schemes in an attempt to pinpoint how colds move from person to person. Does becoming chilled cause increased vulnerability to colds? Apparently not. Volunteers who left warm baths to stand wet and unclothed in dank hallways didn't catch colds more readily than others. Did getting drenched promote catching colds? Not for those volunteers who let themselves get soaked in the rain. Did susceptibility increase or decrease as a result of isolation? Well, it didn't seem to. In 1950 Andrewes marooned a dozen students on an uninhabited island off of Scotland for a summer, then visited them to expose them to cold viruses. The students had had a wonderful time. Fortified with more than two tons of supplies, they had explored the island and its caves, compiled lists of native birds and wild plants, fished, organized cricket matches and concerts, but, alas, they seemed no more or less vulnerable to colds than before. "As it was," Andrewes wrote, "we went away without the answer we wanted. . . . There were also more than twelve people who derived immense enjoyment from a venture planned in the interests of science."

If Andrewes's reaction to the paucity of answers he was finding smacked of good-natured resignation, it was entirely justified, for the common cold seemed determined not to give up its secrets. W. J. Mogabgab of Tulane University and others first isolated a rhinovirus in 1954. David Tyrrell, who joined Andrewes's research group in the mid-1950s, then found a number of other strains of the virus, and discovered the optimum conditions in which rhinoviruses can be isolated and grown. He continues to investigate the disease. But when, in the 1960s, American researchers Jack Gwaltney of the University of Virginia and the University of Wisconsin's Elliot Dick entered the fray, fundamental questions concerning the common cold were still not answered. Though they haven't yet unequivocally solved the mysteries, Gwaltney and Dick have put their own distinct stamp on cold-transmission research, making it one of the most imaginatively investigated fields in virology.

In another time, but not another place, Jack Gwaltney might have been a southern country squire. Fifty-nine years old, straight and tall, slow-speaking and soft-spoken, born and educated in his home state of Virginia, he epitomizes the genteel Virginian — even if this genteel Virginian wears a white lab coat and is involved with tracking the discharge ("streamings," they're called in the trade) from the noses of people with colds. To visit him is to become exquisitely aware of his link to his region and its past. He talks knowledgeably of the nineteenth-century architecture of Thomas Jefferson's "Academical Village," just a stone's throw from his office on the campus of the University of Virginia in Charlottesville. He points to an old poster advertising his father's grain business in Norfolk and pulls out a thick bottle containing his own excised adenoids and tonsils, which his mother put up like a jar of jam almost half a century ago.

Then, just as you're drifting back into the past, lulled by all those southern dropped *g*'s — "somethin', goin', studyin' " — and soft *r*'s — "impo'tant" — he begins to talk of his own ground-breaking experiments, experiments that for a time seemed to settle the question of how colds spread once and for all.

"In 1963 we began a study of a group of insurance company employees," he begins, "here at the eastern regional office of State Farm, about five hundred people. That study went on for fifteen years." The gist was to pin down some basic questions about colds. When did people come down with the disease? All the time, as regularly as eating or sleeping? Or during specific seasons, perhaps prompted by some unique seasonal occurrence that might be managed? Who got colds more frequently, men or women? Adults or kids? How did the virus invade families? And so on and so on.

The State Farm employees were perfect for such an investigation. In contrast to university students, who couldn't always be counted upon ("they didn't show up for appointments; you'd have to find them in the pool hall or somewhere"), the insurance people were reliable and interested. Best of all, they were all housed in one building — "which had a few dividers and looked like a guinea pig cage for studying all these people."

Gwaltney and his colleague William Jordan handed out elaborate "Symptom Cards" on which the State Farm employees could keep track of their daily health. These cards tracked two weeks' worth of answers to seventeen questions:

Feel Sick?	Scratchy Throat?
Feverish?	Hoarse?
Chilly?	Cough?
Headache?	Sputum?
Muscle Ache?	Post Nasal Drip?
Sneezing?	Diarrhea?
Runny Nose?	Stayed Home?
Stopped-Up Nose?	No Symptom?
Sore Throat?	

After six years the researchers had enough information to suggest some interesting facts about the common cold. For example, it turned out that young boys suffer colds more often than young girls, but once adolescence hits the ratio flip-flops and women are hit more frequently than men (making colds one of the few diseases that plague women more than men). Why? Perhaps because women are around kids more, for it became clear that kids play the largest role in spreading colds. They mix cold viruses around in school, bring them home, pass them to parents and siblings. Could that have anything to do with the heavy invasion of colds in September, when school begins? Perhaps so.

But the basic question remained: how, exactly how, were these colds transmitted? Up until then it had been assumed that colds spread through the air. Then Gwaltney came up with a clue that led in another direction, a clue that, interestingly enough, was the result of a mistake. While taking samples from the noses of people suffering colds as well as from those of healthy people, so as to compare them to see the effect of cold viruses upon infected cells, he and associate Owen Hendley accidentally infected some people in the healthy group. After initially taking samples from infected people, the researchers had washed their instruments in alcohol. But the disinfectant hadn't killed all the viruses — rhinoviruses are tough — and when they placed the instruments in healthy noses, some live rhinoviruses passed on.

"That got us thinking that maybe some infections could be directly introduced into the nose," Gwaltney recalls. "If so, the question was, what normally goes into people's noses? That made us focus on fingers."

The question became, are rhinoviruses normally found on the ends of fingers? To find out, Gwaltney took cultures from the noses and hands of University of Virginia students who were suffering colds; sure enough, rhinoviruses showed up in both places. So it was possible that people could stick virus-contaminated fingers into noses and cause infection that way.

The next order of business, then, was to see if rhinoviruses could be transferred from one person's hands to those of another and, if the virus was plentiful in the environment, whether it could be picked up by casual touch. Again, experiments bore out Gwaltney's suspicions. After rubbing or blowing their noses and thereby spreading viruses to their own hands, cold sufferers could indeed pass the viruses to someone else simply by touch. They also deposited rhinoviruses all over the place — on doorknobs, counters, telephones, magazines, books. If someone with a cold reads this book, for example, its pages could easily pick up and harbor rhinoviruses.

The results echoed those of an earlier test done at the Common Cold Unit in England. (In that experiment, fluorescent liquid was rigged to drip from beside the nose of a volunteer to simulate streamings, then was traced in a darkened room.) It was clear that rhinoviruses could be anywhere, and that, they could live for hours on various surfaces. When healthy people came along and opened a contaminated door or picked up a contaminated phone, played a game with contaminated cards or turned the pages of a contaminated book, they picked up the cold viruses. Soon after, they might unwittingly inoculate themselves by touching their noses or rubbing their eyes. It began to look as though colds were primarily transmitted by touch.

Tests and observations that Gwaltney and Hendley compiled between 1973 and 1978 further backed up the theory. While the prestigious British medical journal *Lancet* doubted that the spread of humankind's most frequent disease could be the result of something so ignoble as sticking one's finger into one's nose, Gwaltney and Hendley offered the fruits of secret observations of medical students and Sunday-school children. Yes, indeed, there was a great deal of rubbing and picking of noses going on in these groups. And one of their key experiments showed that while only 5 percent of healthy volunteers caught colds after being exposed to coughs and sneezes of cold sufferers, fully 73 percent became infected after having hand-to-hand contact with infected people and then rubbing their noses or eyes.

But all that, as persuasive as it may have been, was not enough. "We showed that all the links were there and feasible," explains Gwaltney, "but none of that really *proved* that that's what happens under natural conditions."

So far they had amassed circumstantial evidence, but it would never hold up. Gwaltney was really up against it, for the only way to prove that colds spread through touch — and prove it in the real world, mind you, not (as before) in the lab, where variables could be controlled — was to devise an experiment in which you prevent touch-transmission in one group and allow it in another, then compare the incidence of colds in both. If, all else being equal, there were far more colds in the group that transmitted through touch, then you might feel reasonably certain that colds primarily spread via that route. The difficulty was that the experiment necessitated finding a means of blocking the transfer of rhinoviruses from hand to hand in one of the groups.

Block rhinoviruses. If Gwaltney knew how to do that, there would be little reason for all these tests — just as if someone had known how to block polioviruses, there would have been little need for a vaccine. It was a catch-22 situation: to see if colds spread through touch, you had to prevent them from spreading through touch. Good luck.

Undaunted, Gwaltney and Hendley came up with an iodine solution that succeeded in killing the virus. If volunteers applied this solution to their hands, perhaps it would block the spread of virus. The only trouble was, the iodine solution smelled bad and stained the hands brown. The success of an experiment of this kind depends on giving one group the substance to be tested and the other a placebo, with neither knowing the difference, but it would be pretty tough not to know the difference if only one group had brown hands. So Gwaltney and Hendley devised a placebo solution containing food dye that resembled the real thing. Hoping for the best, they gave fifty volunteer mothers the iodine solution and fifty other volunteer mothers the brown placebo and set about measuring the results.

"It was a tremendous amount of work," Gwaltney says. "We had two teams with two nurses each that went into fifty homes each week." For four years, from 1983 to 1986, during every September — the peak season for colds — the teams monitored the experiment, making sure that the mothers filled out the Symptom Cards accurately every day, and that they were applying the hand solutions faithfully. What the researchers found was that the mothers who used the iodine solution had 40 percent fewer colds than those who used the placebo. Perhaps the intervention was working. Perhaps most colds were indeed transmitted by touch.

Still, however, the researchers weren't satisfied. Gwaltney suspected that the people involved were able to differentiate between the virus-killing solution and the placebo. If that was so, their assessments might be biased. For it's a well-known fact among researchers that people have a tendency to report what they think they should report. In this case, if the mothers who used the virus-killing iodine solution suspected it, they might unthinkingly ignore possible cold symptoms, figuring, after all, that they shouldn't be getting sick.

"We made an honest-to-God *very* serious effort to prevent any bias on the part of the mothers and nurses, but how well we succeeded we don't know. We asked them, do you think you're on the active or on the placebo substance? Because the active tended to dry their hands, and the staining lasted a little longer, a number of them thought they were on the active rather than placebo. So I clearly think there was some bias. Whether that was enough to account for the forty-percent difference . . . ?"

So, it was back to the old drawing board. Finally, Gwaltney came up with a new virus-killing preparation, one that didn't stain the hands and, in fact, resembled commercial hand lotion so much that he hoped it would not be distinguishable from the real thing, which he planned to use as a placebo. In September of 1988, again in cooperation with State Farm, he started another study. To mostly women in a hundred families, Gwaltney's team gave bottles of the virucidal hand cream; to those in another hundred families, they gave the placebo, bottles of ordinary commercial hand cream. The volunteers' instructions were to rub the cream on their hands every two to four hours — when they awoke in the morning, any time during the day that they washed or wet their hands, as well as at regular, prescribed intervals.

"I told them what I really believed," Gwaltney says, "that going back to 1963, this was the most important study we'd ever done at State Farm." And when the results are in, they may provide yet more substantial evidence that colds spread mostly through touch (as well as the possibility of a commercial cream to prevent the disease).

But for now, the primary mode of transmission of the common cold remains in doubt. "I think it's entirely possible that colds may spread by more than one route," says Gwaltney. "They may spread both through the air and by the hand route. I've thought about this a tremendous amount. I hope I live long enough to get the ultimate answer — I'm not sure I will."

Elliot Dick might not agree with Gwaltney's assessment. But that would be no surprise, because when it comes to colds, Dick and Gwaltney seldom agree on anything. These two notable cold researchers are also notable antagonists — albeit friendly ones. And the focus of their disagreement is the heart of the issue, nothing less than how colds are transmitted. Whereas Gwaltney leans toward transmission by touch, Dick considers it much more likely that colds spread through the air, with the rhinoviruses floating from nose to nose suspended in invisible moisture droplets wafting in the breeze.

Sixty-four years old, Dick exudes vitality and enthusiasm. Shortish, gray-haired, and inclined to big belly laughs, Dick has the feel of a cheerful, old-fashioned Midwest doctor about him, the sound of the country twanging in his voice. "You know, I've lived long enough to have started right at the beginning," he says, "when John Enders, back in 1949, grew viruses in cell culture. All during my training I've watched viruses come from almost nothing to their flowering now, when we're really looking at them."

That training, which led to a PhD in bacteriology from the University of Minnesota, included a fellowship at Tulane University with the colorful W. J. Mogabgab, the discoverer of rhinoviruses. Hooked by the common could, Dick began his rhinovirus research at Wisconsin in 1961, tracking the spread of colds in families, and later testing antiviral drugs on chimpanzees. Soon enough, however, frustrated by the expense and trouble of working with chimps, Dick turned to humans.

"Humans beings are really marvelous to work with because you can talk to them," he laughs. "You can explain to them what's going on." And here the advantage of studying the common cold is most apparent, for whereas other diseases — polio is a prime example — are simply too dangerous to do any but final vaccine tests with humans, rhinoviruses pose almost no serious risk at all. A researcher can do a multitude of tests with humans without worrying about the inherent dangers.

So Dick began to work with humans. And by the mid-1970s, he had embarked on a series of evocative experiments that not only harkened back to Sir Christopher Andrewes's island isolation test but turned him in the direction of investigating the possibility that colds spread primarily through the air, not by touch. If Andrewes's uninhabited island off the coast of Scotland was a dramatic setting, Dick's was even more so: Antarctica.

Twenty-two hundred miles south of New Zealand on an island at the edge of the Antarctic ice shelf sprawls the United States' largest Antarctic base, McMurdo Station. The base, which consists of more than a hundred buildings and an outlying airport, serves as a supply depot for the support of scientific projects at the bottom of the world — studies in meteorology, geology, and marine biology, plate tectonic work and fossil explorations, and investigations into the nature of the Antarctic glacier. In 1976, when Dick began his studies there, McMurdo station accommodated precisely 200 people, all men, 64 of whom had spent the winter working at the base. The rest flew in for a thirty-six-day stint in September and October.

What they encountered was an icebound paradise that alternately pummeled the senses and entranced them. Temperatures ranged from $-38°$ C to $-12°$ C ($-36°$ F to $10°$ F). Winds could reach seventy-six miles an hour. But sometimes it was bliss. "A clear, calm day at $-20°$ C ($-4°$ F) is lovely," Dick wrote. ". . . walking along the shore of Winter Quarters Bay on such a day, gazing on the continually changing light patterns in the Trans-antarctic Mountains across McMurdo Sound, is one of life's great experiences."

McMurdo offered, then, a world unto itself, a self-contained community that opened to the outside but once a year. It offered a perfect opportunity for Dick to track the spread of colds among the hardy souls who wintered over as compared to the new arrivals, and to see, as Andrewes had on his island, whether isolation made people more susceptible to becoming infected. Once the summer men carting their complement of rhinoviruses arrived, would the relatively germ-free winter-over men suffer more than their share of colds? That's what Dick wanted to find out.

As it happened, he found out more than he bargained for. He discovered that rhinovirus colds do not spread easily after all. Despite the seemingly ideal conditions for transmitting disease, colds didn't spread any more readily at the cloistered station than they do in a typical family setting. He found that the colds brought in by the arriving men did not spread to the incumbent winter-over men any more widely than they did among the healthy incoming men themselves. Here was reinforcement for Andrewes's findings that isolation does not lead to increased susceptibility to colds. And he found, to his surprise, that the infections, relatively few as they were, were not dispersed equally throughout the base. Men living in the main housing unit suffered colds at a rate of only one-fourth as many as those living in the two peripheral residence buildings.

This was perplexing. Just as the incoming and winter-over groups came down with colds at a similar rate, one would have expected that the infections might have been scattered somewhat equally throughout the base. Why weren't they?

Dick reduced the possible reasons to two. First, it was true that the main residential building was less crowded than the other two. The more people jammed into one space, the greater likelihood of disease spreading — that stood to reason. But there was one other difference: the main housing unit had a much better ventilation system. In the outlying residential buildings, movement of air was intermittent, but in the main building the flow was constant — air was initially sucked in from outside, warmed, and then circulated through the building. "In a sense, these uncrowded dormitory rooms were hung in a giant wind tunnel," Dick wrote, "where much of the air for ventilation moved through the walls, presumably rapidly removing virus aerosols."

Ventilation! Could that be the key to controlling the spread of colds? But if rhinoviruses were transmitted primarily by touch, what difference would ventilation make? You could open a building to a gale wind, and it wouldn't affect cold viruses sticking to fingers and lifted to noses. If, on the other hand, rhinoviruses spread through the air — that is, by an aerosol route — now that was a different story.

Three years later, in 1979, after studying the incidence of other respiratory viruses at McMurdo Station, Dick returned to the problem of rhinoviruses and how they spread. This time, rather than *chronicling* their spread, he decided to try *preventing* it. This would be an intervention experiment, an attempt to break the pattern of colds at McMurdo by introducing a heretofore unknown virus-fighting tool: virucidal facial tissues. Or, as they soon became known, Dr. Dick's Killer Kleenexes (D2K2, for short — the movie *Star Wars* had only recently been released).

They were nothing more than ordinary facial tissues soaked in iodine, which, as Jack Gwaltney was about to find out in his own transmission experiments, does a good job of killing off tough rhinoviruses. As it turned out, they had an enormous impact at McMurdo.

Dick waited until the influx of disease accompanying the incoming summer people had a chance to catch hold, and then he and his team went to work. "We handed out tens of thousands of these virucidal tissues," he says. "We asked people to use them every time they blew their noses, wiped their hands, wiped their faces. Everybody sort of ritually used the tissues, anytime, anywhere. Even in the bars they used them."

To imagine the hardy denizens of this polar camp interrupting a beer to reach for a tissue at the slightest tickle of a sneeze — it seems incongruous, even humorous, especially when that tissue is soaked in iodine, which not only stains but smells bad. But that's exactly what happened. And the results were dramatic. For the first four days after the men began using Dr. Dick's Killer Kleenexes, an interval that Dick feels represents the incubation period for colds already contracted, the rate of new colds at the base peaked at ten a day. Over the next five days, it dropped precipitously until the incidence was no more than zero to two a day — this in contrast to the previous years, in which the incidence was still rising during the comparable span of time. And a budding flu outbreak subsided just as swiftly. It looked as though the virucidal tissues might indeed be something special.

Dick and his team went back to Wisconsin armed with all sorts of exciting prospects. Foremost among them were the possibility that virucidal tissues might be an effective way of controlling the spread of colds (but certainly not a practical way if they continued to be soaked in smelly, staining iodine), and the theory that colds may spread primarily through the air rather than by touch. To come were two landmark experiments that, to Dick's satisfaction anyway, proved that both cases were so, and a grand strategy for reducing the spread of colds by consciously applying the accidental approach he had first discovered at McMurdo Station: better ventilation.

In the early 1980s, Dick began a series of two tests, nicknamed "Antarctic Hut" experiments, for which he has become at least as well known as for his earlier work in the real Antarctic. These experiments involved a technique that might be called Dr. Dick's Marathon Poker Games; for in these "huts," actually rooms containing four poker tables and a desk for a monitor, volunteers carried on twelve-hour card games, as a forum for testing whether the spread of colds might be curtailed by virucidal tissues, and whether they spread through the air or by touch.

In the first of the two experiments, Dick filled each of the four poker tables with three healthy volunteers and two people suffering particularly bad colds. He repeated the arrangement in three other similar "huts," so that altogether there were twelve healthy and eight rhinovirus-infected cardplayers in each room — eighty people in all. In two of the rooms, he instructed the cold sufferers to use ordinary cloth handkerchiefs for blowing their noses and catching sneezes and coughs, much as they would naturally. In the other two rooms, he passed out virucidal tissues. But these were new, improved versions, tissues that had been treated with citric acid rather than iodine; they were white and odorless but killed rhinoviruses no less efficiently. He then turned his volunteers loose for their all-day, twelve-hour, marathon poker game, monitored by graduate students who recorded every cough, sneeze, or clearing of the nasal passage, and interrupted only by lunch, dinner, a thirty-minute exercise break in the afternoon, and time out for the researchers to collect streamings.

The "huts" were heated or air-conditioned by units that recirculated rather than brought in new air from outside, thereby keeping all that rich, virus-contaminated air indoors. The poker games were, in themselves, another effective way to facilitate all types of virus transmission — aerosols generated by coughs, sneezes, nose blowing, and conversation, and person-to-person contact from people's touching the cards and one another. The idea was to mimic as closely as possible the many possibilities for virus transmission under natural conditions.

But colds didn't spread at all in the Killer Kleenex rooms, an outcome that, Dick says, was "absolutely astonishing." While in the two huts where the cold sufferers were using cloth handkerchiefs, fourteen of the twenty-four healthy poker players came down with colds — a rate of almost 60 percent — in the virucidal-tissue groups *not one* of the healthy volunteers took sick. Pretty convincing results.

But that test, while offering a way to prevent the transmission of rhinoviruses, didn't solve the problem of determining *how* they spread. In an experiment beginning a year later, in 1984, Dick took on that challenge. In this case, while most features of the experiment were similar, there were two huge differences. First, there were no Killer Kleenexes here; the sick volunteers, all of whom were suffering particularly messy colds, all used cloth handkerchiefs as they might normally. And second, while half of the healthy players simply played cards as before, the other half wore either large plastic collars, three feet in diameter, around their neck or rigid orthopedic braces on their arms. These restraining devices completely prevented them from touching their faces and thus spreading the viruses by touch. "If any of the restrained recipients needed his nose blown or scratched," Dick wrote, "assistance was given by a monitor."

Imagine the scene: some players sniffing, sneezing, blowing, and coughing, others shackled as though just released from some diabolical emergency ward, all trying to keep their minds on a twelve-hour card game. And imagine the condition of the cards at the end of the session. "They were a mess," Dick says, "the *grossest* kind of thing."

But after the twelve hours was up and the rooms were vacated, the experiment was not yet over, for Dick had rounded up twelve more individuals to play another marathon game in a separate room. These healthy volunteers would use the sodden cards, the sticky poker chips, the slimy pencils — all the paraphernalia that had been in use since early in the morning. Immediately beginning another twelve-hour game, this an all-nighter, these people made sure to touch their hands to their faces every fifteen minutes (if they forgot, graduate-student monitors reminded them). And they continually received a new supply of freshly contaminated cards, chips, and pencils; for at the same time, Dick was running yet another poker game, this one for cold sufferers only — what you might call a feeder operation.

There was method to his madness. In the daytime session, if the restrained poker players were to catch cold from the infected volunteers, they would have to do so through the air only — they were simply unable to transmit anything to their faces. As for the other healthy players, they were on their own. If they became sick, it could be from viruses transmitted through the air or through touch. But if the all-nighters caught cold, it would have to be from the virus-contaminated cards — in other words, through touch only. There was nobody breathing cold viruses into the air here.

The next morning, Dick and his associates began the month-long process of tallying the results. Almost as many restrained as unrestrained volunteers came down with colds — ten of eighteen as compared with twelve of eighteen. This suggested that the viruses had indeed spread through the air. And of the all-nighters who played with the heavily contaminated cards, not one caught cold. If rhinoviruses could indeed pass by touch, they certainly failed to do so this time, not even once. Dick's conclusion was clear: "These results suggest that contrary to current opinion, rhinovirus transmission, at least in adults, occurs chiefly by the aerosol route."

Case closed? Will we have boxes of Killer Kleenexes on our supermarket shelves and few colds in our future? Kimberly-Clark Corporation, the maker of Kleenex, did indeed come up with a version of virucidal tissues called Avert, using the citric-acid mix devised by Dick. The tissues were test-marketed in upper New York in 1986 and in Georgia in 1987 — but they simply didn't sell very well. One reason may have been that they cost much more than ordinary Kleenex; another may have been that they were displayed in the medicine sections of stores rather than the tissue sections. And yet another reason may have been that not everyone was convinced that they work. Kimberly-Clark has since taken Avert off the market.

One of the doubters was — you guessed it — Jack Gwaltney. In Gwaltney's studies, the tissues were successful in stopping viruses only about 10 percent of the time. So, the jury is still out on virucidal tissues — at least as far as Gwaltney is concerned. But lest his investigations be a total loss, "the thing that came out of this is that we're using the active ingredients in those tissues in our hand preparations. So it wasn't a total failure in terms of moving us ahead." One man's poison is another man's nectar — the Gwaltney/Dick debates carry on.

For his part, Elliot Dick is still convinced of the effectiveness of the virucidal tissues and, with ten thousand boxes stored on the Wisconsin campus, plans to continue using them. But for now he is moving on to investigate ventilation as a means to control the common cold. He plans to test filtering systems that suck virus-contaminated air away from people and circulate fresh air in its place — much as was done in the main living unit of McMurdo Station in Antarctica.

"We're going to see if we can stop transmission by simply moving air around a room," he says. "I think the air is probably going to have to come up from the floor. The person who's blowing his nose is going to want to have the stuff go up and through the ceiling and then be filtered, rather than just be floating around. Well, that's going to take a little engineering, and it'll be expensive engineering." Dick laughs heartily. "But if it works, it'll be worth it."

*   *   *

How colds spread continues to be a source of disagreement, but the same is no longer true for the nature of the virus that causes most cases of the disease. Now we can say with certainty exactly what a rhinovirus is — how large it is, what it looks like, what it's made of, what lurks beneath its surface, and, most important, how it infects. In fact, now, thanks to Purdue University crystallographer Michael Rossmann, we're able to inspect the landscape of rhinoviruses as closely as the terrain of our own earth. Subscribing to the theory that how a thing is constructed reflects how it works, Rossmann, in collaboration with biophysicist Roland Rueckert of the University of Wisconsin, in 1985 accomplished something that had been considered impossible: he produced an accurate three-dimensional model of a rhinovirus.

Imagine the scale involved. All viruses are tiny, virtually invisible to all but the electron microscope, and rhinoviruses are especially minute. As noted earlier, it would take ten thousand rhinoviruses lined up side by side to span the space between any two words in this book. How, then, could anyone hope to model such an elusive creature? The answer was to take advantage of a remarkable property of viruses, their propensity to crystallize.

Back in 1935, Wendell Stanley was the first to show that viruses could become crystals, a characteristic usually associated with non-living substances such as quartz, or the crystals of ice that form on winter windows. After Stanley led the way by crystallizing tobacco mosaic virus, turning viruses into crystals became one of the staples of the crystallographer's art, a method by which the form of these virtually invisible objects could be discerned.

It is, however, a terribly difficult art. Not only must crystallographers produce uncommonly clear crystals, they must analyze those crystals to discover precisely the nature of the substance they contain. Rossmann is one of the acknowledged leaders in the field. A slight, somewhat stooped, and soft-spoken man, he hides his face behind beard and glasses. Born in Frankfurt, Germany, sixty years ago, Rossmann was educated at the universities of London and Glasgow, and came to Purdue in 1964 because, he says, laughingly, "I got a job." The remark, in its economy, is typical of Rossmann. While friendly and patient with students and interviewers alike,

Rossmann approaches his work with straightforward intensity. Yet he glows with enjoyment when discussing the arcane world of crystallography.

"How does one make viruses crystallize?" he asks, his accent a mix of Germany and England. "When you were in high school, did you ever do any experiments using crystals of salt or sugar? Using saturated solutions? Same way. Or at least that's one way. If you use a saturated solution, after a while crystals come out."

It's not that easy, of course, and Rossmann admits that "it's considered a black art, growing crystals"; but with rhinoviruses, beginning in 1981, that's precisely what Rossmann did. Over the years, he dissolved millions upon millions of rhinoviruses in solution and induced them to precipitate out into hundreds of crystals. The results resemble specks of ice less than a millimeter in diameter, clear jewels suitable for miniature pendants. Rossmann is not easily given to hyperbole, but when it comes to his virus crystals, the reserve cracks. "They're lovely, just fantastic," he exclaims. "Simply beautiful."

They're also replete with viruses. Each rhinovirus crystal contains on the order of a million million viruses — inert, stable, and frozen in time. The viruses line up beside one another and stack atop one another in geometric array, a delicate latticework. And because the millions and millions of viruses in each crystal are virtually identical and all oriented in precisely the same direction, any information Rossmann gleans from the entire crystal can be used to delineate the form of any one virus.

The next step involves X-ray diffraction, a procedure simple in conception but difficult in practice. The idea is to bombard the crystal with X rays and then interpret the patterns those X rays make. The X rays used in crystallography are some four thousand times shorter than the wavelength of visible light, actually about the same size as the space between atoms in many solids. That's why X rays can penetrate substances opaque to the eye.

To uncover the form of rhinoviruses, Rossmann used the most powerful source of X rays available to him, an "atom-smasher" called the Cornell High-Energy Synchrotron Source, one of three such high-powered X-ray instruments in North America. He and his colleagues bombarded the crystals with extremely bright and thin pencils of X rays, over and over, from different angles, replacing the crystals after each shot, as even a short exposure to the X-ray beam damages the delicate structures contained within the bright specks. When the X rays banged into the atoms within the virus crystal, they reflected onto sheets of film, forming a complex pattern of millions of dots that contained a record of the virus's structure.

Eight hundred thousand unique reflections later, the researchers had compiled what was to the untrained eye nothing more than a series of meaningless prints containing patterns thick with dots. Now it was Rossmann's task to use these patterns to work backwards to the structures that produced them. For this, he turned to computers. Before the advent of computers, this kind of X-ray diffraction analysis required extensive mathematical calculations and vast experience. The need for experience remains, but today computers have eased the crystallographer's burden, making possible a heretofore unknown precision in amassing and interpreting the results of such experiments.

Rossmann's computers deciphered the diffraction patterns and translated them into images that the eye could begin to make sense of. These images were in the form of video displays that contained swirling patterns, each of which represented a cross section of an individual virus — imagine looking at successive slices of a cut tomato — down to almost atomic detail. For a three-dimensional view of how these shapes come together, Rossmann converted the computer images into large transparencies. When he stacked these transparencies one atop the other — much like stacking the slices of the tomato to reconstitute its shape — and viewed them over a light box, the shapes fit together to reveal the tiny virus's structure.

Five years ago, Rossmann stood at the light table in his laboratory at Purdue and, in what other crystallographers have called "a tour de force of modern X-ray crystallography," became the first human to view the terrain of a rhinovirus — indeed, of any virus that infects humans. "It was April 6, 1985." Rossmann laughs at himself — the date is engraved in his memory. "It was terribly exciting, maybe the most exciting day of my scientific life." To celebrate, he and his colleagues jubilantly went out for Chinese food.

It was an unprecedented achievement, an event unimaginable

only a few years before. If by first crystallizing a virus Stanley vindicated Beijerinck and d'Hérelle, and if by first making viruses visible the early electron microscopists corroborated Stanley's achievement and the insights of the pioneers that proceeded him, imagine the giants leaning on Rossmann's shoulder as he became the first person to look at a human virus in detail. It was as though he were in a helicopter flying low over the earth — such was his relative perspective toward the viral image he had created — only this planet was normally invisible, a resident of submicroscopic rather than interstellar space. This planet could infect and reproduce. Yet now, as Rossmann peered at it, it might as well have been a mass of rock and earth, for it exhibited many of the same physical characteristics.

Based on information already gathered through the electron microscope, Rossmann knew that the shape of this particular rhinovirus, number 14 of the 100 varieties, was an icosahedron — a kind of twenty-sided soccer ball of a virus. But no one had been able to visualize how the tangled proteins in these shells were organized, how they twisted and folded up and interacted with one another. Now, poring over his atomic-detail, three-dimensional map, Rossmann and his colleagues discovered that the surface of a rhinovirus is made up of four recurring proteins that form protrusions and depressions, much as though this miniature planet were covered with mountains that rim deep, circular canyons.

Beyond simply exploring, however, Rossmann wanted to find out the function of this rugged topography. For a virus has only one face to show to the world: its surface. All its prospects for survival and reproduction depend on how efficiently its surface interacts with the environment surrounding it. For example, the immune system is able to identify a virus as an intruder because the viral surface stands out as foreign to the body; antibodies then attach to specific areas on that surface, neutralizing the viral invasion. These spots are called *immunogenic sites*. If a virus isn't able to manipulate its surface — change its spots, as it were — so as to evade the body's immune system long enough to lock on to an inviting cell, its days are numbered and its prospects nil. Likewise, if it isn't able to infect that cell through the union of other specific sites on its surface with specific sites on the cell's surface — the cell's receptors — it's not in much better shape. Cells are studded with receptor proteins, which appar-

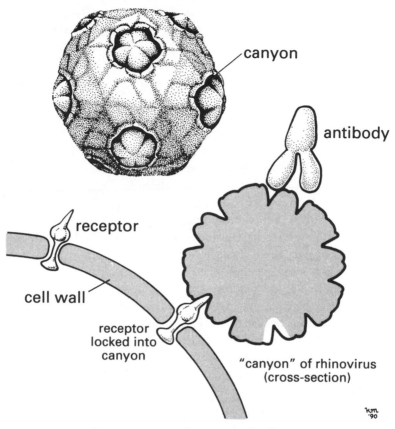

The mountains and canyons of a rhinovirus

Rhinovirus, the cause of the common cold, has "protein mountains" surrounded by deep moats, or "canyons." The canyons lock onto receptors on the surface of cells in the nasal lining. Luckily for the virus, however, the canyons are too narrow for antibodies to fit in and thwart infection (lower drawing).

ently make possible myriads of cellular interactions (their functions are as yet largely unknown), and specific viruses require specific receptors. A cold virus and a flu virus, for example, may infect the same cells, but they do so by attaching to different receptors on those cells. And a virus that can't attach can't infect.

Rossmann peered at the mountains and canyons of the rhinovirus and wondered about their function. It didn't take long for him to come up with likely possibilities. The mountains, Rossmann surmised, are actually the places where our immune system's antibodies, like so many gigantic landers, bind on to the virus in an attempt to prevent infection — the so-called immunogenic sites. "Antibodies bind to the surface of the virus," he explains, "in particular the part that is accessible." And what could be more accessible to an earth lander than the top of a mountain?

But a virus is simply loath to stand still for such docking, as it means death. "The surface of animal viruses is usually in constant change," says Rossmann. "The virus defends itself against the immune system by mutating and escaping antibodies." In this case the mutating involves the mountaintops — their shapes change from one generation of virus to the next so that our antibodies can't consistently recognize them. Even if we are able to fight off the virus today, we may well succumb to its offspring tomorrow.

So much for the mountains — actually changing, shifting immunogenic sites; that left the canyons. What might be their purpose? In the first place, *canyon* may be too mild a word. These are incredibly deep rents in the surface, moatlike gashes fully one-twelfth of the virus's diameter — a scale far beyond any chasm on earth. Even the mile-deep Grand Canyon, for which Rossmann named these depressions ("We even talk about the north wall and the south wall"), doesn't come close to suggesting their relative size.

When Rossmann leaned over his light table and peered into these canyons, he was looking for the solution to an enigma: how could a virus whose surface was constantly changing to escape detection by antibodies retain the ability to invade a stable cell? There must be some part of the virus that remained constant so that it could, time and time again, attach to a particular receptor on the human cells it invaded. This viral-attachment mechanism, wherever it was, would have to bind to its particular receptor and at the same time elude the body's protective antibodies. Quickly the thought came to Rossmann: what if the docking mechanism was buried deep in the canyon?

"The obvious explanation of how to solve this paradox was that, on the one hand, the virus must maintain at least a part of its surface as constant and yet, at the same time, must not allow antibodies to bind to it," Rossmann explains. "It simply hides this region in a canyon where antibodies cannot reach it."

And what Rossmann knew of the nature of antibodies gave support to his theory. Antibodies are relatively fat, unwieldy Y-shaped molecules. Their arms are so broad, in fact, that while a receptor on a cell might readily slip inside a canyon like a card in a slot, antibodies are much too large to fit. They simply can't get to this most vulnerable part of the virus, its receptor attachment site.

Pretty slick — if you're a virus, that is. You disguise your surface while keeping your attachment mechanism safe and out of harm's way. This bit of viral legerdemain means that any successful preventive treatment must take one of three tacks: block the stable attachment sites in the canyon, shield the target cell itself from infection, or interfere with the virus once it gets inside the human cell.

Rossmann and his colleague Tom Smith have recently experimented with the last approach, testing the workings of an already existing series of drugs, and finding that they may be able to prevent colds by paralyzing the virus in the cell, thereby keeping it from taking over the cell's genetic machinery and reproducing. These drugs actually sink into the canyon, past the attachment site, and then through a "pore" that opens into a cave below the canyon floor. The researchers suspect that, like the empty core of a rubber ball, this cavity adds to the flexibility of the virus's structure; by filling in the cave, the drugs stiffen the virus so that it can't disassemble and loose its genetic material in order to reproduce.

What is more, because the roof of the cave is at the same time the floor of the canyon, by stiffening the structure of the cave, the drugs change the conformation of the canyon floor, thereby often preventing the virus from attaching to the cell in the first place. So, the drugs not only look promising as a way to prevent rhinovirus infection, they are providing a means to uncover yet more viral secrets. Sterling-Winthrop Research Institute in Rensselaer, New York, manufacturer of the compounds, is testing them in humans now.

If these drugs work, the ramifications could go beyond rhinoviruses. "Since many, many viruses seem to have basically the same structure," Rossmann says, "maybe we can use the same target in other viruses — HIV [the AIDS virus], for example. We're now trying to crystallize a component of HIV, where we anticipate there will be a pocket like this. If we find it, then we can try to design drugs that will fit into it. There are so many properties of a virus that we can explain just by looking at the structure. It's just fantastic."

It's also frustrating. "I always think we're not moving fast enough," Rossmann says. "You know, in doing scientific work you have long periods of trying things and trying things — success is very difficult to obtain. You never know what problems you will encounter. One day you get success, and you'll be very happy about that for about five minutes. Then the next minute, it's back to work."

It is a credo that Richard Colonno might well share. In 1982, three years before Rossmann mapped the rhinovirus, Colonno, of Merck Sharp & Dohme Research Laboratories near Philadelphia, began to tackle the problem of the common cold by concentrating on the receptors of the cells lining the nasal passages where infection takes place. If he could block the receptors, thereby denying the virus access to the cells, he could prevent the disease. It was a logical approach, theoretically promising — but no more than that.

A friendly forty-year-old from New Jersey who prefers to be called "Rich," Colonno was breaking new ground. "At the time, blocking virus infection by blocking receptors was a novel concept," Colonno explains. "Now, of course, it's popular with AIDS researchers. But with the common cold it made sense because a cold is localized in the nose, plus on top of that it's an extremely benign disease."

His point hints at the paradoxical advantages and difficulties of dealing with the common cold. First, the advantages: because it's a mild disease, experimenting with humans carries little of the risk associated with other ailments. Yet a difficulty ensues because of the very innocuousness of colds: any successful treatment must be absolutely safe and without side effects. An effective therapy for colds puts great pressure on the researcher — it must be state of the art.

Another advantage accrues from the site of a cold, the nasal passage. As Colonno explains: "You help your situation tremendously when you can do a topical treatment. Since the virus had done us the favor of growing on an exposed organ of the body, the nose, there was really no reason, as with treatments for other diseases, to fill your entire body with a medicine to get the little bit you need into your nose."

On the other hand, getting an effective topical treatment into the nose can be a nightmare "because the nose is so efficient in clearing drugs out, along with virus and everything else — that's its job." Thus, some drugs that have seemed in the lab to show promise in controlling colds have been failures when tested in people: they were simply washed away before they could do any good. Those that were not cleared out precipitously tended to have side effects.

A prime example involves interferon, one of the body's own virus-fighting substances, which works by blocking viral replication. Jack Gwaltney, among others, has experimented with using interferon in a nasal spray to prevent the common cold — with some success. In one study, the incidence of colds in families using the interferon spray went down by 40 percent as compared to families not using the drug. But the side effects caused by the spray were sometimes worse than the cold itself. These included dryness and stuffiness in the nose and blood-tinged mucus owing to pinpoint areas of hemorrhage and ulcers caused by the drug.

So the weight of precedence was against Colonno. Nevertheless, he decided to try.

From the beginning, there were some very large snags. The first involved finding the receptor. In Colonno's words, "If you have a hundred different rhinoviruses, how many different cellular receptors do they bind to in a person's nose?"

It was a good question. The only pertinent work that had been done previously tested but three strains of rhinovirus and found that while two bound to the same receptor, the third preferred another.

"You could take that one of two ways," Colonno says. "Either it meant that there were just two receptors for every three viruses, or that we were going to get this whole array of different receptors."

So Colonno's first order of business was to find out how many receptors he would have to deal with. If a manageable number, great — his approach might be feasible. But if the hundred varieties of rhinovirus attached to a multitude of receptors, forget it. Then, assuming the problem could be solved, he could go on to the second order of business: finding a way to block those receptors. It was a tall order.

His first task, determining how many different kinds of cellular receptors were involved, proceeded comparatively well. He upped the ante, testing 24 rhinovirus varieties — a sampling large enough to give a trustworthy indication of the cellular preferences of all 100 — by what's called a "competition binding" experiment. "We saturated a thin layer of cells with a solution containing just one variety of rhinovirus," explains Colonno. "That meant that a number of receptors soon had viruses sticking to them. Then we came back with a second virus and simply asked, 'Does the first one block the second one from attaching?' If it does, then they share the same receptor. But if the second one attaches normally, then it's clearly to a different receptor." By repeating the process for all 24 varieties, Colonno determined that 20 of them bound to the identical receptor. He could then project that more than 80 percent of all rhinoviruses bound to the same receptor site. It looked as though his approach might be feasible after all.

Now Colonno needed to block that receptor. But how? What mechanism could he turn to that would act toward the receptor as an antibody does toward a foreign invader, binding to it and thereby blocking its function? Why not an antibody itself? What if he could introduce a nonhuman antibody onto the scene? Such an antibody would most certainly consider the human receptor as foreign, something that it ought to neutralize as quickly and thoroughly as possible.

That's exactly what Colonno did. He injected laboratory mice with fragments of human cells, which, in the mouse body, were foreign intruders. The tactic worked; in time the mice did their part, producing antibodies by the thousands. Now Colonno could go on to the next step, by a process of elimination figuring out precisely which antibodies targeted the receptor in question. For the fragments of cells he had injected into the mice contained all sorts of

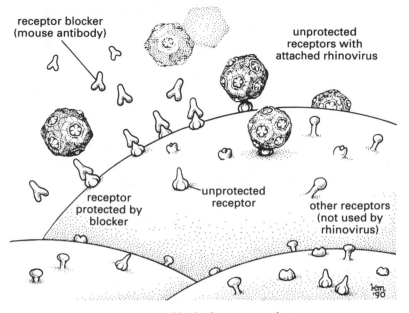

receptor blocker
(mouse antibody)

unprotected
receptors with
attached rhinovirus

receptor
protected by
blocker

unprotected
receptor

other receptors
(not used by
rhinovirus)

An attempt to block rhinovirus infection

One way to prevent colds is to keep rhinoviruses from locking onto
cell receptors. Richard Colonno is experimenting with antibodies that
cover the receptors, thus making them inaccessible to rhinoviruses.

receptors; his rhinovirus variety was only one. The mouse had therefore made a host of antibodies to protect against a multitude of
provocateurs. Colonno had to investigate the antibodies, one by one,
to determine which was the kind he was looking for.

Thus ensued what became famous around Merck as "Colonno's
Brute Force Method." In eight thousand different tests, he and his
team covered a thin layer of human cells with varying mouse antibodies, then infected them with rhinovirus and left the whole business in an incubator overnight. If any of the antibodies bound to the
cellular receptor preferred by the virus, they would protect the cells,
blocking the virus's point of entry. If not, the virus would charge to
the attack unimpeded, leaving behind a collection of dead cells.

For eleven months Colonno's team repeated the test, checking the
cultures each morning and finding nothing but dead cells. Finally, in

May of 1984, he was ready to throw in the towel but decided to do one last batch. "We looked at one another," he recalls, "and said, 'Look, we haven't even seen a hint of protection. If it doesn't work this time, there's something wrong in the whole design.' It came to that. Sometimes," he laughs, "you have to threaten the experiment."

The next morning, Colonno opened the incubator to find that of the eight hundred final tests, *one* displayed a batch of healthy cells. He couldn't believe his eyes. "The first thing I thought was, it's a fluke. Somebody didn't add virus to the cells. So we tested it again the next day, and it repeated positive. That got everybody excited, and we started jumping up and down. We clearly had something."

Had something, indeed. In subsequent experiments, Colonno and his colleagues tested the effectiveness of the antibody by applying it to a plateful of human cells and then adding to the mix ten billion rhinoviruses. If the antibody successfully blocked the receptors, the viruses wouldn't be able to infect the cells. But what a challenge he was presenting his newly found antibodies. All it would take for infection was that one virus latch on to one cell — that's all, just one to one. And here was Colonno smothering the cells with rhinovirus, to the tune of about three thousand viruses to each cell. At best he might hope that his antibodies provided partial protection.

But not one cell became infected. The antibodies provided complete protection, a result virtually unheard of in biological experiments. This was no ordinary antibody — it was a powerhouse of an antibody, a superantibody that didn't just bind to the cell's receptor but actually knocked rhinovirus off in its rush to grasp hold. It bound so tightly to the receptor that Colonno could hardly get it off; this suggested that it would fare well in the flow of the nasal passage. And Colonno discovered another interesting thing as well: almost every cell in the human body (with the notable exception of red blood cells) contains the identical receptor. If rhinoviruses were hardy enough to survive anywhere but in the nasal passage, we might face the fantastic prospect of suffering a cold through every inch of our bodies.

Colonno had finally arrived at the spot he had been aiming toward since he began three years before. He had identified the cellular receptor preferred by at least 80 percent of all rhinoviruses, and, with his mouse antibodies, he had found a way to block that receptor, denying the virus access to the cell. But he had arrived at this momentous point in the laboratory only, with viruses and antibodies vying for the cell's attention in glass dishes rather than in human bodies. The next step was to test the approach where it counted, in human beings.

So in November of 1985, Colonno enlisted the help of Jack Gwaltney and his associate Fred Hayden to test the antibody on humans by applying it directly into the nose. Expectations ran high. Here was a chance to do what had never been done before: prevent the common cold. "We were all hyper," Colonno recalls. "Based on what the antibody had done in the lab, we actually expected that the placebo group would get sick and the control group remain perfect." He laughs. "I was very naive at that time."

The experiment was a huge disappointment. The antibody seemed to offer no protection. Almost everyone in the study came down with a cold.

"We outsmarted ourselves," Colonno says now. "We knew how much antibody it took to protect cells on a glass plate in the lab, but how that related to your nose was completely unknown. The results were devastating." He suspected that the dose had been much too low and administered in too many separate applications. So in a second test, four months later, in March of 1986, the researchers gave volunteers eight times as much antibody. This time, although most of the people again came down with colds, the results were heartening.

"There were two things that were very promising," says Colonno. "The first was that the people treated with the antibodies developed cold symptoms one to two days later than those in the untreated group. The second was that the symptoms were only half as severe." All of which suggested that the antibody was doing precisely what it had done in the lab: blocking receptors and knocking the virus off cells. No drug other than interferon had ever before made an impact on the incidence of the disease. And in this case, unlike the interferon studies, there were no side effects.

So after four years of tedious and frustrating work, Colonno finally stood on the threshold of seeing his antibody become an effective means of ameliorating the common cold. Yet he was, in a way, about to begin all over again. For the antibody had served its purpose, that of demonstrating that a receptor blockade was possible; now it was to be retired. As a mouse antibody, it was a protein foreign to humans; there was always a possibility, slight though it might be, that upon repeated exposure a person's immune system might actually attack it — in other words, produce an allergic reaction in some of the very people the antibody was helping. It was time to go on to the final step, the arduous task of finding molecules that would mimic the antibody's blocking ability while presenting absolutely no risk at all to humans.

To do so, Colonno began distilling the antibody down to its essential functioning structure. "Using enzymes," he explains, "you can clip off the end of the Y-shaped antibody — the part that identifies it as mouse rather than human or anything else. The two arms will work as well as the whole antibody. But the odds of having an immune response against these arms are very remote."

What remains to be determined is how much antibody you can snip away before the odds of producing the desired antibody response become remote. Colonno has broken down the antibody to just one arm, which, he has found, blocks the receptor as readily as before but, in the continual flow of the nasal passage, probably won't be able to stay in place. His task now is to find the antibody configuration that is large enough to work well, small enough that it prompts no harmful immune responses, and capable of being produced in large quantities and sold at an economical, attractive price.

"A cure for the common cold isn't realistic because by the time you've got a cold it's too late to stop the infection," explains Colonno. "But preventing colds is another matter. There's real hope now that this will happen."

The common cold, it turns out, is a tricky customer, but little by little it is giving up its secrets — and not only its own secrets but those of viruses in general. We have learned a great deal about how viruses are constructed, how they infect, how they spread, and how the body responds, all from common cold research.

Nothing to sneeze at, truly. For a lowly, irritating, much-maligned disease, the common cold is proving itself to be quite a noble source of submicroscopic information. Although no one would miss the cold if it miraculously vanished from the earth, we may count ourselves lucky that such a viral disease exists, one that handles us pretty darn gently while offering as many clues to the nature of viruses as any of its other, more dangerous cousins. We may treat it with contempt, but it certainly has amazing stories to tell us.

Twelve

. . . .

The Quick-Change Artist

. . . .

As their lungs filled . . . the patients became short of breath and increasingly [blue]. After gasping for several hours they became delirious and incontinent, and many died struggling to clear their airways of a blood-tinged froth that sometimes gushed from their nose and mouth. It was a dreadful business.

It was the fall of 1918, the place Philadelphia, and the writer a third-year medical student at the University of Pennsylvania named Isaac Starr. The Allies' final, victorious offensive on the western front was beginning; the end of World War I was a little more than a month away. By all rights, the United States should have been rejoicing. But Starr's recollections paint a country in the clutch of a killer whose manner of causing death was as terrible as its outcome.

. . . not long afterward there was shouting from the street, and we discovered that Mike the piano mover was poised on the window ledge ready to jump. Gathering the medical cohorts we converged on him, diverted his attention, rushed him, seized his arms and legs, carried him triumphantly back to bed, and strapped him in. But a little later there was another commotion on the ward; Mike, delirious, had turned the bed over on top of himself and was moving it up the ward on his back. He lasted only a few hours after this.

Then there was the Jewish family whose 18-year-old daughter was desperately ill. She was flushed with fever, and to my eye she was very beautiful. Father, mother, brothers, and sisters had gathered around, and they would not leave her. After suffering for a few days, it was she who left them.

Stories like these became commonplace that fall, as the epidemic gained momentum. It had begun innocuously enough in March when a mess cook at Fort Riley, Kansas, complained of chills and fever — the familiar symptoms of flu. By the end of the week, 522 others, men in their twenties and thirties, an age when resistance to disease is usually at its highest, had come down with the same ailment.

Soon other overcrowded camps were struck by similar complaints, but no one seemed overly worried. Flu was common enough around military bases, and, besides, there were more important problems to solve — such as how to get two million fresh troops to France in time for the offensive that everyone hoped would end the war. Eventually 46 people died at Camp Riley. The deaths were attributed to complications brought about by pneumonia, not flu, and the army continued training and shipping its men across the Atlantic. But when they arrived in France, they brought with them a lethal stowaway: the flu.

By fall it was clear that this was no ordinary influenza. Besides the familiar symptoms reported at Fort Riley, people were now complaining of labored breathing and copious sweating. As in Isaac Starr's account, flu victims began turning blue from lack of oxygen. Purple blisters popped out on their skin; their noses started to bleed profusely. Some were plagued by racking coughs, some spit out pints of yellow-green pus. People began to die, relatively slowly at first, then quickly and more frequently, then violently, sometimes within a few days of becoming ill.

In Europe, fueled by the huge influx of American troops, the flu spread from France to England, then to Spain, where it killed eight million people and became known as Spanish flu (and sometimes, with no affection whatsoever, as the Spanish Lady). It traveled eastward, where it further decimated the beleaguered German army and became known as the Blitz Katarrh, and on to Russia, India, China, Japan, Africa, and South America. The disease even reached Alaska — in Nome, 176 out of 300 Eskimos died — as well as Samoa, where it killed 25 percent of the population. No place in the world was safe from it.

Back in the United States, the epidemic escalated relatively gently, killing "only" 2,899 in August and 10,481 in September, but then it struck with a vengeance, annihilating 195,876 in October and over 100,000 in the remaining two months of the year. During the third week of October, in Philadelphia alone, there were 4,600 deaths from influenza. And there was nothing anyone could do about it, for who had ever seen its like before? The surgeon general, Rupert Blue, a veteran of fighting outbreaks of bubonic plague in San Francisco and yellow fever in New Orleans, declared the obvious — that this was a fierce new killer, unlike flu bugs of the past — and set about recruiting the relatively few doctors not engaged in the war effort to staff a Volunteer Medical Service Corps to deal with the disease. Congress unanimously voted $1 million to fight flu, and Blue, suspecting (accurately, as it turned out) that the disease spread through the air, wrestled with the idea of imposing a massive quarantine on flu victims. Eventually he abandoned the notion, but did call upon states to ban large gatherings.

Nothing stemmed the tide. While in almost all major cities theaters and other public places were shut down, Blue's suggestions were often ignored, as 200,000 people crowded together at a war rally in Philadelphia and 150,000 did the same in San Francisco. Millions jammed into public buildings to register for the draft, and in November, when the war finally ended, millions more filled the streets of the nation in celebration, kissing one another, breathing on one another, spreading flu virus like crazy. (The virus, in fact, played no small part in the cease-fire. As some German generals suggested, not entirely tongue in cheek, it wasn't the Americans arriving in Europe that ended the war, but rather the virus they brought with them.) And the epidemic continued.

Yet who knows whether the anticrowd regulations would have had any effect. Flu is neither entirely preventable nor curable. Even today, armed as we are with vaccines, flu continues to bedevil us, and of course in 1918, a flu vaccine was nothing more than a dream. No one was even sure what caused the disease, as the virus would not be discovered until 1933.

So around the world the pandemic ran its course. While devastating, and incredibly widespread, it was blessedly short. By the beginning of 1919 it was virtually gone, vanished as mysteriously as it had come. Yet the wake it left was cavernous. In the United States, twenty-five million people came down with the disease, a fourth of the entire population; more than 500,000 died. By comparison, no war — not the Civil War, not World War I, not World War II or any afterward — has ever claimed so many American lives. Worldwide, an estimated two billion people were stricken, and at least twenty million — some say twice that number — died, all in little more than ten months. There had never been as much loss of life in a comparable period of time in the history of the world, nor has there been since.

But that's not to say that otherwise the flu bug has left us alone. Every couple of years, to one extent or another, it lays us low; some people predict that we're likely to experience major pandemics every decade or two. Other, more publicized illnesses notwithstanding, influenza may be the most dangerous infectious disease we know.

"I always remind AIDS researchers that many more people die of flu than AIDS," says Peter Palese, of the Mount Sinai School of Medicine in New York. "Between twenty thousand and fifty thousand die of flu in the US every year, and even more in pandemic years. Flu is not so dramatic a disease, but clearly in terms of numbers over the years, AIDS is peanuts."

But why is flu such a problem? In contrast to AIDS, there are vaccines against flu. And in contrast to the common cold, another recurring viral disease, which is caused by some two hundred different viruses, there are only three types of flu virus, designated A, B, and C. Two of these viruses, type B and type C, cause relatively few cases of flu (although type B is associated with Reye's syndrome, a frequently fatal disease of children). The brunt of the illness, therefore, is the work of only one virus, type A. Why should it be so hard to get this one type of virus under control?

It's a question that Peter Palese has asked time and time again. Chairman of the Department of Microbiology at Mount Sinai, director of an elegant laboratory on the sixteenth floor overlooking the East River and the Triborough Bridge, Palese is involved in mapping the genes of flu viruses so as to discover precisely which gene produces which part of the virus. Discover that, and it may be possible to block the production of new viruses.

In one sense, he's picked a virus that is especially amenable to study. It was, after all, discovered in 1933 and has been the subject of investigation ever since. "With influenza we're lucky that so much has happened for over fifty years," he says. "No other animal virus has that length and breadth of study. And now we have the molecular techniques to look at its genes."

In another sense, however, Palese couldn't have picked a more difficult subject, for the flu virus simply refuses to stand still. As we saw in the last chapter, rhinovirus has come up with some survival strategies that seem infernally clever, but flu virus puts it to shame. From its bizarre form to its propensity to profit from the many *mistakes* it makes in attempting to reproduce and carry on, it's simply a fascinating creature. It has occupied Palese for almost twenty of his forty-five years, ever since he came to New York from his native Vienna. He still carries a Germanic accent, punctuating his remarks with a rhetorical "yah?" "It's a very nice virus for study," he says. "It's a model for dealing with other viruses, yah?"

But the obvious motivation for investigating the virus is to be able to do something about the disease. "We certainly hope that we can design in the future better vaccines and learn better how we can develop antiviral drugs," says Palese, "all based on a molecular understanding of these viruses."

At the molecular level, the flu virus is unique. Midsize as viruses go (by comparison, rhinovirus is about a third the size, vaccinia three

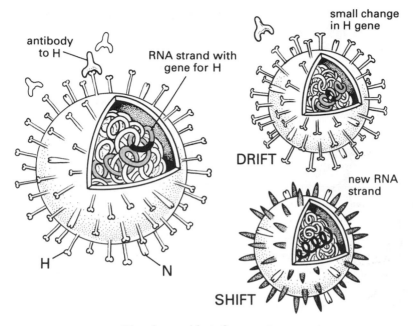

The changeable influenza virus

Flu viruses have two proteins — Ns and the more numerous Hs — on their surfaces, both of which may be recognized by our antibodies (left). But crafty flu is always changing its proteins to thwart our antibodies. In the mechanism of "drift" (top right), small changes in the virus's genes cause small changes in the surface proteirs, which may keep antibodies at bay for a time. In "shift" (bottom right), an entirely new gene strand appears, causing the surface proteins to look totally different, thus evading antibody surveillance. Most new, virulent strains of flu virus arrive by means of this mechanism.

times as large), flu virus is a collection of genes enclosed in a tough sphere made of protein. Nothing so unusual there, at least by viral standards, but the protein shell is itself cloaked in a fatty membrane punctuated all around with slender, jagged projectiles. It resembles nothing so much as the business end of a medieval mace, that daunting ball of iron studded with spikes that was used to bludgeon through an opponent's protective coat of armor. In the case of the virus, however, the coat of armor it means to breach is the wall of the human cell. It is the most fearsome-looking of all viruses and one of the most efficient.

The virus enters the body much as does the rhinovirus, through the nose and mouth. Then, again like the rhinovirus, it attacks cells in the nasal passage, hooking to receptors on the surface of those cells and infiltrating to take over the cell's reproductive machinery and force it to make new viruses rather than new cells. Unlike the rhinovirus, however, the flu virus will infect cells lower down the respiratory tract, in the throat, windpipe, and lungs.

Within a couple of days, you'll start to suffer the familiar miseries of "the flu": pain in the chest and throat, cough, fever, and possibly headaches, overall aching muscles, shivering, a feeling of lethargy. Happily, however, most bouts of flu don't last very long. Virus-fighting antibodies may appear in the blood as soon as four days after the onset of infection (but much later if it's your first exposure to the virus). Within another few days, if all goes well, you should be feeling pretty good once again, and by two weeks all should be back to normal. But if all doesn't go well — and such is often the case during influenza pandemics — the flu infection, while itself not necessarily life-threatening, can pave the way for other infections that are. The primary complication, of course, is pneumonia, which can be the result of the flu virus's penetrating deep into the lungs or, by destroying the protective layer of mucus and cilia, blazing a trail that bacteria follow to infect the now exposed respiratory tissue. It is to prevent such secondary infections that doctors often prescribe antibiotics during bouts with flu. The antibiotics do nothing against the virus, they do a great deal against bacteria.

While such a process may explain the basic illness, it doesn't explain why flu strikes in recurrent waves of pandemics and epidemics against which previous infections don't immunize us. The answer involves two essential components of the virus, *hemagglutinin* and *neuraminidase*. These convoluted terms provide an essential entrée into the arcane world of the flu virus. (For the sake of convenience, I'll often use the abbreviations H and N.) They are the names of the proteins that constitute the two kinds of spikes that dot the surface of the macelike virus. Jutting out as they do from that surface, they are the first parts of the virus that our immune system encounters. Just as advance scouts may identify opposing armies by their flags of battle, our body identifies the flu virus as a foreign invader by recognizing these two proteins.

But these viral battle flags are more than just identifying markers; they also perform vital functions. In the case of N, its role is only now being determined — more on that later. What H does is well known: it hooks up to the receptors on the cells in our body. Once that bond is forged, the virus can invade the cell and go about its business of making new viruses.

As we've seen, the body is not defenseless. It is often successful in stopping the virus before it has a chance to start by producing antibodies that make a beeline to the hemagglutinin spikes and stick to them like magnets to metal, preventing them from making contact with the cellular receptors. Prevented from binding, the virus is helpless. Soon the scavengers of the immune system, the white blood cells, engulf the virus and destroy it.

Too often, however, the antibody defense to flu simply doesn't work. Why? Because routinely the flu virus changes its battle flags by changing the configuration of its hemagglutinin. Confronted with these altered spikes — with what is in effect a viral disguise — the body's antibodies can no longer recognize the virus and prevent infection. It's a very clever and effective ploy. As a quick-change artist, the flu virus simply has no peer (with the possible exception, as we'll see, of HIV, the AIDS virus).

As an illustration, imagine that the H spike resembles a key and the receptor on the cell a lock. The key fits into the lock, and, once engaged, it opens the lock much as the spike opens the cell, readying it for viral invasion. Now imagine making a wax mold of that key. Since the impression in the mold is identical to the shape of the key, the mold can blanket the key perfectly, immobilizing it and preventing it from being able to slip into the lock — that's the action of an antibody against the virus. But if you encounter another key that fits the lock, this one with a slightly different configuration, much as a master key is slightly different from an individual key but still does the job, the impression in the wax mold is now off just enough that it may no longer be of any use in trapping the key. You must make another mold if you want to be sure to encase the new key perfectly and prevent it from opening the lock.

The flu virus makes just this kind of change, subtly altering its hemagglutinin spikes so that the body's antibodies can no longer neutralize them. By the time we're able to produce new, effective antibodies, it's too late. The infection is well under way. Flu virus makes these changes as a matter of course, and the body simply can't keep up. The result is a new flu outbreak every couple of years, and vaccines aimed at a particular version of the virus that risk being obsolete almost before they pass from the syringe into your body. They must be modified again and again to keep up with the virus's alterations. "I'm convinced that we'll never control flu virus by vaccination because the virus is too smart," says Robert Krug, Chairman of the Department of Molecular Biology and Biochemistry at Rutgers University, who is investigating precisely how the virus replicates so as to devise a means of interfering with the process.

But this pattern of continual subtle changes, called *antigenic drift,* is not the only trick the virus throws our way. It occasionally undergoes an even more dramatic change, this one called *antigenic shift.* If drift produces a virus with a subtly altered face, something on the order of growing a beard where before there was none, or getting a nose job, antigenic shift creates a completely new hemagglutinin protein. It is a change on the level of major plastic surgery, an entire face reconstruction. Now the virus looks brand-new, with all new flags — that is, with surface spikes able to elude the immune system and infiltrate respiratory cells with newfound alacrity. Now the body can't *begin* to recognize the invader. Now the result can be widespread, severe illness on the order of the 1918 Spanish flu, the 1957 Asian flu, and the 1968 Hong Kong flu.

In trying to understand the virus, here is where the value of metaphors reaches its limits. It's all well and good to talk about disguises and plastic surgery, as though the virus were some miniature intelligence that intended these changes in order to outfox the immune system. But the truth — at least, what is considered to be the truth — is quite different and, in its way, much more amazing. For these changes, these mutations — both minor, as in the case of antigenic drift, and major, as with antigenic shift — are simply the result of mistakes, of accidental, random disruptions among the virus's genes. Disrupt genes and you disrupt hemagglutinin spikes. Minor disruptions may be enough to produce a virus that can temporarily slip past the body's antibody sentinels — thus antigenic drift. But major disruptions, events that completely change genes or replace them with others from elsewhere, lead to antigenic shift and a virus that, completely able to escape antibody surveillance, may invade with a vengeance that can cause a pandemic.

And wouldn't you know it, it just so happens that flu virus is uniquely set up for such devastating mutations. It's simply not a very stable virus. "There's an intrinsic property of this virus," says Palese, "namely, that it makes a lot of mistakes as it copies its genes."

The reason involves the fact that when you descend past its surface spikes and lipid membrane and past its protein shell into the interior of the virus, you encounter genes of RNA rather than DNA. RNA is simply more prone to errors in replication than the more stable DNA. And added to this tendency is the astounding fact that unlike other viruses, flu does not simply contain just one strand of genes (as do rhinovirus and poliovirus), nor two intertwined strands (as do vaccinia, Epstein-Barr virus, and other herpesviruses), but eight separate strands of RNA, all curled together in the heart of the virus. Together they make up the whole of the virus's complement of genes, an aggregation called a *genome*.

Now things get complicated. The ordeal for our infected cells begins when a flu virus, using its hemagglutinin spikes, binds to the receptor of a respiratory cell. At this point neuraminidase, about which we know much less than we do about hemagglutinin, comes into the picture. Surprisingly, it is able to destroy the very receptor to which the virus binds. "You'd think, isn't this crazy — it destroys the receptor," says Krug.

When it comes to talking about the flu virus, Krug, a short, earnest, fifty-year-old, has a fondness for colorful language. "Well, first of all, there's less neuraminidase than hemagglutinin, and it obviously doesn't destroy the receptor in a normal infection," he says, peering over the top of his half-glasses. "Otherwise, there are several hypotheses as to what the neuraminidase does. One is that there are lots of similar receptors all over the place before the virus gets to cells. Maybe neuraminidase blasts a path so that the virus will not bind to anything before it gets to the cell." In other words, the neuraminidase may run interference for the virus, clearing its way of any obstacles that might hinder its binding to the proper receptor in the cell.

Once the virus makes fast, the cell fights back. It engulfs the virus and, intending to destroy it, ferries it into the cell interior in a special chamber called an *endosome*. Within the cell, the virus-containing endosome eventually becomes a slightly different cellular chamber called a *lysosome,* which is the cell's garbage dump, as it's filled with enzymes and acids that dissolve waste and foreign material.

Things then get a little hazy, but at some point the virus undergoes a dramatic transformation. It may be that it happens while this enclosing cellular chamber is still relatively neutral, still ferrying the virus into the cell; it may happen when the chamber begins working on the virus, starting to digest and dispose of it; but suddenly the virus's versatile outer membrane begins to change. The membrane actually unfolds like a blossoming flower and fuses with the membrane enclosing the chamber, thereby allowing the virus access to the interior of the cell. The virus then dumps its innards into the heart of the cell. And those innards are, of course, the virus's genes, all eight strands of them.

Once free and loose within the cell, the genes make their way to the nucleus, where they usurp the cell's own reproductive machinery and begin directing the construction of new viral genes and proteins. Soon these fresh viral ingredients migrate back to the edge of the cell where the virus first entered, separate into appropriate proportions (a pinch of genes here — eight strands will do — a dab of N and H there, a protein coat enclosing all) and, submitting the cell to a final indignity, tear from its lipid membrane an outer cloak to protect the new viruses in the hostile world outside. The young viruses then break away from the cell, finally on their own, identical — usually — in every way to the original invading virus.

Now the neuraminidase may play another role. The scrap of cellular membrane that the emerging virus has annexed for a cloak may contain all sorts of cellular features that would only hinder the virus — receptors, for instance. The N cleans the virus of those receptors, making it sleek and streamlined, ready to go out and infect. "Basically," says Krug, "the N is thought to function in helping the spread of the virus. It clears a path to the cell, and it also cleanses the virus, making it battle-ready."

Within seven to ten hours, hundreds of new viruses escape the hapless cell. Soon it will die.

Sometimes, however, the replication process is somewhat less than perfect — that's where the mutations come in. As one strand of genes begins to produce another, somewhere along the line one single element called a *nucleotide* — out of fourteen thousand nucleotides in the genome — may fail to attract its complementary nucleotide. This seemingly insignificant mistake, a mistake that in the lab occurs in only one out of every million viruses, is all the virus needs. The result can be a new strand of genes that gives rise to a new virus containing subtly altered H spikes. And that means a virus that, next time it infects, may be able to elude antibodies built up against the original invader. Thus, antigenic drift.

And sometimes something more dramatic occurs. If a cell happens to be infected by more than one virus at a time — it happens rarely, but frequently enough to have dramatic consequences — a strange intermingling can take place. One infecting virus may be a familiar version of flu, but the other may be a much different strain. "What is the source of this other virus?" asks Krug. "That's been the topic of a real long discussion. The idea is that there is an animal reservoir for flu virus. Pigs, for example — swine influenza, remember that? A pig virus could infect somebody at the same time that he or she is infected by the strain that's already around."

So instead of the genes of just one virus loose inside the cell — and that means eight separate strands loose in the inner workings, remember — now added to the mix are genes from the second infecting virus, another eight strands. It doesn't take much imagination to recognize the confusing possibilities. Now *sixteen* different gene strands are mucking about inside the cell, and they can easily become mixed up, shuffled together, and, when finally separated into groups of eight for packaging inside the new viruses, reassorted into entirely new combinations. You may end up with a new virus containing seven gene strands from one of the infecting viruses and one strand from another — seven human strands and one swine gene strand, say — or some other equally original concoction; there are 256 possibilities. If that wayward gene happens to be the one that produces hemagglutinin, look out.

When this new virus infects a new cell and replicates, the result will be viruses with brand-new hemagglutinin spikes — not just a sightly altered H spike, as in antigenic drift, but a completely different molecule. And that means real trouble for an immune system that has a hard enough time dealing with the subtly modified spikes produced by drift. Now, with a previously unknown variety of H confronting the body, all bets are off. As Krug explains: "The same cell gets infected by both viruses, there's reassortment, and a new virus picks up the pig H gene. Therefore you get a virus that is really totally resistant to the prevalent antibodies."

Thus the flu epidemics that regularly arise — but that *cannot* be anticipated. It's possible to track the mutations of the flu virus, but it's not possible to predict them. "It is a chance business," says Palese. "It's like a lottery — you know every month a ticket will win, but you can't predict which one." For the mutations are simply random, the result of the mistakes of a mistake-prone virus, and there's no way to know which will turn out to be the kind of antigenic drift or shift that will enhance the virus's ability to infect, or which will be a useless mutation that actually inhibits its ability to survive. "The making of changes has basically a neutral effect," Palese explains. "If the virus makes too many mistakes, and introduces too many lethal changes, then the organism can kill itself. On the other hand, a certain amount is quite helpful to the virus in terms of its survival. There has to be a balance. A little bit of change is quite good, flexibility is quite good, but if the flexibility is too much, it arrives at a stage where it just can't produce functional proteins. So it's a fine line where an organism like this can go."

Those mistakes that result in successful infection and reproduction remain, at least until the next mistake; those that are not successful disappear. And the flu virus goes merrily on its way making error after error, mutation after mutation, continuing to cause disease on a regular basis. You have to hand it to the little creature.

Over the years, researchers tracking these changes have come up with a notation system for categorizing the major fluctuations. For example, the virus that caused the great pandemic of 1918, the so-called Spanish, or swine, flu virus, is referred to as H1N1, signifying a variety of (H)emagglutinin and (N)euraminidase that have both been labeled type 1. That virus, in subtly different guises caused by drift, circulated for almost forty years, until in 1957, the Asian flu virus emerged to replace it. This virus was designated H_2N_2, as both its hemagglutinin and neuraminidase spikes were substantially new. Eleven years later, in 1968, antigenic shift struck again and yet another subtype appeared, the Hong Kong flu virus. This virus, H_3N_2, sported new hemagglutinin spikes but retained the neuraminidase from its predecessor. Today, although it has undergone numerous subtle changes as the result of drift, and though it no longer is as dangerous as it was initially, the Hong Kong flu virus still circulates. Altogether, flu researchers have identified thirteen distinct H subtypes and nine distinct N subtypes, a number of which have never seen the inside of a human body (why that is so in a moment).

In the midst of this cataloging sequence — as though the virus were loath to allow its behavior to be pinned down too easily — researchers received a major shock. In 1977 there appeared yet another type of virus, the Russian flu variety. This virus didn't play by the rules. Palese and his Mount Sinai colleagues discovered that rather than exhibiting new hemagglutinin and neuraminidase, it presented spikes virtually identical to a previous strain, a variety of the original 1918 swine flu virus that was still making the rounds in 1950. Moreover, not only were this new virus's spikes the same as those of the 1950 strain, so were its genes. It was virtually the same virus and so was given the same designation, H_1N_1.

The discovery destroyed the researchers' notion that, as changes in flu virus are the result of mistakes, new viruses will continue to turn out different from one another. For here was substantially the same virus that had plied its trade way back in 1950. "That was very surprising," says Palese, "because it had never happened before, one, and secondly because it was not clear how a virus that changed so rapidly could survive from a period of 1950 to 1977 unchanged genetically."

How, then, could it have happened? Could the virus have actually mutated back to a previous form? "That I would exclude," says Palese. "It would be very difficult for it to undo all the nucleotide changes which had accumulated over time."

Could it have been *our* fault, the result of inadvertently allowing an old virus that had been stored in the laboratory to escape back into circulation? "Many people are suggesting that," says Palese. "And there would be some possibilities in addition to a man-made introduction, such as that the virus was frozen in nature and somehow reintroduced into the human population. But it is difficult to understand how that could have happened."

So, we're left with the sobering if not entirely unexpected realization that despite more than fifty years' worth of tracking the virus, it is still capable of producing some unexplainable surprises.

Yet perhaps nothing should be put past a virus that spends much of its time in the belly of wild ducks. Flu virus travels the world in wild ducks and other waterfowl, such as various small shore birds and gulls. These feathered, mobile incubators function as ideal carriers for the virus. While virtually unaffected themselves, they spread it throughout the world, where it moves through horses, pigs, even seals and turkeys, and, of course, humans.

Perhaps no one knows that better than Robert Webster. A tall, slim fifty-seven-year-old New Zealander, Webster chronicles the worldwide travels of the flu virus from his vantage point as chairman of the Department of Virology and Molecular Biology at Saint Jude Children's Research Hospital in Memphis, Tennessee. That's not to say he follows the virus from his office, however. Webster and his team have gone on hunting expeditions around the globe, tracking the virus in ducks from Iceland to Australia, Canada to Peru, and, of course, to Russia and China, the point of origin of so many flu viruses.

Why ducks? Two reasons. First, flu viruses, all of them, get along famously with ducks. Every variety of flu virus sporting every variety of hemagglutinin and neuraminidase is found in the feathered fowl. Second, as the ducks provide a safe, hospitable home for the viruses, the viruses, as mentioned, return the favor — they cause no disease in the birds. As in humans, they do infect the respiratory tract, but only a little, preferring to do most of their multiplying deep down in the intestines where they do no harm. That arrangement suits the virus, for it offers it an ideal means to spread to other ducks, through feces and therefore through the water of the innumerable lakes and ponds where the ducks live and breed. "In August you can go to

Canada and take samples from all the local lakes," says Webster, "and you can isolate any kind of influenza virus you like, none of which causes any disease at all in the ducks. And recently we've found another reservoir of influenza — little shore birds that migrate north in the springtime from South America. We think that these birds maintain the virus in the spring, ducks in the fall, some sharing viruses and some not."

How does ducks' and other wild waterfowl's providing a mobile haven for flu viruses help to explain why so many of the epidemics of the disease seem to come from China and Russia? The enormous expanse of Asia is home to huge numbers of ducks and other fowl. And despite the enormous numbers of people in these areas, many tend to live in rural settings, where they often come into contact with the wild birds. So, the scene is set: a virtually limitless supply of birds carrying untold millions of flu viruses, and millions of people who live in proximity to this "avian reservoir," as Webster likes to put it. All that is needed is that the viruses be able to pass from bird to human.

That they do. And like airplanes landing to refuel partway to their destination, their route involves an intermediary step, a stay at a sort of halfway house before mustering the courage to go the rest of the way. That cozy haven, Webster suggests, is supplied by pigs.

As we've seen, the 1918 Spanish flu pandemic was caused by a swine flu virus, but it wasn't until the swine flu incident of 1976 that pigs and flu became linked in the popular consciousness. It was in January of that year that a flu outbreak killed a soldier at the army base at Fort Dix, New Jersey. The virus taken from the dead soldier turned out to be a descendant of the virus that caused the 1918 pandemic. Epidemiologists feared another pandemic, and, accordingly, a vaccine was quickly produced, and a nationwide inoculation campaign launched, with the hopes of heading off the problem before it got out of hand.

As it turned out, however, the outbreak didn't even travel out of Fort Dix. Why was it a false alarm? The answer, according to Webster, is that the virus was able to travel from pig to human but lacked the ability to spread from human to human.

"In 1988 we had another swine flu incident in the northern

United States that went almost unnoticed," he says. "Investigators from the Centers for Disease Control reported that a pregnant woman died of flu in Minneapolis. The virus isolated was swine virus. All the genes in that virus came from one known to infect pigs. She had gotten infected at a county fair. There had been a little flu epidemic in the pig population, and all of the 4-H children who had pigs at the fair got infected. Some of them transmitted the virus within their families, but that's as far as it went." This swine flu virus, too, was unable to travel farther.

So, what pigs commonly seem to provide the flu virus is a rest stop between ducks and people, a kind of "mixing vessel" where the virus can mutate and acquire mammalian characteristics until the day that, like the 1918 virus, it will be able not only to travel from pig to human but to continue its spread in people thereafter. The 1976 variety was, in effect, an impetuous adolescent, strong enough to make the leap from pig to human but not quite ready to go beyond that effort. Meanwhile, other viruses continue mutating in pigs, preparing for the day when they try again. It is not a happy thought.

When the virus *is* able to make the transition to successful, recurrent human infection, the results can be catastrophic. Studies of H3N2, the Hong Kong flu virus that appeared in 1968, show that while six of its eight genes came from the strain that preceded it (a descendant of the Asian flu virus of 1957), the other two genes probably came from a duck virus. And one of these two intruding genes was the all-important one that produced the new hemagglutinin spike, the H3 in the notation. Webster surmises that the Hong Kong virus emerged when the descendant of the Asian flu virus circulating in 1968 and a strain of flu virus from a duck simultaneously infected a single cell in a single pig or human being, probably in China. Two of the duck virus genes became mixed up with the Asian virus genes, all of which ended up packaged together in one virus. The result: antigenic shift and a new flu epidemic. From such humble beginnings do great catastrophes grow.

So that's evidence provided by research. Are there any eyewitness accounts of flu viruses passing from bird to animal to human? The answer is yes, and Webster had a front-row seat to one of the occasions. In 1980 he was at the Institute of Experimental Pathology outside of Reykjavík, Iceland, investigating a particularly virulent strain of flu known as H7H7. This virus had been devastating seal populations along the North Atlantic coast of America, inflicting the worst flu-caused lung damage Webster had ever seen and killing up to a third of the seals. The virus was genetically related to another murderous influenza strain, fowl plague, which could kill a chicken in forty-eight hours by destroying its central nervous system. This was, then, hardly a virus to become exposed to if you could help it.

Unfortunately, one of Webster's colleagues couldn't help it. As Webster was examining a sick seal that had been experimentally infected with the H7N7 virus, the animal suddenly twisted around and sneezed directly in the face of a lab technician who was attempting to hold it steady from behind. It was a very worrisome moment — after all, the seal harbored an especially virulent virus. If the sneeze passed the virus to the technician, the man might soon find himself in dire straits. Webster badly wanted to know if such transmission could take place, but he certainly hadn't planned to find out at the expense of the technician.

In just forty-eight hours, Webster found his concern to be at least partially justified. The technician came down with a bleary-eyed case of conjunctivitis, and in the infected eye Webster found high concentrations of the seal virus. The virus *had* passed from the seal to the man via the sneeze and had infected him. Fortunately for all concerned, the virus did no apparent damage beyond inflaming the technician's eye.

And that's not all. Genetic tests at Webster's lab later showed that the virus had originated in birds. Here, then, the proof that a flu virus from the avian reservoir, harmless among the birds themselves, could make its way to other animals (in this case, a seal rather than a pig) and then to humans, causing severe disease along the way.

Webster had been finding evidence of such viruses in birds for years, but this time the scientific community sat up and took notice. It was one thing to have vast quantities of flu virus multiplying without incident in aquatic birds — it was another for these same viruses to cause killing disease in animals and readily make the leap into humans as well. What if the disease had proved as fatal in the lab technician as in the seal? That was the prospect that terrified Webster, and that, projected onto the human population as a whole, terrifies him still.

He illustrates his concern with a homely barnyard parable. "In the spring of 1983, influenza virus made its way into chicken houses in Pennsylvania," he says. "We were fortunate that an astute veterinarian isolated virus from what was then a very mild infection. It was H5N2. In October, the virus became virulent, and every chicken was killed."

This virus, too, came from waterfowl, where it caused no disease whatsoever. And, in the beginning, in the chickens it acted in much the same fashion, causing nothing more than a mild infection. Then, within the space of half a year, something happened to change it into a killer. "Nothing more than the hemagglutinin gene changed between April and October," says Webster. "That single mutation was enough to change that benign virus to one that was lethal."

Then he drives the point home. "How does one cope with such an epidemic? The Pennsylvania agricultural department stopped it by the standard method of eradication, which was to kill the birds, dig a deep hole, and bury them. That's the way it's been done throughout the world for centuries. They destroyed seventeen million chickens, at a cost of sixty-one million dollars to the taxpayer. But we can't help asking ourselves, what would we have done if this virus had occurred in humans? If you try to eradicate it that way with humans, you don't do so well, do you?" He laughs, then abruptly turns serious once more. "That's the point. The chicken population in Pennsylvania is like the world as it is at the moment. There are thousands, hundreds of thousands, millions of chickens just waiting to be infected."

So, our ancient battle with the slippery flu virus is far from over. "All the genes of all influenza viruses in the world are being maintained in aquatic birds," Webster says, "and periodically they transmit to other species like pigs. The 1918 viruses are still being maintained in the bird reservoir. So even though these viruses are very ancient, they still have the capacity to evolve, to acquire new genes, new hosts. The potential is still there for the catastrophe of 1918 to happen again."

*   *   *

That being so, what can we do about it? The obvious approach is to develop vaccines, by far the most effective means yet devised for coping with viral attacks. And there *are* effective flu vaccines, preparations that can reduce the incidence of the disease between 60 and 80 percent. But they all have an Achilles' heel: they're primarily effective against only one strain of the virus. As the vaccine consists of viruses — usually killed, but sometimes live and attenuated — from one particular strain of flu, when another strain comes along with a different configuration of hemagglutinin spikes, the vaccine may be able to do little. So while the vaccine may protect against the viruses circulating during one flu season, it may fail to stop next season's strain. For a flu vaccine to be effective at all, it must be thrown into circulation early on in the course of an epidemic, before the virus has drifted or shifted so much that the vaccine no longer protects.

That necessity puts great pressure on those whose task it is to recognize new flu virus strains quickly. The World Health Organization maintains a network of centers in over a hundred countries throughout the world — facilities where experts isolate and examine any new occurrences of influenza in their locality. Whenever one of the centers identifies a new strain, it immediately sends it to one of the two WHO Collaborating Centers for Reference and Research on Influenza, located in Atlanta, Georgia, and London, England. Scientists there scrutinize the viruses sent to them and, if they corroborate the findings of the far-flung centers, put out the word that a new flu epidemic may be around the corner. Vaccine manufacturers can then begin work on a vaccine so as to have it ready for the coming season.

While this approach has resulted in a number of effective flu vaccines, it has several limitations. The obvious one is that sometimes — the 1976 swine flu incident is a prime example — the expected epidemic might never actually come to pass; you've expended all that effort and spent all that money on a vaccine that may be unnecessary. When it comes to flu, the crystal ball is almost invariably clouded. Another weakness to this approach is that finding a new virus variety and analyzing it is tough enough, but designing, manufacturing, and distributing a new vaccine is a daunting task in the best of times. When a flu epidemic is imminent, it's yet more difficult. The manufacturer must get the virus to grow prolifically in the lab, then prepare the vaccine and test it for safety and efficacy, then distribute it to the people who will administer it, who in turn must actually give the vaccine to the people who need it. Then, after all that, a week or more is usually required before the person vaccinated develops immunity (it depends on which viruses a person has already been exposed to). The whole process can take several months. By that time the virus may have gone on its way, infecting huge numbers of people and concurrently already beginning to drift toward its next mutation. So while the outcome of the effort can sometimes be successful and satisfying, it can also be terribly frustrating.

That being the case, it would seem as though flu fighters might adopt the approach of those who are investigating other quickly changing viruses, such as HIV and rhinovirus. That approach is to find a spot on the virus that does not change — a spot, say, that plays a role in docking with the unchanging receptor on the target cell. If such a spot could be found, perhaps it would be possible to use it to produce an all-purpose vaccine, a vaccine that would be effective no matter what flu virus strain was circulating during any given season.

To find such a spot, however, necessitates knowing a great deal more about the virus's hemagglutinin spikes. A huge step in that direction was provided in 1981 by Harvard University's Don Wiley and Ian Wilson. Using crystallization and X-ray diffraction techniques similar to those used by Rossmann to decipher rhinovirus, they produced a three-dimensional map of a hemagglutinin spike. As a result, H is now one of the most completely understood biological molecules. It turns out that this spike has three legs that perch on the surface of the virus; they look something like a tripod. At the end of the legs are clublike projections, each of which contains four "hot spots." These hot spots, really short segments of amino acids, are the antigenic sites that the body recognizes as foreign and against which it prepares an antibody defense. The receptor-binding mechanism seems to be located in a cleft between the legs.

As suspected, these hot spots are highly unstable; they change constantly as a result of antigenic drift. And all that's necessary for a virus to elude the body's immune system is a change in a single amino acid — a *point mutation,* as it's called. In other words, to produce a newly infective flu virus the entire H molecule doesn't have to change; nor does a significant part of it. The alteration involved here is subtle to a degree that's hard to conceive. In effect it would be as though changing the color of your contact lenses made you no longer recognizable as the same person.

With Wiley and Wilson's map as a guide, however, researchers can now try to pinpoint the area on the H that binds to the cell, with the hope of making a vaccine that uses this purportedly conserved and unchanging structure. "So far," says Peter Palese, "all of these attempts have not been very successful."

The problems stem from the fact that to be successful a vaccine must prompt the body to make antibodies that can prevent a dose of the real disease. To do that, the vaccine must present a safe version of those parts of the virus that the body recognizes as foreign and dangerous, the antigenic sites. The antibodies produced to fend off this pretender will then be there to protect against the real thing. But these conserved areas on the flu virus may not be antigenic at all. Since they seem to be sunken in a cleft between the tripod legs of the hemagglutinin, much as the binding site of the rhinovirus may be deep in the canyon, they may not be accessible to the immune system and therefore will fail to provoke it to produce antibodies. And if a vaccine doesn't raise antibodies, it won't do a thing to protect against the disease.

So with the possibility of an all-purpose vaccine looking dim, Palese is focusing his attention on yet another difficulty involved in stopping flu — diagnosis and analysis. Even in the best of situations, flu fighters find themselves running as hard as they can simply to keep up; so any mistake or delay in diagnosis of the disease or analysis of the virus can irretrievably throw a wrench into the works.

"At this point the diagnostic procedures used by physicians are miserable," he says. "If you are really sick, and you go into a doctor's office, he will look at you, and if you have a runny nose and fifteen hundred other people have been in his office in the last five days, all with runny noses, he will say, 'It's influenza.' But in many instances that's wrong."

In contrast to ailments caused by bacteria, for example, viral diseases allow no quick and definitive diagnostic procedures. If you go to the doctor with a sore throat and cough, a throat culture and a simple chemical test will within minutes reveal if you're suffering strep throat, which is a bacterial problem, or something else. If the verdict is something else, your doctor is likely to announce that you have an unspecified virus, perhaps flu, perhaps some other respiratory problem, and in any case there's little to do about it except wait it out. But Palese expects all that will change.

"In the future there will be fast tests to assay sputum," he explains. "You'll take a little bit of spit and, by genetic-engineering technology and new applications of molecular-biology methods, identify within a half-hour whether or not it is indeed influenza. Then we can choose therapeutic agents. And we do already have antivirals that are effective against influenza, yah?"

The most familiar of these is amantadine. One of the oldest antiviral drugs, it is also one of the most effective, both in preventing flu and, if given early enough, lessening the miseries of the disease once you've gotten it. The drug works by blocking the flu virus from penetrating the cell — how, no one is quite sure. But because there are some side effects, such as nervousness and insomnia, amantadine is used primarily for people who are at high risk either of contracting flu or suffering complications once infected — people with heart and lung problems, for example, and the elderly. And amantadine has two other, even more important drawbacks: it only works on some influenza strains and not on others, and the virus's frequent mutations may easily neutralize the drug.

On the drawing boards, however, are other anti-influenza drugs. Robert Krug is among those pursuing one particularly promising avenue, and the trials and tribulations he's experienced offer eloquent testimony as to the difficulties this formidable virus presents, not only to physicians and flu sufferers, but to researchers as well. Krug's approach involves looking at precisely how the flu virus replicates inside the infected cell. In general, all viruses reproduce in a similar manner, but on a closer level each exhibits its own, unique style. As is only appropriate, the iconoclastic flu virus is no exception to that rule. It goes through a very interesting series of steps that Krug calls "cap snatching," and against which he's readying his own "MX missile."

As we've seen, viruses reproduce by relaying the instructions contained in their genes to the protein-making machinery of the cells they infect. Cells reproduce in much the same way; the blueprints from the genes in their nuclei drive the cells' protein factories. For both organisms, therefore, the problem is the same: how do you get that genetic information out to the factories? What's needed is some kind of messenger service to carry the instructions into the field. And, as discussed, that's precisely what they get — a blueprint messenger service in the form of mRNA, a special messenger form of ribonucleic acid. The genes within the cell or virus thus must first make mRNA before they can go on to make the proteins that constitute new viruses and new cells.

With the aid of enzymes that act as catalysts for the process, that's exactly what takes place. Once produced, mRNA gets right to work, ferrying the precious genetic information out of the nucleus of the cell to the cell's cytoplasm, where await the protein factories called ribosomes that then churn out custom-made proteins.

For some reason, however, flu virus simply can't make its own messenger RNA. It has the necessary genes, and it carries the necessary enzymes, but it just can't do it on its own. It must borrow part of the cell's own mRNA-creating capability. Actually, *borrow* might be the term preferred by a viral politician; *stealing* is more accurate. "What the virus does," says Krug, "is go into the nucleus and steal one end of newly synthesized host-cell mRNA. The virus actually chops off this end, which includes a section called the 'cap,' and uses it as a primer to initiate the synthesis of its own messenger RNA. It does its own genetic engineering." This cap is a segment on the very end of the cell's mRNA strand — it "caps" off the strand. Though not completely understood, the cap's functions probably include stabilizing the mRNA and helping it more effectively to produce protein. Stealing it is a step almost no other virus needs to take.

That got Krug to musing: "So you would think that if you could block this step, you'd be able to stop all flu viruses. But that's assuming you could devise some sort of inhibitor that would get into cells and would be specific for this particular step. People have tried, and nothing has come of it." Suddenly his face brightens. "But it turns out that there may already be an inhibitor of this step that's naturally occurring. It's something called the Mx protein. The Mx protein — it's a biological missile, believe me. It's a protein that's induced by interferon."

We've encountered interferon before, most recently as one of the means used to try to control the common cold. It's part of the armamentarium of the immune system, a protein produced by infected cells that puts all the cells in the area on virus alert. "What happens," says Krug, "is that interferon goes into a cell and causes the production of special proteins that somehow set up an antiviral state. People have worked on this for twenty-five years, and they still have found no clear-cut mechanism to explain how these interferon-induced proteins cause an antiviral state. And they don't understand the state itself. The Mx protein is probably the best way to find out."

Krug goes on to tell an interesting tale. In the late 1970s, researchers in Switzerland, headed by Otto Haller, were investigating the properties of flu virus by injecting it into mice. In the course of their study, they found that whereas in general mice, like people, readily succumb to the virus, there was one strain of the rodent that wasn't affected at all. It scurried around as lively as usual, while all the other mice were exhibiting the same kind of miseries as those suffered by flu-infected people. But why? Upon examination the researchers discovered only that these flu-resistant mice contained a gene that the others didn't have. They called this gene Mx, for myxovirus, the family to which flu virus belongs.

All well and good, but did the gene have anything to do with preventing flu infection? To find out, the Swiss researchers extracted cells from the mice with the Mx gene and cells from the mice without it, then ran an extensive series of experiments in which they studied the genetic output of each of the cells under a variety of conditions. They found that when the cells containing the Mx gene were exposed to interferon, and only then, they produced a protein seen nowhere else. The researchers dubbed it the Mx protein. This protein seemed to stop flu virus in its tracks.

Then, in collaboration with Krug, they found that the protein did indeed prevent influenza by blocking production of the virus's messenger RNA. The protein didn't have a similar effect on any other virus they tested, just flu. Other viral infections, they found, were inhibited to one degree or another by interferon whether or not the Mx protein was present. But without the Mx protein, interferon did almost nothing to stop a flu infection. This was a protein specifically directed against the influenza virus. It was as accurately aimed and as effectively deployed as an MX missile.

It was an exciting discovery; but it is one thing to observe the protein working in mice and in the laboratory and another altogether to make use of it as an influenza treatment in humans. Simply dosing people with the flu-fighting protein is out of the question, as the body would produce antibodies against it as a foreign invader. "It's the mechanism we're interested in," says Krug, "in figuring out how it works. If you could [discern the mechanism], you could design a drug that could work the same way. But first you have to prove how the Mx gene works." He suspects that it might interfere with the virus's practice of snatching the primer caps from the infected cell.

Krug shakes his head wearily. Even if he should prove that the protein acts by preventing cap snatching, he would be only partway there. "And this Mx protein," he says, "the fact that it inhibits flu may have nothing to do with its normal function. It may have a lot of other functions. How it acts just isn't clear, but it's a very good lead."

So now he and the Swiss group, collaborators no more, are racing each other to be the first to come up with the answers concerning the Mx protein and, from them, a new flu treatment (in this effort Krug is involved with the pharmaceutical company Merck Sharp & Dohme). "On a scale of one to ten, we're at five to seven," Krug says, laughing ruefully. "All we really know is that this is a very specific protein, specifically aimed at influenza virus. And we know that in cells expressing the Mx protein, flu mRNA synthesis does not occur. But we don't know how the protein does that. We don't know if it acts alone. We know that it doesn't need other interferon-induced proteins, but does it need other host-cell proteins that are already there? Does it act directly, indirectly? We don't have a clue. Once we know the mechanism, we'll jump higher on the scale. But that's a big jump, and it's been hard."

Krug spreads his hands and looks over his half-glasses. "Science is hard. Having the right ideas and designing the right experiments is very hard. In science you can work for six months, a year, two years, have all the right ideas, do all the right experiments, and get nothing. It can be frustrating, really tough." He laughs. "Somebody could find the mechanism next week. But who knows. I can't predict that. Somebody will get the answer — I hope it's us."

Science is especially tough when you're working with a virus that has made a career out of staying one step ahead. But for all the difficulties it presents researchers — and all the illness it brings to the rest of us — the flu virus may provide crucial insights into how our own cells reproduce and grow. That's the beauty of working with viruses: in some ways they're very much like stripped-down cells, and they exhibit in simplified form many of the same processes cells go through. If viruses, then, are tough to understand, cells present difficulties of an even higher order. It can be a relief for a molecular biologist to gear down to the minimalist world of viruses, no matter how frustrating the chase.

Flu viruses also cast light on the workings of other viruses. On how genes replicate, for example. Together, the eight gene strands in the flu virus contain about 14,000 nucleotides; each gene strand contains only a fraction of that number — the largest 2,400 nucleotides, the smallest only 890. A single gene strand in retroviruses, the family to which HIV belongs, can be close to 10,000 nucleotides long. In our own cellular chromosomes, the nucleotide count can reach into the billions. Better, then, to uncover the workings of flu's eight relatively small strands and then use them as a model for the more complicated processes of the larger organisms. As Krug underscores, "flu viruses are very nice molecular models for normal cell functions."

And that's as good a place as any to leave the flu virus — in the midst of being appreciated by a hardworking virologist, someone who spends most of his time dealing with the amazing creature. As fascinating as it may be, and as much as it has to teach us, we may feel grateful that the flu virus doesn't spend most of its time working with *us*. A bout with the flu once every year or two is plenty, thank you.

Thirteen

· · · · · · · · · · · · · · ·

The Virus That Turns You Yellow

· · · · · · · · · · · · · · ·

Saul Krugman leans back in his chair in his office at New York University Medical Center. "I was there," he says, "in the South Pacific. I treated many of the patients. I was vaccinated for yellow fever, too, but I was lucky." Now seventy-nine years old, with graying hair receding on his forehead, thick glasses, and a lean, almost gaunt frame, Krugman is a renowned specialist in infectious disease. As a young physician in 1942, though, he could do nothing but look on in dismay as recruit after recruit fell ill.

The war was going badly. During the first half of the year, Douglas MacArthur retreated from the Philippines, vowing, "I shall return." Bataan and Corregidor fell. Sick and starving American and Filipino POWs were forced to make what has become known as the Death March, sixty miles to Japanese internment in the interior.

Three years earlier, in 1939, the surgeon general of the United States had called on the Rockefeller Institute to mass-produce the new yellow-fever vaccine for the army. By the spring of 1942, the Rockefeller labs had ready seven million doses of the vaccine. They were delivered and administered quickly. The war effort was discouraging enough — a disease epidemic was the last thing the army needed. Unfortunately, an epidemic was the very thing it got.

But not of yellow fever — the vaccine was effective against that dread disease. These recruits were coming down with an ailment that was much more widespread. At the time it probably was exceeded only by the common cold, flu, and measles as the most prevalent viral malady. And, even more disturbing, its cause, and treatment, were as yet unknown. The disease? Hepatitis.

Popularly called *yellow jaundice,* hepatitis was, and is, a disease that attacks the liver — thus the term *hepatitis,* Latin for "inflammation of the liver." It comes on as if a case of the flu, but it lingers, with a fever that reaches over 100 degrees Fahrenheit and a marked loss of appetite. People lose their taste for meat, rich foods, cigarettes, alcohol. Accompanying all is a feeling of terrible tiredness, and headaches, chills, and vomiting. After a week or so of fever, the patient might start to feel somewhat better. At that moment the jaundice appears — a sickly mustard yellow skin tone that persists for another couple of weeks at the least.

Although most people eventually recover from hepatitis, it's a particularly debilitating disease while it lasts and can result in long-term, even fatal, complications. It wasn't known at the time, but the variety of hepatitis that later became designated hepatitis B, if acquired early enough in life, leads to cancer of the liver. In some parts of the world — Africa, China, Taiwan, and Southeast Asia, in particular — liver cancer may be the most frequent kind of cancer death, and most of the cases are directly attributable to a strain of hepatitis acquired at birth.

The army reacted quickly. Within months the surgeon general stopped the use of the Rockefeller vaccine, pending an investigation. The damage, however, had been done. In 1942 alone, almost 30,000 soldiers came down with hepatitis; 84 of them died. From 1942 to 1945, from all causes, there were 200,000 cases of hepatitis in the US Army, with 352 deaths. (On the other side of the front lines, five million Germans, civilians, as well as military personnel, are said to have contracted the disease.)

The army seemed to have averted one catastrophe, only to suffer another. How could a vaccine designed to prevent yellow fever cause an epidemic of hepatitis? It turned out that the otherwise effective viral vaccine was contaminated with a different virus, one that caused hepatitis. But no one knew that then. "You have to remember that this was back in the forties," Krugman says. "When they prepared the vaccine they used human blood serum as a stabilizer, but there was no way to test it for a hepatitis virus. Nobody even knew about such things. The first time that it would have been possible to detect the virus would have been somewhere around the late 1960s."

So, nothing could be done. The many thousands of soldiers fell ill, and the war dragged on. In time the Rockefeller Foundation laboratory resumed production of the vaccine, this time bottled in water instead of blood serum, so as to preclude the possibility of blood contaminants making their way into the mix. By 1947, after the war was won, the lab had supplied the army and the rest of the world with twenty-eight million doses of yellow-fever vaccine. Finally, belatedly, the effort was a great success. By 1960, a hundred million people worldwide had received the vaccine, and the once horrible scourge of yellow fever was reduced to a sporadic disease.

Not so hepatitis. It emerged from the war more widespread and mysterious than ever — especially because what had been perceived as a disease of squalid and overcrowded living conditions now, as a result of the contaminated vaccine, showed itself liable to be borne through blood. It wasn't a disease of the environment only; it could infect anyone, anytime, anywhere. But what caused the disease remained unknown. Animal experimentation was of no use because animals suspected of being susceptible to hepatitis from exposure to humans, such as chimpanzees, simply didn't respond when infected experimentally. The answers to the fundamental questions surrounding hepatitis — what caused the disease, were there different strains, was there a way of preventing it? — boiled down to one: no one knew.

Such was the situation a decade later when Saul Krugman, then a professor of pediatrics at NYU Medical School, received a call from the Willowbrook State School for retarded children. The institution was plagued with hepatitis, among numerous other health problems. Was there anything Krugman could do to help control these infectious diseases?

Situated on 375 acres on Staten Island, Willowbrook had been designed and built to care for 3,000 mentally retarded children from the greater New York area. In 1955, when Krugman came to the school, its twenty-four buildings housed upward of 4,000 children — the most severely retarded, most handicapped, and most helpless of all of those sheltered in the New York State system. Over half of the children weren't toilet trained. Over half couldn't feed themselves. Almost half couldn't even get around on their own. Another third suffered convulsive seizures. And if all that weren't difficult enough, the overcrowded conditions made caring for the patients all but impossible. During the day, when the children were crowded together in the dayrooms, they spent much of their time attacking each other, soiling and abusing themselves, and destroying their clothing. Many of them were prone to put everything they picked up into their mouths. Often those things included filth and feces. At night their beds had to be pushed together so there would be room for all the children to sleep. Except through one narrow aisle, the only way for attendants to reach the children was by climbing over the beds. The conditions were simply intolerable.

Krugman, then, found himself entering a nightmare, one that promised nothing so much as the specter of an unchecked epidemic. In the years before he came to the school, over a thousand cases of hepatitis had been recorded (the rate was actually higher among the adult attendants than the children themselves), and these were just the obvious, jaundiced cases. His own epidemiological work soon convinced him that within months almost all newly admitted children would themselves surely become infected. Clearly, something had to be done. But what?

The action Krugman took became a watershed in the fight against hepatitis. In a few short years it rid the Willowbrook School of its epidemic, provided a context for the discoveries about the nature of hepatitis that have come since, and led to a vaccine that has prevented millions from suffering the disease. At the same time, it was so controversial that Krugman's efforts were censured on the floor of the New York Senate, and a bill was introduced to prevent such an experiment from ever happening again. But it all began with a decision that to Krugman was little more than good common sense.

From the beginning, Krugman's mandate was clear: rid the school of hepatitis. But how do you eliminate a disease that doesn't respond to treatment? Answer: you prevent it. And when it comes to viruses, prevention means a vaccine.

The only catch is that to produce a vaccine for a viral disease, you first must be able to identify the virus. That done, you must be able to grow it and work with it in the lab. In the case of hepatitis, no one had ever tracked down the agent. No one even knew for sure that it was a virus that caused hepatitis — it had simply been assumed so because no other infectious agent had ever been found.

So Krugman's hands were tied. Hepatitis was flourishing all around him, but he could neither treat the disease nor isolate the offending agent and deal with it apart from the havoc it caused. He made a fateful decision: in order to stem the hepatitis epidemic at Willowbrook, he would perform experiments that involved artificially inducing the disease in the children themselves.

"It was inevitable that all newly admitted children would come down with hepatitis, they couldn't escape it," Krugman explains. "So we realized that the only way we could learn something about the disease which would eventually lead to prevention was by studying small groups of children and actually exposing them to the same Willowbrook strain of hepatitis that they were going to live with anyway as soon as they came into the institution."

It was a bold and risky decision. To perform medical experiments on human beings was — and is — controversial, but in the case of consenting adults it's commonplace. To perform medical experiments on children enters into a gray area of medical ethics, but it, too, is often done; the development of the polio vaccines is a prime example. But when those children are retarded, incapable of understanding what is being done to them or appreciating the reasons why, the ethical justification becomes more muddled — at least in the eyes of some. It is inevitably an emotionally-charged situation.

Krugman saw no other choice. He decided to go ahead, but with one condition: "Only children whose parents gave consent would be included," he says. "Children who were wards of the state or children without parents were never included in our studies." And the children he worked with would live in a specially equipped and specially staffed ward. They would be isolated from the host of other diseases rampant at Willowbrook, maladies such as measles, dysentery, respiratory infections, and infections caused by parasites. Even if nothing came of the experiments, at least these children would be better off than otherwise. In 1956, he began his experiments. The first series would last seven years.

During that time, Krugman made substantial inroads against the disease, eventually cutting the obvious, jaundiced cases by as much as 85 percent among children and staff alike. By giving injections of gamma globulin, the part of blood plasma that is rich in antibodies, he proved that antibodies built up from another's exposure to hepatitis provide temporary immunity to someone not yet afflicted. The "borrowed" antibodies in the gamma globulin offer a "passive" immunity, acting as an internal bodyguard to weaken the effect of the disease for up to about six weeks.

Krugman also proved that once exposed to hepatitis, you were thereafter immune to the disease. He fed groups of six to fourteen children doses of diluted blood serum from other Willowbrook children who had already contracted hepatitis, then, after they had recovered from the induced disease, fed them another dose of infected serum. Sure enough, the children remained healthy the second time around.

But soon certainty turned to mystery. As he continued to monitor the rates of infection, Krugman began to notice something that was both alarming and surprising: almost one in every twelve children who had contracted the disease suffered a second attack within a year. Krugman was dumbfounded. He had devoted seven years to proving, among other things, that a bout with hepatitis provided immunity. Yet here were these kids getting infected again.

What could be the cause? Perhaps it was the result of an especially heavy exposure to the virus — all too likely at a facility as contaminated as Willowbrook — a viral bombardment that somehow overwhelmed the immunity the body had built up following the first attack. But that just didn't ring true. Krugman began to think that the cause of this uncooperative disease might be more complicated than anyone had yet suspected. Could this second attack be the result of another virus altogether?

If so, it would explain a great deal. Krugman thought back to his days in the Pacific during World War II. What about the many soldiers he had treated for hepatitis caused by the yellow-fever vaccine? How these thousands of troops came down with the disease had been a mystery. Back then, hepatitis was considered a hazard associated with overcrowded, unhygienic conditions — Willowbrook-like conditions. These soldiers, however, contracted the disease through the blood, as a result of a contaminated vaccine. Could hepatitis actually be two distinct diseases?

Krugman decided to find out. In September of 1964 he began another series of experiments. They would last for three more years.

The first step — which Krugman called the First Trial — involved giving a group of eleven newly admitted children a mix of blood drawn from a large number of hepatitis victims at Willowbrook. The mixture therefore presumably contained all of the hepatitis viruses at large in the school. If there was an unknown virus among them, Krugman would track it down. This first dose he would *feed* to the children. After an incubation period of thirty to sixty days, blood tests revealed that ten of the eleven children had come down with the disease. Krugman drew a sample of blood from one of the children, a boy, and labeled it MS-1.

Six months later, long enough for the children to purge the virus from their bodies and recover from the disease, Krugman and his team once again gave the children the same virus mixture, this time by *injection*, and then waited eagerly for the results. This was the Second Trial. If the children became infected once again, Krugman would know that indeed something strange was going on.

And, sure enough, within two months four of the children again came down with hepatitis. Two others who had not been in the First Trial were also infected. It was another blow to the commonly held notion that contracting hepatitis brought about immunity to further infection — unless this new infection was the result of a different virus, an unknown virus. By this time Krugman strongly suspected as much, for he had been careful to expose the children to the virus *only* by injection during the Second Trial, thereby assuring that if they did become infected once again, it would not be from ingesting the virus. This second infection, then, was blood-borne, as was the hepatitis spread by the yellow-fever vaccine during World War II. And, as before, it took longer to show up and longer to run its course than the earlier, oral hepatitis infection — a finding that reinforced Krugman's growing suspicion that this was a different strain. Krugman then took another dose of blood from the same boy whose blood he had drawn during the First Trial. He labeled this sample MS-2.

Now Krugman and his team were armed with two different blood samples, MS-1, and MS-2, which they thought to harbor two different types of hepatitis viruses. It was time to find out for sure. Thus began the Third Trial. They admitted a new group of fourteen children to the unit and injected eight of them with the MS-1 blood sample, the blood from the orally induced hepatitis infection. In about a month seven of the eight injected children came down with hepatitis, and soon thereafter every one of the six uninjected children, simply from contact with the others, came down with the disease. The experiment confirmed what the team already knew: the MS-1 virus, the original hepatitis virus, was highly infective. From worldwide spreading as the result of unhygienic conditions, to the epidemics at filth-ridden Willowbrook, to this recent transmission from no more than everyday exposure to other infected children, this was a terribly contagious disease.

Now the stage was set for the culmination of the experiment, the stage that would either confirm or deny Krugman's conviction that there were two distinct hepatitis viruses at large at the school. He began the Fourth Trial.

Krugman admitted another group of fourteen children, all of whom he kept isolated from the Third Trial group. He inoculated nine of these Fourth Trial children with the other, MS-2, blood sample. Would this sample contain a different virus that caused hepatitis, or were the two attacks nothing more than variations of the identical infection?

The answer was not long in coming. In about a month and a half, seven of the nine inoculated children became infected, but, in contrast to the Third Trial results, only two of the five noninjected children came down with the disease, both as a result of intimate contact with inoculated children who may have had open sores or cuts that allowed the exchange of bodily fluids. Conclusion: there was indeed a separate second strain of hepatitis active at the school. It was highly infective when transmitted by inoculation but, in contrast to the MS-1 strain, less contagious otherwise. These studies also indicated that the MS-2 strain could be spread by contact.

It was an important breakthrough. It made sense out of a hodgepodge of conflicting faces of the illness, solved the mystery of the World War II hepatitis epidemic, and allowed investigators to go after two specific, identifiable culprits. These two kinds of hepatitis viruses caused similar but distinct illnesses, and they were spread in fundamentally different ways — one by mouth, the other by blood. In time Krugman's MS-1 and MS-2 designations came to be known as hepatitis A and B, terms that had been proposed soon after the war as a way to differentiate what were then suspected to be two different manifestations of the disease. Krugman's findings delineated the strains once and for all.

The discovery was the accomplishment of his career, and as he made ready to publish his results, Krugman justifiably thought that his work would receive notice. Receive notice it did. In fact, it caused a sensation — but not for the reasons Krugman anticipated.

On January 10, 1967, a New York State senator named Seymour R. Thaler stepped onto the floor of the senate chambers in Albany. Excited and angry, admitting that he had "searched his soul and conscience," he exclaimed: "The medical profession has presumed to act as God over the health and lives of the medically indigent. I have the documentary proof. I have undergone a terrible inner conflict on whether to bring to the attention of the public that thousands of patients are being used daily as medical guinea pigs."

He went on to accuse Krugman and his team of injecting children with live hepatitis virus simply because they wanted to start a hepatitis research program. And he linked the Willowbrook experiments to alleged medical abuses elsewhere, accusing Harlem Hospital of removing limbs of congenitally deformed children and performing a hysterectomy without a patient's consent for no other reason than to demonstrate the procedures to interns and residents, and charging Bellevue Hospital with causing the deaths of five alcoholics and derelicts after taking liver biopsies for its research program.

Thaler then introduced a resolution calling on the State Investigation Commission to "examine into the conduct, practices, and procedures of the medical establishment."

Reaction was swift. Dr. Jack Hammond, the administrator of Willowbrook, defended the Krugman program, stating: "We're not using the youngsters because they are mentally retarded, but because hepatitis is a particular problem at Willowbrook. We have the consent of the parents of every child enrolled in the program."

New York State medical officials agreed, pointing out that Krugman's program had all but eliminated hepatitis at Willowbrook. An association of staff physicians at the city's hospitals went further, charging Thaler with employing "smear and scare tactics." His accusations were simply "malicious, irresponsible, and untrue."

Thaler countered by calling the hepatitis research at Willowbrook "unethical, immoral and illegal." He went on to say that "it violates the legal and moral rights of the children, who are being used as guinea pigs," and he introduced a bill to ban research on children. It sought to nullify written consent by parents or guardians except in the case of an emergency. This was too much for Saul Krugman. Calling the bill "a disaster, a real disaster," Krugman explained that if such a law had existed in the early 1950s, polio would still be running rampant and hundreds of children would be dying of measles every year. The vaccines for these diseases had been tested on children and simply "could not have been tested any other way."

"That business with State Senator Thaler was very difficult," Krugman says now, with obvious pain. Over two decades later, the incident continues to haunt him. "It happened when he ran for re-election. Politicians have to get publicity, so he invited the press to join him and went out to the Willowbrook School and held a press conference and [made accusations], completely, of course, out of context." He shakes his head. "It was difficult."

Soon after Thaler's outbursts, in May of 1967, Krugman published the results of his studies confirming the existence of two distinct strains of hepatitis, A and B. Again reaction came quickly, but this time it was unequivocal: Krugman was hailed as a hero. He was praised not only for his results but for his *methods*. The *Journal of the American Medical Association (JAMA),* where Krugman's article appeared, stated, "These recent studies of Krugman . . . would have been impossible without the judicious use of human beings in carefully controlled experimental studies."

The editor of the *New England Journal of Medicine* went even further: "By being allowed to participate in a carefully supervised study and by receiving the most expert attention available for a disease of basically unknown nature, the patients themselves benefitted . . . How much better to have a patient with hepatitis, accidentally or deliberately acquired, under the guidance of a Krugman than under the care of a [rights-minded] zealot."

Case closed? Not quite. Although Thaler's bill quietly died in the senate, he had opened Pandora's box. Fanned by his criticism, the controversy surrounding Krugman's experiments would not die. Critics contended that although Krugman did indeed receive written permission from parents of the children involved, his being able to experiment on those children was never really in doubt. Krugman had simply made these parents an offer they couldn't refuse. "Coercion" may have been too strong a term, but what practical alternative to Krugman's proposal did parents of a newly admitted child have? Despite the risks of the experiment itself, wasn't life in Krugman's isolated, clean, well-staffed ward preferable to the overcrowded and understaffed chaos of the school as a whole? And from time to time regular admissions were closed because of the overcrowding, making Krugman's unit the only game in town. The choice was simple: sign the consent form or look for another institution for your child. In 1972, when Krugman received an award from the American College of Physicians, a cordon of police surrounding the podium was necessary to protect him from 150 protesters denouncing the ethics of his efforts.

"What of Krugman's contention that his research was an experiment in nature?" write the authors of *The Willowbrook Wars,* a 1984 book critical of his efforts.

The claim ignores the fact that the underlying problem was not ignorance about a disease but an unwillingness to alter the social environment. Had Krugman wished to, he could have insisted that hygienic measures be introduced to decrease the spread of the virus. Should the facility resist carrying out the necessary cleanup, he might have asked the Department of Health to close the place down as a health hazard, which it surely was. Furthermore, Krugman had at hand an antidote of some efficacy. His own findings demonstrated that gamma globulin provided some protection, and yet he infected control groups with the virus and withheld the serum from them in order to fulfill the requirements of his research design.

But if his methods remain a source of controversy, his accomplishment in discovering the two strains of hepatitis stands. In fact, what *JAMA* called "an important contribution to our knowledge of hepatitis" was about to become more than that. It was about to facilitate the other watershed discovery in the fight against hepatitis. And the impetus for that unlikely finding was a little vial of blood drawn from the veins of an Australian aborigine half a world away.

In 1963, while Krugman was tracking down hepatitis A and B, Baruch Blumberg — "Barry," to those who know him — a geneticist at the Institute for Cancer Research at Fox Chase Center in Philadelphia, was trying to figure out why some people get certain diseases and others don't. A stocky, powerful, and outgoing native of New York City, with a PhD in biochemistry from Oxford University and a medical degree from Columbia, Blumberg had in 1949 traveled to the South American country of Surinam to study infectious diseases caused by parasites. Surinam was a country populated by an extremely diverse mix of people — Africans, Javanese, Hindu Indians, Chinese, native Indians, and, of the relatively small population of whites, a group of Sephardic Jews who had come north from Brazil. Blumberg was struck by the fact that these people from different backgrounds responded differently to prevalent infectious diseases, especially one called elephantiasis, which causes painful swelling in the legs, breasts, and testicles. But why? Why wouldn't the parasitic worms that cause elephantiasis infect everyone similarly? What was it about different populations that caused differing susceptibility to disease? Perhaps it was a difference in genetic makeup.

When Blumberg returned to Columbia medical school in New York City the following year, he brought with him a brand-new interest: population genetics. He decided that in order to figure out if genetic differences lead to differing susceptibilities to disease, he must first find out what those genetic differences are. Knowing that, he could then go on to the next step of aligning genetic variations with incidence of disease.

He had opened the door to a vast and intricate world. No two people are exactly the same in every way. Essentially, everyone is unique — and that uniqueness is reflected in variations in our genes. So Blumberg set himself the herculean task of discovering and chronicling those differences.

His method was to focus on proteins that circulated in the bloodstream. Different genes create correspondingly different proteins. Chronicle the distribution of any one protein among the peoples of the world, and you delineate the distribution of a particular genetic characteristic. Find enough different proteins among enough different blood samples, and you begin to draw a map of genetic distribution worldwide. Then, by investigating the prevalence of certain diseases among those genetic populations, you start to understand the role genes play in getting sick or staying well.

So Blumberg traveled the world collecting blood samples (he has by now 350,000 samples stored away in deep freeze). And he decided to look at the blood of people who had received a number of transfusions. "The notion was," he explains, "that with all their transfusions these people would have received blood containing proteins they hadn't inherited. If some of these proteins were antigenic [that is, able to provoke an immune response], then the transfused people would develop antibodies against them. Then we could use the antibodies to find the proteins."

The work went well. In the blood of patients with an inherited anemia-causing blood disease called thalassemia, Blumberg found antibodies that led to the discovery of a heretofore unknown family of proteins. Then in 1963, while studying the blood of a group of hemophilia patients, he and his associate, Harvey Alter, a National Institutes of Health hematologist, found a new set of antibodies. These reacted against the blood of an Australian aborigine. There must be an antigen in the aborigine's blood that provoked these antibodies. Blumberg called it the Australia antigen.

Curious. Whereas his earlier-discovered antibodies had reacted against blood from a large variety of sources, showing that the newly discovered protein was widespread, this time the antibodies reacted against this one sample of blood only. Why? Why would the blood of a hemophilia patient from New York display antibodies that only reacted against the blood of an aborigine from Australia? Blumberg and Alter decided to find out.

They began to look for the antigen and the antibodies in blood samples from different populations with various diseases. In time they found the antigen in other blood samples, but, in the United States at least, it was rare — it showed up in only one of every thousand blood samples. In blood from various Asian and African populations, however, it turned up much more frequently. Then they discovered something even more intriguing: Australia antigen often appeared in the blood of leukemia victims.

Now this was interesting indeed. What if the presence of the Australia antigen indicated leukemia? What if people with the antigen were more likely to develop leukemia than those without? There may be no more powerful urge in all of medicine than to discover a marker for cancer, and, as leukemia is cancer of the blood, Blumberg began to think that he and Alter might have happened on to just that. Yet the Australia antigen showed a strong relationship to other diseases involving blood transfusions — leukemia was simply the most frequently associated. The antigen also showed up in people from populations in which leukemia was not particularly prevalent but in which people suffered frequent bouts with infectious diseases. It was, for example, highly prevalent in the infectious-disease-ridden tropics.

Nevertheless, in 1965, Blumberg, who had by this time moved to Fox Chase Cancer Center, published the results of his study under the title "A 'New' Antigen in Leukemia Sera." He then went about trying to track down the antigen in people who had not yet developed leukemia but were at a high risk of coming down with the disease. Among those were a group of children with Down's syndrome. For reasons that are not yet clear, leukemia is more likely to attack Down's syndrome children than it is other people. Blumberg found that about a third of these children had the antigen.

Then another unexpected thing happened. In early 1966, a teen-ager with Down's syndrome, who had originally shown no sign of the Australia antigen in his blood, was tested again. This time the antigen turned up. Why? Had he suddenly contracted leukemia?

"We brought him into our clinical research ward and observed him," recalls Blumberg, "and one of the things we looked at clinically was if anything had gone wrong with his liver. The reason for that is that many proteins are made in the liver, and this antigen was a new protein. So the notion was that there may have been some liver abnormality."

Yes, the young man's liver had changed since he had first tested clean for the antigen. But this liver abnormality didn't indicate leukemia — far from it. It was proof that he had come down with the disease famous for causing "inflammation of the liver." He had come down with hepatitis.

Hepatitis. Could there be a relationship between Australia antigen and hepatitis? Blumberg — a geneticist, not an infectious-disease specialist — hadn't emphasized the connection before, but now he undertook an analysis of the blood of hepatitis victims and, sure enough, discovered that many of them had the Australia antigen in their blood during the course of the disease.

How could that be? The answer involved the fact that leukemia victims commonly receive multiple transfusions, as do hemophiliacs. Saul Krugman had shown that there was a distinct hepatitis virus that was transmitted by blood, and what more effective way to cause infection than by transfusion? Therefore leukemia patients and people with hemophilia were prime risks for hepatitis. And Down's syndrome children, who have faulty immune systems to begin with, almost invariably have lived in institutions for the retarded (all the children Blumberg tested were institutionalized), where hepatitis is almost a given. It was hepatitis that these people had in common. It must be.

Krugman sent Blumberg samples of his MS-1 and MS-2 hepatitis-infected blood, so that he could see if the serum contained the Australia antigen. He also sent samples to the New York Blood Center, the National Institutes of Health, and Yale University. Testing revealed that Australia antigen *was* present in hepatitis blood, but only in the MS-2, or hepatitis-B, strain. The Australia antigen was indeed a marker for the presence of the hepatitis B virus. In time it became officially known as *hepatitis B surface antigen.*

The World Health Organization called the discovery "the most spectacular advance in the seemingly insoluble problem of human hepatitis." Here, finally, was a recognizable marker for the disease — and, more than that, for hepatitis B in particular. Though it was still impossible to track down the virus itself, here was the next best thing. Blumberg's discovery corroborated Krugman's work, while Krugman's findings made it possible to understand the true significance of Blumberg's achievement. Without Blumberg's Australia antigen, there would have been no marker for hepatitis; without Krugman's delineation of the two strains of the disease and his resultant blood serum samples, there would have been no context in which to test the antigen. It would have been a discovery in a vacuum.

And were it not for Blumberg's bulldog nature, there may not have been any discovery at all. His uncovering the significance of the Australia antigen is yet another example of Pasteur's dictum that "chance favors only the prepared mind." For facing what was no more than a curious development, Blumberg had the presence of mind and perseverance to stay with it.

"I think Blumberg's strong point in this whole thing was his tenacity, his ability to hypothesize and follow through," says his early collaborator Harvey Alter. "This is where he's terrific. As soon as he sees something, he goes into hypotheses: What could this be? What should we do? He has a wall covered with previous hypotheses and how they came out. This thing could've been dropped very easily from the beginning, but he kept pursuing it and pursuing it — and he came up with this big payoff."

Blumberg won the Nobel Prize for Physiology or Medicine in 1976 in recognition of his discovery of the hepatitis B surface antigen. Now that it was possible to track down the disease with accuracy, perhaps even identify the virus, specialists could turn their efforts toward developing a vaccine to prevent its infection and spread.

*   *   *

Thanks to an explosion of research following Blumberg's discovery, a great deal became known about hepatitis, information that made it clear that the disease was more insidious and dangerous than had ever been suspected. The hepatitis B virus showed up in more than 150,000 Americans each year — over 3,000 of whom died. That was disturbing enough, but worldwide, untold millions of people were infected by the virus. Studies culminating in epidemiologist Palmer Beasley's investigations of cancer in Taiwan showed that if contracted early enough in life, hepatitis B was a primary cause of liver cancer. And, perhaps most alarming of all, not only could the virus be transmitted from mother to child at birth, it could be carried in the blood, no longer causing illness, but there to be passed on. Estimates are that three hundred million people worldwide are silent carriers of the hepatitis B virus.

This new ability to track down the hepatitis B virus showed further that infection could be transmitted by the exchange of bodily fluids: sex could do the trick, as could the passing of blood products via a kidney dialysis unit, for example, or simple transfusions. Accordingly, US blood banks revolutionized their policies. In 1970 they began the world's first widespread testing of donated blood, looking for the Australia antigen, an effort that in time virtually eliminated transfusions as a cause of hepatitis B. This became the basis for routine blood screening, a practice that we now take for granted, and that in our day has come to include screening for the hepatitis B and C viruses (hepatitis C, a more recently discovered virus, is discussed later in this chapter), the AIDS virus, the human T-cell leukemia virus, and syphilis, which was being tested for even before hepatitis B. The result has been a smaller population of suitable blood donors and a much safer supply of blood.

But for all its effect, just what was this Australia antigen, this so-called hepatitis B surface antigen, anyway? Electron micrographs had shown scores of small particles, some round, some sausagelike, proliferating in blood samples that contained the antigen. Were these the virus itself? Or were they something else, some part of the virus, say, that announced the presence of the real thing?

That question, too, was about to be answered, for in 1970 an English scientist named D. M. S. Dane finally identified the hepatitis

B virus. It was a round particle approximately forty-two nanometers in diameter, small as viruses go but fully twice the diameter of the strange particles the electron microscope had already discovered, and related to these particles by the fact that they both contained the Australia antigen. But while the smaller particles prompted the body to develop antibodies, they lacked the virus's core of genetic material that causes infection and so were harmless. They were antigenic — that is, capable of causing an antibody response — but not infectious. And they outnumbered the viruses themselves to the tune of as many as ten thousand particles to every one infectious virus. There might be as many as a billion to a trillion particles per milliliter of blood. It was a phenomenon no one had ever seen before.

As it turned out, the hepatitis B virus is unique in that it causes its commandeered body cells to manufacture this staggering amount of noninfectious virus particles, all of which proclaim the presence of the live virus. It was these particles that had stimulated antibody reactions in Blumberg's experiments and that he had named the Australia antigen, the particles that have come to be called the hepatitis surface antigen. Just why the virus misfires by making so many surface antigen particles is unknown — if, indeed, misfiring is what the virus is actually doing. Producing so many decoy particles might instead be a sophisticated viral survival strategy. "From the virus's standpoint, it could be very useful," says Alter. "By sending out these dummies to pick up the radar, to suck up antibodies, it clears the way for the real bomb. I don't know if that's why the virus does it, but it's conceivably a way that the virus protects itself."

The irony, however, is that if the particles are a protective mechanism, they could very well make possible the next step in the fight against hepatitis — a vaccine. For it quickly became obvious to a number of researchers that because these particles prompted the body to produce antibodies, perhaps it would be possible to harness them to confer a long-lasting immunity. One such researcher was Baruch Blumberg himself, who in 1969 — even before the virus was discovered — together with his associate Irving Millman, applied for a patent for a vaccine based on the use of surface antigens. At much the same time, virologist Maurice Hilleman of Merck Sharp & Dohme Research Laboratories was casting intrigued glances in the direction of hepatitis B virus. " 'Boy, this is interesting,' I thought, but you couldn't grow the virus in cell culture so the conventional vaccine approach was out. If you're a virologist and you're interested in vaccines, you think about a vaccine on the day you're able to grow the agent. Then one day I got the notion, 'Who knows, maybe there's enough antigen in blood plasma to make a vaccine.' " So Hilleman, too, began to look into the possibility of a vaccine that used hepatitis surface antigen, and he soon launched studies on how best to concentrate and purify the antigen.

Work proceeded slowly. Blumberg's lab was not set up for developing a vaccine (in fact, his "vaccine" was purely conceptual: the patent simply described a number of possible approaches, many of which were not feasible), and Hilleman had many other projects that had higher priority. What was needed was a prod, a spur, some event to convince those involved that while an antigen-based vaccine might seem theoretically promising, it was practically attainable as well. That event was provided by Saul Krugman.

On March 24, 1971, the *New York Times* ran this headline on page 1: "Immunization Is Reported In Serum Hepatitis Tests." Lacking resources, technology, expertise, and funding for vaccine research, Krugman had simply taken matters into his own hands.

"It was a very, very exciting time," he says, "but I wasn't really trying to develop a vaccine. Actually, all we did in our little laboratory, our little kitchen, so to speak, was boil hepatitis B blood serum and water." Krugman was interested in determining how little heat it would take to inactivate hepatitis viruses, both A and B. To find out, he added a small amount of water to serum containing the newly discovered hepatitis B virus and, in effect, rolled up his sleeves, tied on his apron, and boiled the mixture for one minute on the kitchen stove at Willowbrook. He then injected the mix into selected Willowbrook children.

Again, as has been so often the case in the fight against hepatitis, the results were surprising. In most cases not only did the injections *not* cause the children to come down with hepatitis, but they provided immunity to the disease. Boiling the serum and water mix for one minute was just long enough to destroy the virus's capacity to cause the disease but at the same time preserve the ability of its millions of surface antigens to stimulate the production of antibodies. Krugman had concocted a "homemade vaccine."

"I don't like to call it a vaccine," he says, "because it really wasn't a vaccine. Our finding was serendipitous. It demonstrated that a plasma-derived vaccine might indeed be developed. What was needed then was for the vaccine manufacturers with their highly sophisticated technology to follow up our lead."

And so they did. Robert Purcell of the National Institutes of Health set to work developing a vaccine for research purposes, and Maurice Hilleman stepped up his own efforts. Now seventy-one, Hilleman has developed more vaccines for human use than anyone else. They include vaccines against measles, mumps, rubella, chickenpox, pneumococcal pneumonia, influenza, and meningitis, but none of them took as much effort as the vaccine against hepatitis B. "That's a story that I could talk about for a week," he says. "The trials and tribulations of making the vaccine, starting from scratch — that was an absolute nightmare."

A native Montanan, tall, with a ring of silver hair and bushy eyebrows that shelter deep-set, mournful eyes, Hilleman looks as though he carries the weight of the world on his shoulders. And during the development of the hepatitis B vaccine, that was exactly how he felt. "Krugman reported that he had been doing these experiments in people, boiling this stuff to measure heat sensitivity. He said that these preparations were partially protective. So I thought, well, that demonstrated the principle that yes, by George, antibodies *did* mean protection. So I said, 'it's full speed ahead,' because obviously there must be enough antigen in that plasma to be able to induce immunity — let's go for it!"

Thus ensued one of the most intense vaccine-developmental-research projects ever attempted — and the most unusual. For this was to be a vaccine made from the blood of hepatitis victims themselves. The very people who suffered the disease would offer protection to those who might get it — a delicious irony there. And the irony was deepened owing to the fact that these were the same people who had been rejected as blood donors as a result of the new blood-screening procedures. Now this infected blood would be used to prevent infection, a variation on the old theme of taking a little hair of the dog that bit you.

The reason for using this unorthodox approach was that the hepatitis B virus resisted conventional means of vaccine preparation. In contrast to poliovirus, for example, which Salk had grown in the laboratory and then inactivated, and which Sabin had grown and then weakened, hepatitis B virus refused to be cultivated in the lab. But while it was uncooperative in this way, the virus was more than cooperative in another — it's voluminous production of surface antigen particles in the blood of those infected. So Hilleman would build upon what Krugman had done, trying to make the blood serum approach feasible, on a scale that boggled the mind.

"I got from Krugman a little bit of the plasma from the stuff that he had used," Hilleman recalls. "There was an *enormous* amount of antigen. That gave me a number to work with — a few micrograms of purified antigen per dose would probably immunize. But how to make it safe?"

That was the crux of the matter. It seemed likely that the vaccine would work, but derived as it was from human blood — and infected blood at that — how could Hilleman purge from it contaminants ranging from hepatitis B virus itself (Krugman's "homemade vaccine" had still contained some live virus) to heaven only knew what else? The specter of World War II's contaminated yellow-fever vaccine reared its ugly head, as did the Cutter polio vaccine incidents of the mid-1950s. The sucess of the hepatitis B vaccine depended on its being absolutely safe, safe to a demonstrable degree never before attained.

Hilleman's solution to the problem was to subject the blood plasma to a sequence of purification procedures so powerful that nothing, infective or otherwise, could possibly live through the experience. This vaccine was going to be "deader than dead." It was overkill to the extreme, as the surface-antigen-containing plasma had to pass from the veins of the infected donors through a Rube Goldberg–like series of torture chambers before being deemed suitable for clinical testing.

First, blood was drawn from volunteers who were known to be carriers of the virus — primarily gay men living in New York City, a group particularly plagued by the disease (more on that later). Lab technicians then spun the blood in a centrifuge so that the relatively solid red cells separated from the infectious liquid plasma. (The red blood cells were later mixed in a saline solution and injected back into the donors' bloodstreams so as not to deplete their supply.)

Then, once the plasma had been trucked to the Merck laboratories near Phildelphia, the steps necessary to separate, purify, and inactivate the surface antigen particles began. These were performed by Merck technicians protected head to toe by sterile gowns, masks, and gloves. For the plasma was so infectious that as little as a twentieth of a billionth part of one liter of the stuff could infect an entire roomful of people. In fact, for optimum safety much of the operation was done by remote control.

Hilleman's team separated the surface particles from the virus by spinning the plasma in a centrifuge, thereby separating the lighter and smaller particles away from the larger and heavier virus as well as from much of the protein in the blood. They then loosed upon the solution digestive enzymes, which devoured what impurities might have been left. The team even physically altered the remaining particles through treatment with urea that, in Hilleman's words, "took these surface antigen particles and unrolled them and then let them snap back together — that kills virus." The particles then passed through a pipe into a tank filled with formaldehyde, the same chemical used to kill virus in the Salk polio vaccine and others.

Finally, after all this — perhaps as much as seventy million dollars had been spent — by late 1974 Hilleman had a vaccine. Tests in chimpanzees showed it to be uncommonly effective and perhaps the safest vaccine ever made. But the battle was not yet won. Developing the vaccine and testing it in animals was one thing; now Hilleman faced the necessity of trying it out on human beings. And that wouldn't be easy, because who would want to try out a blood-derived vaccine, no matter how safe it appeared in animal and laboratory tests?

Recognizing the problem, Hilleman conducted the first human-safety trials of his new vaccine among the executives and staff of Merck itself. "Where in the hell was I going to go?" he recalls. "I had to go to people who weren't going to be [otherwise] exposed to hepatitis B — one case of hepatitis B among the volunteers, and that would have been it for the vaccine project. So where were you going to get a population that would volunteer, number one, and number two that didn't appear to be used as guinea pigs? Boy, that was a mess. So I got Merck executives."

Trying out the vaccine on his own company's employees not only solved the problem of finding volunteers, it was also a way of putting his money where his mouth was: if Merck people were willing to have Hilleman's new vaccine injected into their arms, it must be all right. But it wasn't only Merck executives who volunteered. Two other interested parties did as well — Saul Krugman and his wife, Sylvia.

Krugman was then sixty-four. He had finished his work at Willowbrook and so no longer handled hepatitis-B-infected blood on a regular basis. Moreover, he and his wife had often tested negative for hepatitis B antibodies. They were clean.

So in November of 1975, eleven persons — Saul and Sylvia Krugman and nine Merck executives — received the first inoculation of the new hepatitis B vaccine. (Maurice Hilleman was not among them, for the reason that as architect of the vaccine, and therefore perhaps the one who handled more contaminated blood than anyone else, he was too much of a risk. "I was in the lab, and we lab people handled blood all the time. If one of us had been infected with hepatitis B, it could've destroyed the vaccine. No, my arm is out there for anything that I can do, but you do not destroy your experiments by being a hero.") Krugman recalls: "I had the nurse give me the first injection. Then I gave everybody else theirs."

Six months of surveillance — the maximum incubation period of hepatitis B — then followed, during which all the subjects submitted to numerous blood tests. At the end of that time not one person showed evidence of hepatitis B infection. The vaccine was safe. If it contained any hepatitis B viruses at all, they were thoroughly inactivated ones.

Now it was time to see if the vaccine *prevented* hepatitis B infection. Hilleman, Krugman, and others conducted new tests among several hundred health professionals who volunteered to receive injections of the vaccine. Again, the vaccine was a success: it was safe and produced antibodies in those inoculated, which suggested that it would be effective in preventing the disease. At the same time, researchers were able to figure out optimal doses and times for giving the vaccine. Now, with the decade of the 1970s almost over, the stage was set for the most ambitious and most difficult tests of all — those that would determine whether the vaccine would prevent hepatitis in a large human population. And again the team faced the same old bugaboo: who could they find for their tests?

There simply was no precedent for the situation. By contrast, in the case of the polio vaccines, parents readily volunteered their children for testing because they feared the disease so much that the risks involved with the vaccines seemed much the lesser of evils. Hepatitis, though, struck no such fear into the general population. Although by the beginning of the 1980s hepatitis attacked 200,000 people per year in the United States (many more than had ever contracted polio annually), and although 10,000 to 20,000 of those became chronic HBV carriers (adding to the 800,000 carriers already in the country), and although almost 4,000 people died from hepatitis B every year, chances were, if you polled the general public, hepatitis would rank nowhere near the top in any list of most feared diseases. Its symptoms, while dramatic, usually disappeared in time, and its most grievous manifestations — cirrhosis of the liver, cancer of the liver, and a huge, silent population of infectious carriers — were long in coming or had hidden impact.

But health professionals knew about hepatitis. They knew that it cost the US economy $750 million a year in hospitalization expenses, medical fees, and time lost from work. And they knew that while these figures were staggering enough, they represented only a fraction of the global impact of the disease. Some two hundred million people were thought to be infected with hepatitis B at the time. The ramifications of such an army of sick people were simply incalculable.

That was worldwide. In the United States, only two groups besides health professionals knew much about the disease: male homosexuals and intravenous drug users. For them, it was hardly a disease of little moment — it was the major medical problem in their lives. Now, in retrospect, it seems as well an eerie prelude to AIDS, the more fearsome disease that turned out to be lurking just around the corner. So with the 1970s almost past and the portentous 1980s looming, Hilleman turned for help to these two groups. The battle against hepatitis was about to be augmented by what has been called the finest field trial in the history of medicine.

In 1973, the Gay Men's Health Project in New York City was asked to participate in a study to determine the incidence of hepatitis in various populations — medical workers, blood donors, Chinese-Americans, gay men. The Health Project itself was an expression of the concern gay men felt at the increase in disease that accompanied the flowering of the gay liberation movement. Ever since 1969, when a police raid on a gay bar in Greenwich Village had provoked a riot that became known as the "Stonewall Rebellion," many New York gays had been living a more open and varied sexual life than ever before. In bars, bathhouses, or simply by meeting people along Christopher Street and vicinity in the Village, gay men were finding opportunities for new relationships, from random sexual encounters to long-term alliances. It was an exhilarating time, but along with greater freedom came greater incidence of disease. Gonorrhea, syphilis, and hepatitis B were the most serious, and of those hepatitis was by far the greatest threat. Antibiotics were effective in treating syphilis and gonorrhea; not so, hepatitis B. So when the New York Blood Center asked the Gay Men's Health Project to participate in its hepatitis study, the members of the group responded quickly, donating blood samples and answering a questionnaire concerning their sexual history.

The results of the study shook the gay community. As compared with only a 5-percent prevalence of exposure to hepatitis B virus in the general population, more than 50 percent of the gay men showed evidence of either current or past infection. A sexually active homosexual man was a hundred times more likely to come down with hepatitis B than a sexually active heterosexual man. And it was clear that, unchecked, the disease would continue to spread. Something had to be done.

Enter an unlikely hero. Wolf Szmuness ("*shmew*-ness") had arrived unannounced at the doorstep of the New York Blood Center in 1969. A small, fifty-year-old man with a large head, pockmarked face, and startling shock of yellow hair, Szmuness, accompanied by his wife, Maya, had only months earlier arrived from his native Poland armed with fifteen dollars in his pocket and a painfully acquired expertise.

He had undergone a long journey. Born to a middle-class Jewish family in Warsaw, Szmuness had lost his parents in the Holocaust and had himself fled to Russia ahead of the advancing German army. For the next two decades he studied and practiced medicine in Siberia and later did so in the Ukraine, where his wife nearly died from hepatitis caused by a blood transfusion. Szmuness was to devote the rest of his life to combating the disease.

In 1959, twenty years after fleeing Poland, Szmuness returned, settling in the town of Lublin to continue his work in public health, epidemiology, and hepatitis. But in 1967 the Six-Day War between Egypt and Israel broke out, galvanizing already widespread anti-Semitic feelings in Poland. Szmuness was ordered to protest Israel's actions at a large rally. This he refused to do. When he arrived to work the next morning, letters of dismissal were awaiting him.

Two years later he was in New York City and, supported by his wife, who found a job sewing neckties, Szmuness made the rounds of the city's medical establishments, but to no avail. Finally, on the verge of taking a job at a factory, he was hired by the Blood Center as an assistant. It was the beginning of a blossoming as unexpected as it was swift. Four years later, in 1973, he directed hepatitis testing in New York's gay community. A year after that the Blood Center created a new Department of Epidemiology for him to head. Simultaneously, he rose from a lecturer at Columbia's School of Public Health to full professor. But as impressive as were these feats, they were only a prelude to the culmination of Szmuness's lifework: the testing of Maurice Hilleman's new hepatitis B vaccine.

The vaccine had been shown to be safe, but in relatively small groups; and it had been shown to be effective in experimental studies in chimpanzees. Before it could be licensed for general use, it had to be proved similarly advantageous in people who were at high risk of infection. Moreover, since it was made from hepatitis-infected human blood, it had two strikes against it already. Any hint of defect, no matter how minor, might scotch the vaccine for good. This trial would have to be extraordinary. Szmuness, therefore, was determined to test the vaccine on an optimum population, a population that would provide the clearest, fairest, most unequivocal answers. For it was likely that if the trial failed there would be no others.

But what kind of people would be needed for such a trial? What was the ideal test population? To begin with, there must be enough people to provide statistically valid results. Close to a thousand people would be necessary. And these should be people at high risk for the disease. If they weren't likely to come down with hepatitis, how could Szmuness be sure that the vaccine was protecting them from the disease? These people should be healthy and young, and they should constitute a reasonably homogeneous and identifiable group. And they should be educated and accessible. If Szmuness couldn't get to them to inject the vaccine, and continue to get to them for follow-up testing, the trial simply wouldn't work.

As it turned out, he didn't have to look far. Just across town from the Blood Center's East Side offices were precisely the people he needed. They were numerous, young, contracted hepatitis B ten times as often as people in the general population, were educated and concerned about the disease, and Szmuness had worked with them before and knew them to be accessible and reliable. Yes, the gay male community of New York City would be perfect for such a test. Now it would simply be a matter of convincing enough gays to participate.

Agreeing to have your blood tested was one thing — allowing a relatively unproved vaccine to be put into your body was quite another. No matter what vested interest a group may have in seeing hepatitis B a thing of the past, it might not be easy to persuade people to become guinea pigs for months at a time. This was to be a randomly selected, double-blind, placebo-controlled trial. Half the volunteers would receive vaccine, half a placebo, and no one — not Hilleman, not Krugman, not the administering staff, not Szmuness himself — no one would know which was which. Secrecy would be assured by means of a code — the vaccine and placebo groups would be identified by number only — and only two people, the officers responsible for the safety of the trial, would have access to information about how to break the code. Everything was set. Now all Szmuness had to do was recruit his test volunteers.

It would be hard to imagine a more unlikely group of characters. Here, in the great battle to control hepatitis B, were thrown together Hilleman, the tall, dour virus cowboy from Montana; Krugman, who had done the controversial early studies that pointed the way to this very moment of promise; Szmuness, the refugee from the horrors of Hitler and anti-Semitism; and the gay men's community of New York City, a group reviled, oppressed, and misunderstood. Ironically, gays now had the opportunity to help provide for general societal protection against a disease that was a particular hazard for themselves (just as they had provided, with their own blood, the basis of the vaccine in the first place). But they had to be convinced. In 1978, Szmuness, along with Dr. Cladd Stevens of the Blood Center and a particularly dedicated staff, went out to do the convincing.

In her 1985 book *Quest for the Killers,* writer June Goodfield quotes the recollections of one of Szmuness's staff people, Allison Brennan.

> By the spring of 1978, when we began the massive screening to find the 850 suitable men we mounted a massive PR campaign and started with the Bloodmobiles. These are enormous vans which travel all round the city, going to the blood donors rather than have them come to the Blood Center. But now we took the Bloodmobiles all over the place — to the gay bars, down to the pier on the West Side where the gay men congregate, and caught a lot of people who were just going to sunbathe. We would haul them into the Bloodmobile and invite them to take a blood test.
>
> . . . we had a picture of Dracula in the Bloodmobile to inspire people and make them feel comfortable, and one of the volunteers had a set of vampire teeth that he brought along too. . . . We had to ask the guys lots of intimate questions too, not only about how old they were and what illnesses they'd had or were having, but when did their active gay life begin and how many sexual contacts they'd had over the past year. These could run between 200 and 700 a year — the average was about 160.

Afterward Szmuness's team gave the men a brochure, "What You Should Know about Hepatitis B in Gay Men, and the Hepatitis B Vaccine," and asked them to return in a week. If the blood test turned up no sign of hepatitis, they would be asked to register for the trial.

Szmuness contacted the gay community leaders and invited them to the Blood Center to discuss the disease and the upcoming vaccine trial; he went to gay churches and synagogues, gay bars, gay bathhouses. Volunteers began to trickle in, but not nearly enough. The team increased their recruiting effort.

"So we created the next pamphlet with a lively headline, 'Is Variety the Spice of Your Sex Life?' " Brennan recalled.

> This implied that if you want to keep up sex and not worry, you had better get this test for hepatitis B, because there's nothing gay about being sick. . . . We just grabbed the people, opened up a clinic wherever we were and hoped that they would go from us on the street into the clinic. The main thing was to get that blood test. Meantime, of course, we were still going to all the clubs and all the usual organizations, and people from these groups were joining in as volunteers. So there was this kind of reaching out everywhere and it was really quite wonderful and extraordinary.

In the end, they screened more than 10,000 people — 1,083 signed up. And in January of 1979 the vaccine trial began.

It lasted eighteen months, until June of 1980. During that time, Szmuness and his team gave each of the 1,083 volunteers three doses of hepatitis B vaccine or a placebo and followed up on each dose with blood tests and question-and-answer visits — a mammoth, exhausting undertaking. Through it all, Szmuness, typically edgy in the best of times, fussed and fidgeted like a man possessed. His characteristic assessment of progress was, "This trial is a disaster." He worried about the rate of compliance among the volunteers, worried about the usefulness of the follow-up blood tests (he eventually increased their number), worried about how the trial would be received by the world. "He was constantly nervous, whether things were going wrong or not," remembers Cladd Stevens. "He kept us all nervous."

And in the end, after a few close calls that tempted him to shut down the entire trial, he, the vaccine, and the thousand volunteers triumphed. Of the 1,083 men who began the trial, 1,040 turned up for the second injection — that's 96 percent, an unheard-of rate of compliance. Szmuness had expected that fully 40 percent of the men would drop out by the end — only 15 percent did. Over 90 percent of the scheduled thousands of follow-up visits actually took place. In terms of reliability and cooperation alone, there had never been anything approaching the results of Szmuness's trial.

And the vaccine itself performed equally spectacularly. Only 3 percent of the vaccinated group developed hepatitis B; in the placebo group, 27 percent came down with the disease. Of those who received all three doses of the vaccine, 96 percent developed antibodies against the virus. Overall, the vaccine was more than 92-percent effective in preventing hepatitis B. To say that the trial and the vaccine were a success would be saying too little.

Szmuness and his team published the results of the trial in the October 9, 1980, issue of the *New England Journal of Medicine*, which called it a "landmark study." The same week, *Nature* praised the study as "a milestone in the annals of preventive medicine." On November 17, 1981, the vaccine was licensed by the US Food and Drug Administration. It had been twenty-five years since Krugman began his Willowbrook experiments, sixteen years since Blumberg had discovered the hepatitis surface antigen, nearly twelve years since Dane had found the hepatitis B virus itself. Ten years before, Krugman had brewed his homemade vaccine, and soon after, Maurice Hilleman began working on developing a vaccine in earnest. Now, finally, after all that, the vaccine, having been shown to be uncommonly safe and effective, was licensed for general use.

Soon the architect of the final phase departed. In June of 1982, as though, in a life filled with disappointments and travail, he had been granted just enough time for one crowning achievement, Wolf Szmuness died of lung cancer. He was sixty-three. "I remember when he was ill," says Stevens. "He was angry and very disappointed. He had just gotten there, and now he was dying. He felt it wasn't fair."

While the other leading characters in the drama remain, none actively does hepatitis research any longer. Saul Krugman, now seventy-nine years old, the elder statesman in the field, still is involved in long-term follow-up studies on the persistence of immunity in those people he originally inoculated in 1978; he travels and consults as well. Baruch Blumberg, sixty-four, is now Master of Balliol College at Oxford University. He has put his connection with Fox Chase Cancer Center on hold. Maurice Hilleman has retired from Merck Sharp & Dohme but continues to consult with the company regarding vaccine development.

And the fruit of their labor, the hepatitis B vaccine, has flourished. In the first place, it has continued to prove itself uncommonly safe — especially important in this day of AIDS. Says Harvey Alter: "Hilleman was very careful about making the vaccine. He inactivated it in many more ways than were absolutely necessary. As it turned out, it was a great thing because then AIDS came along and scared everybody to death about taking vaccines. But he had done all the right things to kill the AIDS virus, even if he didn't know it was in there." There has been no AIDS associated with the vaccine, nor has it transmitted any other known viruses or caused any known disease.

In the second place, it has continued to protect against the hepatitis B virus. Well over two million people worldwide have received the vaccine, although in the beginning acceptance was slow because of the fear of a blood-derived vaccine. Old suspicions die hard. And in combination with hepatitis B immune globulin, essentially the same kind of gamma globulin preparation Saul Krugman gave children at Willowbrook, the vaccine works well in reducing the number of hepatitis B cases that are passed from mother to child at the time of birth. One study in Taiwan has suggested that the combination may prevent 95 percent of such cases, thereby breaking the ageslong cycle in which newborns of hepatitis-B-carrier mothers themselves become carriers and go on to infect others.

But much of the hope offered by the vaccine is still no more than that, hope. For despite its demonstrated effectiveness, the vaccine simply hasn't yet had the impact its creators hoped for. While there are some problems associated with the vaccine — for example, the relatively few people whose immune systems are not functioning properly (such as kidney dialysis patients) are not protected well by the vaccine — the most pressing difficulty, that affecting by far the greatest number of people, has nothing to do with antibody response, effectiveness, or safety of the vaccine. In fact, it has nothing to do with the vaccine's ability to prevent hepatitis B at all. It has to do with getting the vaccine to the people who most need it.

In the United States the hepatitis B vaccine has had little impact for the simple reason that the people at highest risk are by and large difficult groups to immunize. Physicians and medical workers, a group often exposed to infected blood, comprise one of those high-risk groups, and although it would seem as though they more than most would be interested in using the vaccine, the opposite is true. As a group, they tend to be conservative, and while most may now be immunized, the going has been slow.

Ironically, the situation among gay men is little better. Despite their crucial contributions in the making and testing of the vaccine, as a group they haven't embraced it. There are fewer cases of hepatitis B in the gay community, but the reduction is more a function of less-hazardous sex practices among gays since the onset of AIDS than the effect of any vaccine. And the remaining high-risk group, intravenous drug users, is for the most part simply inaccessible to existing medical care.

Besides that, says Robert Purcell, the head of the Hepatitis Viruses Section at the NIH, "forty percent of hepatitis B cases in the US are associated with no known exposure to the virus. That means that we probably still don't understand how the virus is spread in the general population. So unless you immunize the whole population, you're just not going to have a very great impact."

Worldwide, the situation is similar, but for a different reason: cost. "The problem," says Harvey Alter, "is that the people who need it the most, Asians and Africans, can't afford it." At approximately ninety dollars for three shots — an initial inoculation, another at one month, and a final injection at six months — the vaccine is simply too expensive. "That's not enormous for us, but when you're talking about three hundred million people already infected worldwide and hundreds of millions needing to be vaccinated, the cost of the vaccine has to come down to pennies."

What will bring the cost down to pennies? While the price of the vaccine has become substantially lower — a Korean company now offers it at less than a dollar a dose, and China is developing its own vaccine — the great hope is that genetic engineering will do the trick. In the United States, recombinant hepatitis B vaccines have pretty much replaced the original plasma-derived preparation. The vaccines are prepared in yeast — that is, scientists remove the gene that makes the hepatitis B surface antigen from the virus itself and insert it into the chromosomes at the heart of yeast cells. With the addition of the foreign gene, the yeast goes on to manufacture not only yeast cells but hepatitis B surface antigen as well, and it does so indefinitely. So instead of being forced to turn to the blood of infected individuals for a source of surface antigen, vaccine manufacturers now have a source that has nothing to do with blood at all. The advantages of such an arrangement are many.

"High yields, simplicity, unlimited source of material," says Hilleman, who began developing the recombinant vaccine before his retirement. "And you get rid of that nagging doubt in people's minds: 'It's a blood product — am I going to get AIDS? Am I going to get other diseases?' This new vaccine will be cheaper as well."

"The hepatitis B story is a success story," says Alter. "If you could make the vaccine available to the world, then you'd *really* have a success story. The potential is there for this vaccine to eradicate hepatitis B, as most of the spread is from mother to infant in Asia and Africa. If you could get infants vaccinated routinely, you could break the infectivity cycle and ultimately stop the disease. The basis for that success is there, but the practical application is not."

And what isn't always appreciated is that regardless of the failures of the vaccine program so far, the hepatitis B vaccine, whether it be the original blood-derived version or the recombinant preparations on the way, is much more than simply a means to prevent infectious hepatitis. It is the first example of something medical scientists have been searching for and the general public has been yearning for ever since vaccines became commonplace. For hepatitis B is not a finite disease. As mentioned, it often leads to other, more dangerous complications — primarily cirrhosis and cancer of the liver. In preventing hepatitis B, the vaccine is essentially the world's first cancer vaccine. And now that it's clear that other cancers —

leukemia, for example — are caused by viruses, the hope is that it will not be the last.

Hepatitis B, however, is only one of the forms of hepatitis. That was determined almost twenty-five years ago when Saul Krugman delineated another strain, *hepatitis A,* that was even more infectious than B, if not so long-lasting. And now at least two other strains of the disease, *non-A, non-B* (at last, mercifully, becoming commonly designated *hepatitis* C) and *delta,* have been added to the mix. Although the efforts to prevent these versions have not been as successful as those aimed at B, some major achievements are worth noting.

Hepatitis A

In contrast to hepatitis B, hepatitis A infects through food and water. It's the form of hepatitis that depends largely on a suitable environment; as sanitation increases, for example, hepatitis A decreases. Also in contrast to B, there are no chronic carriers of hepatitis A virus, nor does it lead to problems down the line, as hepatitis B leads to cirrhosis and cancer of the liver. Hepatitis A is a self-limiting disease, like flu or the common cold, in that it makes its appearance, you get sick, and, although the disease can occasionally be lethal, you generally recover. The whole process might take a few weeks to a few months.

Moreover, hepatitis A is not a threat to blood banks, as is B. "It makes sense if you think about it because there are no carriers of hepatitis A," explains Alter. "So the only way to transmit it by transfusion would be to have donors coming in who were infected but not yet sick, because once they got sick they wouldn't be accepted as donors. It would have to be that narrow period of a week, maybe two weeks max, that they might be infectious but not ill. And their blood would have to go into a recipient who was susceptible, and that would be tough in itself, because about half our population is already immune to hepatitis A — it's a very common childhood subclinical infection."

Nevertheless, hepatitis A is a debilitating disease, and much work has gone into identifying the virus and developing a vaccine. In 1975

Robert Purcell and his NIH associates Stephen Feinstone and Albert Kapikian first found the virus, and in 1979 Maurice Hilleman and his Merck Sharp & Dohme team grew the virus in tissue culture in the lab, albeit with great difficulty. Even today, hepatitis A virus resists laboratory study, as it deigns to grow slowly if at all and produces relatively few offspring. The virus also makes things hard in that as a single-stranded RNA virus, it mutates relatively often. That means that, like poliovirus, it has a tendency when attenuated to revert to virulence. It may be that a killed-hepatitis-A vaccine, rather than the recalcitrant live version, will be the first to be licensed for human use. "A hepatitis A vaccine is just on the verge," says Purcell. "It's about five years behind the development of the new hepatitis B vaccines."

Non-A, Non-B (Hepatitis C)

When Feinstone, Kapikian, and Purcell began to work with the hepatitis A virus in 1975, they derived a hepatitis A blood test. Test in hand, they went to Harvey Alter to evaluate numerous cases of hepatitis in heart surgery patients whose blood Alter had been studying. They thought that these patients' hepatitis had been transmitted through transfusions. Certain that they were not instances of hepatitis B, they presumed the cases to be hepatitis A. They were in for a surprise: not a single one of them was hepatitis A. Some other agent was causing these cases. But what could it be? Unable to find out, and certain only that it was neither hepatitis A nor B, the NIH coworkers tentatively named it *non-A, non-B*. It was an ungainly designation, if an accurate one, but it stuck. So now the world had three varieties of hepatitis to cope with.

As it turned out, non-A, non-B hepatitis not only had its own cause, but it provoked problems of its own as well. It tended to be milder than hepatitis B in its early stages, but while only about 5 to 10 percent of B patients became chronically afflicted with that disease, non-A, non-B became chronic in 50 percent of the people it infected. Approximately 20 percent of those went on to develop cirrhosis of the liver, and recently it has been shown that non-A, non-B may lead to liver cancer as well. In total, today, probably as much as 25 percent of all hepatitis in the United States is non-A, non-B (including 95 percent of all hepatitis caused by transfusions), a figure that translates into as many as 150,000 cases a year. That means about 75,000 chronic infections every year and at least 15,000 cases of cirrhosis of the liver. So in the long term this virus is responsible for a considerable amount of liver disease.

Until 1987, however, when associate director of research Michael Houghton and his colleagues at Chiron Corporation in California positively determined that non-A, non-B was indeed caused by a virus, then developed a blood test to find it, it had only been presumed that non-A, non-B hepatitis was a viral disease. The measures undertaken to find out for sure are a microcosm of the virus-hunting techniques established in past decades.

First, to determine if the agent was transmissible through blood, as they suspected, Alter, Purcell, and Feinstone in 1975 injected five chimpanzees with blood from the sick heart-surgery patients. All five chimps became infected. The virologists were indeed dealing with a transmissible agent.

Next the NIH researchers passed blood from infected chimps to healthy chimps, in effect the same kind of experiment old Beijerinck had conducted with tobacco mosaic virus some seven decades earlier. And, like Beijerinck, they found that their agent could continue to infect when passed serially from host to host. The healthy chimps became infected from the blood of the previously infected animals.

Next the modern world intervened, as Purcell performed a test that wasn't part of the early virus hunter's arsenal. He and Feinstone — as well as Dan Bradley, working independently at the Centers for Disease Control — discovered that the agent was sensitive to applications of chloroform and could be dissolved by solvents, which meant that, whatever it was, it was probably surrounded with a lipid membrane or envelope — a common feature of viruses.

The next test again harkened back to the beginnings of virology, as Bradley and Purcell passed blood containing the agent through a succession of filters to find that it would pass through a filter containing holes 80 nanometers in diameter but only occasionally through holes 50 nanometers in diameter. That information suggested that the agent was 50 to 60 nanometers in diameter, small as enveloped viruses go (if indeed that's what it was) but not unheard of. In fact, one class of viruses, called togaviruses, precisely fit the description — enveloped viruses approximately 50 to 60 nanometers across. They included the viruses that cause rubella, yellow fever, and dengue.

So by relying primarily on time-honored virological techniques, Purcell, Alter, Bradley, and their colleagues had gleaned a pretty good idea of what they were dealing with. It was left to Houghton and his team at Chiron to prove the matter. That they did by using the techniques of a later day, those of modern molecular biology. They succeeded in extracting the genetic material from the non-A, non-B agent, cloned it, and discovered that the agent is indeed a virus, similar to the togaviruses, with a single strand of RNA and a lipid envelope. It is, more specifically, a virus very much like the one that causes yellow fever — a flavilike virus, in technical terms.

Subsequent tests showed that the great majority of people with non-A, non-B hepatitis had antibodies to this very virus — it was indeed the cause of non-A, non-B. And finally, as noted, the term *non-A, non-B* is giving way to a new designation, *hepatitis C,* and the newly discovered virus to *hepatitis C virus,* or HCV. So now the hepatitis virus family includes HCV, HBV, HAV — which, however, like adopted siblings, have little to do with one another in terms of structure, size, or type. They are related only in what is to most people their most important characteristic: they cause similar diseases.

Chiron has as well developed a blood test for hepatitis C. This means that the virus's days of causing disease through transfusions are numbered. Routine blood screening for HCV will become part of the armamentarium of the nation's blood banks. It also looks as though a vaccine may well be possible, for it is from the flaviviruses that we've gotten one of our best vaccines, the yellow-fever vaccine.

Delta

In 1977 an Italian gastroenterologist named Mario Rizzetto noticed something unusual while studying liver cells from patients suffering hepatitis B. In the nuclei of the cells he found evidence of what

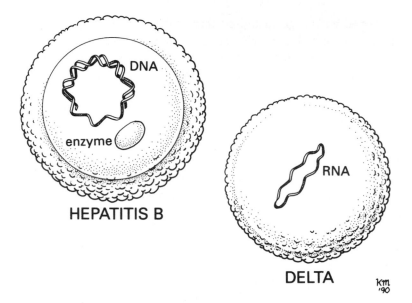

HEPATITIS B

DELTA

km
'90

Hepatitis B virus and its parasitic cousin, the delta virus

Hepatitis delta has been linked to a particularly severe form of hepatitis. Delta is far simpler in structure than hepatitis B virus, having only a single, flattened ring of RNA, while hepatitis B has an incomplete double ring of DNA and enzymes. Externally, however, the two appear virtually identical, since delta slips inside and steals hepatitis B virus's coat as the viruses are assembling inside cells.

looked like a new protein. Certainly it wasn't part of the hepatitis B virus, as it stimulated antibodies different from those that attacked the virus. Something strange was going on here.

More recently Rizzetto and others have established that it was indeed something unusual; circulating in the blood of some hepatitis-B-infected patients was not only hepatitis B virus but something they had never seen before. It was an elongated circle of RNA, covered by a coat of the protein Rizzetto had noticed in the liver cells, all of which was packaged inside another wrapping, that belonging to the hepatitis B virus itself. Somehow this strand of RNA had pirated the protective coat of hepatitis B virus for its own. Rizzetto named this new agent *delta*.

Further investigations uncovered a shadowy picture of microbial grand larceny. Delta infects only cells already inhabited by the hepatitis B virus. Disguised in the virus's outer coat, delta appears to gain entry to the cell using the same surface receptor as the B virus does — but now delta rather than hepatitis B penetrates and deposits its genes inside. Once there, delta somehow forces the now doubly infected cell to produce mostly delta agents, which then in effect elbow their way onto the viral assembly line and usurp the coats intended for hepatitis B viruses. Thus protected, they pop out of the cell to infect again. "A strange bird, this delta," says Chiron's Michael Houghton, one of the researchers who first sequenced delta's genes in 1986. "It's the cuckoo of the viral kingdom, stealing the coat of the B virus just as the cuckoo steals the nests of other birds."

It's a process — and an agent — never before seen, "literally an emerging virus structure," says Houghton, who suspects that there may be other similar agents out there, as yet undiscovered, playing a role in human disease. Meanwhile, delta intensifies the infection and the patient suffers a more severe case of hepatitis. Delta hepatitis is already endemic in Italy and elsewhere in the Mediterranean, and in several regions of South America, including northern Colombia. It is inching into the United States, as well, predominantly among intravenous-drug users and homosexual men, who are traditionally among those most at risk of hepatitis B infection. Many cases of fulminant hepatitis, a particularly severe form of the disease also known by the graphic term *acute yellow atrophy,* are now found in retrospect to have been a superimposed delta infection. In fact, up to half of all cases of fulminant hepatitis previously thought to be caused by the hepatitis B virus are associated with delta.

Because delta infects only in the presence of the hepatitis B virus, however, if one eliminates B, delta will no longer be a problem. The hepatitis B vaccines, therefore, may perform a greater service than anticipated by preventing delta hepatitis. It's simply another incentive for finding a way for the vaccines to be distributed worldwide as quickly as possible.

Still, much of the mystery surrounding delta remains. For instance, where did this defective, parasitic virus come from in the first place? Although no one knows for sure, a possible answer emerges from the fact that there exist in the world of plants even more primitive agents called *viroids*. These viroids, like delta, are tiny circles of naked RNA that, unlike delta, cannot so much as form their own inner coats of protein. Nevertheless, they can be devastating in effect. The viroid that causes a disease called cadang-cadang has killed millions of coconut trees in the Philippines, thereby inflicting heavy economic losses. Other viroids infect potatoes, hops, cucumbers, and avocados.

Houghton theorizes that at some point in the past a viroid somehow fused to a piece of RNA in a living cell and began to make its own protein. From there it evolved into a parasite that happened to package itself into the coat of the hepatitis B virus. "I think that in nature everything is possible," he says. "When a molecule exists the chances are at some point in history it has interacted with another molecule, and something new is going to come out of it. It's going to happen and it's going to be selected for or against." In this case the unlikely combination survived.

But the greatest mystery of all is how delta succeeds in replicating. Viruses can make more viruses only by taking over the cell's reproductive apparatus. In order for this scheme to work, however, the virus must get its genetic material working in the cell, a transfer that usually entails using its own special enzymes. But delta has no enzymes. Its core includes nothing more than its flattened ring of RNA. By all rights, delta should not be able to replicate at all. But it does. How it does, no one knows for sure. It's assumed that viroids use the enzymes of infected cells to replicate. Perhaps the same is true for delta.

The implications are fascinating, for it's clear that delta is a link to the very beginnings of life. As delta may have evolved from viroids, viroids may have evolved from expendable parts of primitive genes — pieces of RNA that could cleave themselves out of the strand during reproduction. And delta, viroids, and these self-splicing bits of RNA may all be direct descendants of primordial RNA.

"More and more people now think that it was RNA rather than DNA that was the first genetic material," says Purcell. "In the beginning, some three billion years ago, it may have just been RNA self-replicating."

*   *   *

The discovery of delta is a reminder that our knowledge of the world of viruses is incomplete. Scientists have recently found yet another hepatitis virus, this once called *hepatitis E,* or *epidemic non-A, non-B,* which causes symptoms much like hepatitis A in epidemics — twenty thousand people at once, for example — as well as in isolated cases, in Asia, Africa, and Mexico. And researchers suspect that there may be yet another agent out there causing those few cases of non-A, non-B hepatitis that can't be laid at the doorstep of hepatitis C virus.

The entire family of hepatitis viruses, in fact, constantly serve notice as to how amazing viral behavior can be. Hepatitis B virus, for example, has been found to undergo an involved, contradictory process — reverse transcription — unknown to all but one other notorious class of viruses, retroviruses, of which the best known is HIV, the AIDS virus. Some scientists now speculate that hepatitis B and retroviruses come from the same primordial ancestor. And even on a more mundane level, the relationship between the two families of viruses is startling. For, as we've seen, hepatitis is blood-borne, is sexually transmitted, and afflicts gays and intravenous-drug users disproportionally. It thus can be seen as a kind of prelude written in a minor key to the crash and tumult that surround AIDS. In the next chapter we'll dive deeply into the realm of AIDS, HIV, and HIV's revolutionary family, the retroviruses.

Fourteen

· · · · · · · · · · · · · ·

The Virus That Works Backwards

· · · · · · · · · · · · · · · ·

"Not long ago in the dark ages of medicine, one could think nearly anything about disease because one knew almost nothing."

So wrote Peyton Rous in 1936. Twenty-five years earlier Rous had discovered that viruses can cause cancer tumors, and in 1936 he had only recently returned to the problem. His initial findings had met with so much resistance and just plain indifference that he had given up investigating viruses and their role in cancer. But in 1933 an associate at the Rockefeller Institute found a virus that causes wartlike tumors in wild cottontail rabbits and gave samples of the tumors and the offending virus to Rous. Inspired by this discovery, Rous once again set about investigating the problem in earnest. He had been a man ahead of his time; now the world was finally beginning to catch up.

Still, however, another decade and a half passed before Rous's finding that viruses can cause cancer received substantial corroboration. In 1951 Ludwik Gross, a Polish refugee working at the Veterans' Administration Hospital in the Bronx, discovered that a virus similar to Rous's original chicken sarcoma virus caused cancer tumors in mice. Here was the first indication that viruses like Rous's could cause cancer in mammals. But these were laboratory strains of mice, and the cancer was the result of a lab experiment. Many people still doubted that in the real world outside the lab infectious agents such as viruses had anything to do with cancer. And, cancer aside, the same people tended also to doubt that these viruses had any significant role in the outside world. Perhaps they were no more than laboratory curiosities.

In the early 1960s all that changed. William Jarrett of the University of Glasgow in Scotland discovered feline leukemia virus, the cause of leukemia — cancer of the blood — in cats. This, too, was a virus similar to Rous's virus. And to those who still doubted that cancer could be caused by an infectious agent came further news from groups headed by Jarrett and Max Essex of Harvard University that the disease readily spread from cat to cat in normal, household environments. Feline leukemia, then, was an *infectious* disease.

Soon enough it became clear that all sorts of animals — monkeys, hamsters, rats, frogs, birds — suffered viral-caused cancers. In 1966, as belated recognition of his trailblazing work, Rous was awarded the Nobel Prize. He was eighty-six years old.

Yet Rous and the other researchers in the field might as well have been in the Dark Ages for all they knew about how these viruses worked. They had a good deal of structural information — the genetic material of these so-called rousviruses was made up of RNA rather than DNA; the protein shell protecting the RNA was itself enveloped by a layer of lipid, or fat, "borrowed" from the cells the viruses infected; and the particles were middle-sized as viruses went, somewhere between small RNA viruses such as poliovirus and rhinovirus and more robust DNA viruses like Epstein-Barr — but how these viruses went about causing cancer, no one knew.

Cancer is a disease of ordinary body cells that for some reason become "transformed" into cancer cells — cells that multiply out of control. In so doing, they no longer exhibit a basic property of normal cells: the tendency upon encountering one another to stick together and stop growing. It's that attribute that provides a stable cellular matrix, with each cell maintaining itself in healthy relationship to the others. Cancer cells, on the other hand, don't seem to care if they encounter one another or not. They simply keep growing and multiplying, over, around, and on top of each other, piling together into a chaotic mass — the tumor. Or, in the case of leukemia, blood cells proliferate out of control, in extreme cases literally clogging up the blood, making it thick and viscous. And in most cases cancer cells continue this unchecked propagation until by one means or another they're stopped or, in the most unfortunate scenario, the host body dies.

The question for Rous and the researchers who succeeded him therefore became, how do these viruses cause this transformation? The possibilities seemed to be two: either the viruses enter the cells and then proceed to infiltrate the cells' nuclei, in effect acting as genes that transform the cells into cancer cells, or, surprisingly, *the viruses don't have to enter the cells at all* — they're already there. For despite the evidence offered by Essex and Jarrett and their colleagues, some were not at all convinced that *infectious* rousviruses caused cancer. In their view, again, these rousviruses remained little more than laboratory artifacts. What counted was a type of virus whose genetic material appeared in the genes of a number of different animals, apparently without any indication of infection. In other words, these viral genes — "endogenous," originating from within, as opposed to "exogenous," originating from outside — lurking in the animal chromosomes from generation to generation, were a remnant of an ancient infection, and they were transmitted from one generation of animal to the next by the normal process of inheritance. Such was the view proposed by Robert Huebner and George Todaro of the National Cancer Institute. They proposed that cancer came about when these genes — they called them *oncogenes* — were turned on by carcinogens in the environment. So in the early 1960s it seemed likely that if rousviruses did have a role in cancer, these viruses, whether endogenous or exogenous, began by changing the genetic makeup of cells.

But how? Viruses were known to kill cells, not alter them so that they continued to live (and live with a vengeance). Besides, while the genetic material of cells was made of DNA, these tumor viruses contained only RNA. If these viruses were somehow to insert their genes into those of cells, whether now or in ancient times, the act would require the viral RNA segment to graft onto the cell's DNA, and everyone knew that that was simply impossible. The two kinds of nucleic acids could not commingle in such fashion. It was an apples-and-oranges scenario.

No, the only conceivable way for these RNA tumor viruses to take up residence in the cells' genes would be if the viral strand of RNA could somehow produce DNA, which might then be able to insert itself into the cells' complement of DNA. But that, too, seemed impossible. In a cell DNA makes RNA for purposes of reproduction, but the reverse simply ran counter to the "central dogma of molecular biology" as elucidated by no less than Francis Crick. This doctrine stated that all begins with the blueprint of life, DNA, which proceeds to make RNA, which, as a messenger, directs the production of proteins, the building blocks of life. DNA to RNA to proteins — that was the direction of life. It flowed "downhill."

And even obvious exceptions to this rule such as other RNA viruses didn't really contradict the dogma at all. For the RNA in poliovirus, for example, is itself messenger RNA. The virus works by instructing the cell to make proteins, in this case enzymes, which help make more RNA, which directs the production of more protein, until finally new viruses stand completed. Poliovirus, therefore, may truncate the central dogma, but it doesn't negate it.

But poliovirus doesn't cause cancer. It doesn't bring about the transformation of cells. There's no question as to whether it inserts its genes into the cell's genes — it doesn't. Yet here were these puzzling RNA tumor viruses that seemed to be doing precisely that. And if they weren't, well then the mystery deepened. For somehow, by some unimaginable process, these viruses were succeeding in bringing about the cancerous transformation of cells.

RNA tumor virus researchers, therefore, were caught in a purgatory between the reality of what they saw happening before them and the riddle of how, given what they knew about the scheme of creation and reproduction, such a thing could possibly happen. Rous himself acknowledged such difficulties when he wrote that "nature often speaks her secrets with a still, small voice out of a dense thicket of happenings. He who would hear and comprehend can have no pride of intellect, no fixed preconceptions; he can only listen intently and ask himself what he may have heard."

Those who were listening intently to the problem of RNA tumor viruses were indeed about to hear something that would challenge their preconceptions. These viruses, they were soon to discover, did work in a heretofore unknown manner, a manner contrary to the "central dogma." And not only did these viruses cause cancer, they were at the heart of another, entirely unforeseen scourge of this last quarter of the twentieth century: AIDS.

In 1960 a twenty-five-year-old postdoctoral student from Caltech named Howard Temin began working as an assistant professor of biology at the University of Wisconsin. His doctoral thesis had been on the Rous sarcoma virus, and upon settling into his less-than-imposing lab in Madison — "My first laboratory was in the basement, with a sump in my tissue culture lab and with steam pipes for the entire building in my biochemistry lab" — he began two experiments that led him to propose a novel and unorthodox answer to the question of how the virus brings about cancer.

The first experiment involved an antibiotic called actinomycin D, which recently had been found to prevent DNA from making RNA by wedging itself between the coiled strands of the DNA to keep necessary enzymes away. Temin decided to try the drug on cells infected by Rous sarcoma virus to see what would happen. He expected that since the virus's genes were made of RNA rather than DNA, the antibiotic would have little effect on the virus — it would simply continue replicating within the cell; meanwhile, the antibiotic would interfere with the normal cellular reproductive cycle of DNA to RNA to protein — that would be shut down. He was in for a surprise. It turned out that the antibiotic shut down the virus as well.

"When I added actinomycin D to cultures of cells producing Rous sarcoma virus," Temin wrote, "I found that the antibiotic inhibited the production of *all* RNA. One would have expected the replication . . . of an RNA viral genome to continue without hindrance. This result was the first direct evidence that the molecular biology of the replication of Rous sarcoma virus was different from that of other RNA viruses."

How could that be? How could an antibiotic that stopped the transcription of DNA to RNA prevent the replication of an RNA virus, which contained no DNA whatsoever? The answer Temin came up with seems perfectly logical in hindsight, but at the time it was revolutionary: central dogma to the contrary, he surmised, the Rous sarcoma virus must produce DNA at some point in its reproductive cycle. An RNA virus it may be, but somehow it must go through a DNA phase.

His second experiment reinforced that hypothesis. Temin loosed the virus onto cells in laboratory dishes, then immediately added drugs that inhibited the production of DNA. The result was that the cells did not become infected. No DNA, no viral replication. The significance was clear: in some manner and to some degree, this RNA virus needed DNA to replicate.

Temin called his unconventional supposition the "DNA provirus hypothesis." Again, the idea was that somehow, at some point in its replicative cycle, the RNA of Rous sarcoma virus made DNA. Then, presumably, the DNA went on to make RNA, which then simply continued with its normal task of making the proteins for new viruses. Rous sarcoma virus, therefore, and by extension all RNA tumor viruses, worked backward, a modus operandi unlike that of any other viruses or, indeed, any other living system.

It was a bold concept, one that not only explained the results of the experiments, but also offered a rationale for how these viruses might transform cells. For now it made sense to conceive of the virus's inserting its genes — its *DNA* genes — into the DNA of the cell, and in that way transforming it into a cancer cell. The provirus theory tied together the loose ends concerning RNA tumor viruses.

But, however ingenious and compelling, it was no more than a theory. And when Temin presented it at a meeting of virologists at Duke University in the spring of 1964, he was rudely reminded of the fact. "At this meeting and for the next 6 years the hypothesis was essentially ignored," he wrote.

Temin was now all of twenty-nine. A slight fellow, with curly, brown hair, a narrow face, thick glasses, and a persistent manner, he simply refused to give up. It's easy to imagine him regaling his fellow virologists, most of them his seniors, in his thin voice, expounding on the promise of his provirus theory, and producing over the years a stream of articles and lectures advancing the theory —

all to no avail. For the theory was simply too unorthodox to be entertained readily by a conservative scientific community. In the absence of any direct evidence, it could hardly have much of an impact. "Teminism" — that's what his provirus theory began to be called, and none too kindly. If he wanted to be taken seriously, Temin would simply have to prove his case.

At the same time, in a laboratory half a continent away, another of the young turks of microbiology, David Baltimore of the Massachusetts Institute of Technology, was trying to make heads or tails out of poliovirus. Although it had been much studied in the years before and after the great polio vaccine effort, no one knew much about how the virus replicated. Earlier, as a researcher at the Salk Institute for Biological Studies in La Jolla, California, Baltimore had discovered that when poliovirus infects a cell, instead of translating its genes into proteins bit by bit, as did other viruses, it converts its whole genome, all seventy-five hundred nucleotides of it, into one long, continuous protein; he called it a *polyprotein*. The virus then snips the lengthy protein into usable pieces as it sees fit. So little by little, he was getting to the bottom of the virus, while simultaneously developing an interest in the reproductive strategies of RNA viruses in general.

Then an unexpected event catapulted Baltimore in a direction that would take him away from polio for the time being to address the same problem Temin was trying to solve: how do RNA tumor viruses infect cells? It was the appearance at the Salk Institute of a Chinese-born postdoctoral student named Alice Huang. Huang was interested in another RNA virus: vesicular stomatitis virus, or VSV, which causes a foot-and-mouth disease in animals. When in 1968 Baltimore moved to MIT as an associate professor, she came also. Now, along with a new position as a researcher at MIT, she sported a new personal title, Mrs. David Baltimore. With the help of graduate student Martha Stampfer, she began to apply Baltimore's poliovirus techniques to the study of VSV, engaging her husband's interest in the process.

Baltimore wondered how VSV was able to infect at all. The virus was armed with a so-called negative strand of RNA, an impotent form that was useful only as a template for making an active, positive strand — the kind that had the ability to infect. That procedure needed the help of an enzyme, or *polymerase* ("pol-*im*-er-ace"), a protein that acts as catalyst for such biological processes. No one had ever found an RNA-to-RNA polymerase in cells, so Baltimore reasoned that it must be in the virus.

Baltimore is short, dark-haired, and bearded — he could be an Eastern European rabbi — with a raspy voice and a particularly direct and concise way of framing questions and answers. Today he is fifty-two years old, the president of the Rockefeller University in Manhattan. In 1969, when he began to investigate the workings of VSV, he was thirty-one. "Whereas a positive-strain virus like polio can come into the cell as messenger RNA and make protein and more RNA as a first act," he explains, "a negative-strain virus can't do that because it's got a senseless RNA as its genome. So it needed a polymerase. But where was it? Either the virus had to carry its polymerase with it or there had to be one in the cell, and we didn't know of any such enzymes in cells. So I went and looked for the enzyme in the virus and found it."

In only a couple of hours he, Huang, and Stampfer found that, indeed, VSV carries its own reproductive enzyme, its own RNA polymerase. They found further that there were a number of such minus-strand RNA viruses that also carry enzymes that help make plus strands. Then Baltimore had another thought: "What other viruses might have enzymes in the virus particle? It occurred to me that RNA tumor viruses might, because from what we knew about them, they had the same sort of problem. They were RNA viruses, and they had to do something to infect cells, but it was murky what happened after that." So Baltimore looked in Rous sarcoma viruses for an enzyme that, like the one in VSV, helped RNA make more RNA. This time he couldn't find any. Then he remembered Howard Temin.

Baltimore had known Temin for years, ever since he had spent a summer at the Jackson Laboratory in Maine. Temin was four years older than Baltimore, a college student while Baltimore was still in high school, and the acknowledged "guru" of the lab. Through the years, Baltimore continued to admire Temin and was well aware of his theory that RNA tumor viruses transcribed DNA. "I knew that Temin had suggested this years before and that the evidence was pretty meager, but I figured I'd test that one too."

"Chance favors only the prepared mind," said Pasteur. Baltimore wasn't particularly interested in RNA tumor viruses. He likely wouldn't have been investigating RNA viral enzymes at all had it not been for the appearance of Alice Huang at the Salk Institute those few years ago. And here he was, slipping in through the back door, about to make his share of one of the great discoveries of modern virology. "Some people, what they set out to do takes them years and years and years of background work and developing systems," he says. "In this case, although everything I had done up to then in a sense went into it, this one just fell right in my lap. The experiment itself was trivial. It took a couple of days' work. When it's right it's right."

Baltimore modeled his experiment on work Nobel laureate Arthur Kornberg had done in the late 1950s to find DNA polymerases. In a procedure similar to that which he had performed to find the VSV enzyme, Baltimore first chemically stripped the protective lipid envelope from mouse leukemia viruses and then added to the now exposed nucleic acid a chemical called TTP (for thymidine triphosphate) that had been made radioactive. TTP is a precursor to DNA, one of the chemical ingredients that makes up the final nucleic-acid form. If the virus happened to contain an enzyme that made DNA from RNA, it would use the TTP in the process. And because this TTP was now radioactive, Baltimore would readily be able to detect it in any subsequent traces of DNA, proof that this was new DNA made by the virus itself. He could in effect watch the process unfold.

Sure enough, after Baltimore allowed the viruses and TTP to incubate together at body temperature for forty-five minutes, he found newly formed DNA containing the radioactive TTP. The virus did indeed carry an enzyme capable of transcribing RNA to DNA. He then repeated the experiment with Rous sarcoma virus, with similar results. Here was demonstrable evidence that these RNA tumor viruses contradicted Crick's central dogma, and a very strong indication that Howard Temin's provirus theory of how these viruses cause the cancerous transformation of cells was correct. Without ever intending to, Baltimore had vindicated the much-maligned approach of his old acquaintance.

So, he picked up the phone and called Temin. "I said to him," Baltimore remembers, " 'You know, there's an RNA/DNA polymerase in mouse leukemia virus.'

"He said, 'Yes, I know that.'

"I said, 'How do you know that?'

" 'Because I found it,' he said.

" '*You* found it?' I said. 'But *I* found it!' " Baltimore laughs. "That was the first time I discovered that he had also been working along these lines. He and Mizutani had evidence of the same thing."

After years of trying, Temin himself had finally come up with evidence for his provirus theory. In an experiment on Rous sarcoma virus similar to that performed by Baltimore, he and his colleague Satoshi Mizutani had arrived at the identical conclusion: the virus contains an enzyme capable of producing DNA from the virus's genetic core of RNA. Without realizing that the other researcher was attacking the problem, each man had independently performed a comparable experiment to come up with an answer. It was on the one hand poetic justice, on the other a striking example of the capriciousness of scientific discovery; for while Temin had been involved in the problem for years and had finally made the discovery that vindicated his long-belittled hypothesis, Baltimore, with no personal stake in the problem, had arrived at the identical point at the same time almost by accident.

Temin and Baltimore wrote up their results and sent them off to the journal *Nature* within two weeks of each other, and on June 27, 1970, the papers appeared together under the general title "Viral RNA-dependent DNA Polymerase." Almost immediately microbiologist Sol Spiegelman of Columbia University not only confirmed their findings but found a similar enzyme in no less than six other RNA tumor viruses. Although *Nature* cautioned that the results were "very preliminary" and "heretical," the journal's reaction was nevertheless hugely enthusiastic. "Central Dogma Reversed," headlined the editorial. "The discovery of the unprecedented enzyme, which obviously has profound implications for the whole of molecular biology, as well as for the mechanism of cancer induction by

RNA viruses, is an extraordinary personal vindication for Dr. Howard Temin. If ever a man was in a position to say I told you so, it is he." Temin's and Baltimore's discoveries, the editorial went on, and Spiegelman's confirmation "are likely to generate one of the largest bandwagons molecular biology has seen for many a year."

The statement could not have been more prescient. Soon thereafter *Nature* dubbed the enzyme *reverse transcriptase,* imparting a certain intriguing flavor to this heretofore unknown enzyme and the process it facilitates. And not long after that, in reference to their proclivity for reversing the usual order of life, these RNA tumor viruses became known as *retroviruses,* another evocative designation that has stuck. Five years later, with Renato Dulbecco of the Salk Institute, who had shown how DNA viruses can cause cancer, Temin and Baltimore shared the Nobel Prize for Physiology or Medicine. Retroviruses were on the viral map to stay.

Retroviruses. Today they may be the most well known viral family; certainly they're the most extensively studied. As a group they encompass a huge variety of viruses and a wide range of consequences, from cancerous tumors, to leukemia, to various immune-deficiency diseases, to no apparent effects at all. And, of course, the most notorious retrovirus of all, human immunodeficiency virus, or HIV, is the cause of the most notorious disease of our day, AIDS.

All retroviruses, while differing in a multitude of small ways, go about their business in generally the same manner. After inserting its innards into the cytoplasm of the cell it infects, a retrovirus begins the remarkable process that gives it its name. Employing its onboard reverse-transcriptase enzyme, and using its single positive strand of RNA as a template, the virus manufactures a complementary negative strand of DNA. This minus strand of DNA then itself becomes a template for a positive DNA strand, with the result that the virus soon contains a double-stranded DNA molecule — a double helix — precisely the same form as the DNA in the cell's nucleus. The viral DNA, by means that aren't yet completely understood, then snakes its way into the nucleus, where, like a latecomer butting into a long line, it inserts itself into the cellular DNA. And there it stays.

Now part of the cell's chromosome, the virus — that is, all that's left of it, its genes — is in the catbird seat. Like any good DNA, it may transcribe messenger RNA, which travels back into the cytoplasm, some of it directing the cells' ribosomes to manufacture new viral proteins, some of it becoming enveloped by the emerging viruses to form their new cores of RNA. Or the integrated viral DNA may exert its influence upon the cellular genome and cause the cell to reproduce aberrantly, erratically, uncontrollably, thereby transforming it into a cancer cell. Or the viral genes may do nothing at all, may simply lie low — for years, perhaps — safe and undetected within the heart of the cell, until prompted once again to become active and produce more viruses, or a transformed cell, or both. What causes that activation isn't always entirely clear. There are many retroviral mysteries still to be unraveled.

But whatever may prompt this unorthodox behavior, almost everyone agrees that it's a masterful strategy. "It's a particularly bright way of existing," says David Baltimore, "because the virus doesn't continually have to find new hosts in which to grow."

That in contrast to poliovirus, for example, or the virus that causes measles. Because poliovirus wreaks such devastation on the cells it infects — it readily kills them — its offspring must pass from one host to the next to continue to reproduce. In any single host the viruses have a lifetime of about a week; after that amount of time the cells they infect are dead, and the viruses themselves can't exist in anything but a live cell. They can't rely on reinfecting the same host because by that time the body readily builds up immunity to the viruses. So if they don't quickly pass on to somebody else, that particular strain of virus dies, never to be seen again.

Measles virus, another RNA virus, has much the same needs. It, too, maintains itself by rapidly passing from person to person — in the case of measles, mostly from one child to the next. So it takes a large population containing many as-yet-uninfected people to maintain measles virus. In small villages that are isolated from the world — Eskimo villages, for example — there is no measles. The population simply isn't large enough to sustain the virus. (But because these isolated peoples have never been exposed to measles, they've never had the chance to build up immunity to the virus. If someone infected with the measles virus should come into the area, the disease could decimate — and *has* decimated — whole villages.)

This necessity for such viruses quickly to find new bodies to infect explains the fact that the most devastating disease epidemics are often short-lived. Influenza is a prime example. Once a new influenza virus arrives on the scene, it methodically makes its way through the susceptible population, and then, with no one left to infect, abruptly fades away. And were it not for the virus's virtuoso ability to mutate into newly infective forms, that would be the end of that.

Retroviruses, on the other hand, are simply in no rush. They come on in, make themselves at home, and hang around for a while. In fact they become so much a part of the household that while the infected cell may alert the body that there's a virus around, the immune system's virus fighters simply don't do enough about it. And the virus — that is, the viral genes — taking advantage of such generous laissez-faire, becomes a semipermanent guest. It's a particularly ingenious and efficient way to conduct an infection.

All of this is good news for retroviruses, but decidedly mixed news for people. Yet how quickly things change. For back in 1970, when Temin and Baltimore discovered reverse transcriptase and retroviruses became an enduring focus of molecular biology, there were no known retroviruses that infected humans, and no known human-retroviral diseases. Chickens, mice, cats, cattle, apes — by that time it was common knowledge that retroviruses infected these and other animals. But there was great doubt in the biological community that any human retroviruses existed. That point of view was largely based on the fact that in animals retroviruses had been shown to replicate extensively, creating huge numbers of new viruses, before they caused cancer. Scientists could see the many new viruses in electron-microscope images. But no matter how carefully they investigated the cells of human victims of the kinds of cancer caused by retroviruses in animals — leukemia, say — researchers found no evidence of the viruses. Their conclusion was that there simply were no such human viruses.

At the time, a young physician from Connecticut named Robert Gallo was studying enzymes that produce DNA in blood cells. Brash and ambitious, Gallo simply didn't accept the conventional wisdom.

"Perhaps," he thought, "human retroviruses exploit a different cancer-causing mechanism, which does not entail extensive viral replication. If so, electron microscopy (a cumbersome tool for the purpose) would never detect the pathogen. A more sensitive assay, though, might."

Gallo determined to find that assay and see once and for all whether there did indeed exist any human retroviruses, as he suspected there did. He was in for an adventure.

Today Robert Gallo is fifty-three years old. Of medium height, graying, with glasses perched above a prominent nose that in profile imparts to his round face a probing, impatient aspect, like the prow of a ship, Gallo is a man in a hurry. He speaks so rapidly that often the last words of one thought tangle up with the first words of the next. He seems most at home in the midst of turmoil — the telephone ringing, people waiting, experiments brewing, interviews beckoning. The typical life of any top-level scientist? Well, yes, but more intensely — with Gallo, always more intensely.

For Gallo is working in an area, retroviruses, that is today not only the leading field within virology — the one attracting the most researchers and the most money — but a field that, because of AIDS, has received more widespread and impassioned attention outside scientific circles than any before it. HIV, the retrovirus that causes AIDS, may be the most widely known virus in history. Because of AIDS, the word "virus" is on our lips more frequently than ever before, carrying with it an emotional jolt the like of which has rarely, if ever, been experienced before. In the United States, you must go back to the early 1950s and polio for a medical phenomenon even remotely comparable, and earlier than that nothing comes close. But while polio also elicited visceral fear and apprehension, it struck far fewer people than does AIDS; polio is transmitted through mundane processes — water droplets, for example — rather than through emotionally charged practices such as sex or IV-drug use; and, in the early fifties, it was publicized by reporters dependent primarily on radio and the printed word. Television, and particularly television news coverage, was only in its infancy then. Today, thanks to TV, the intimacy, lack of social inhibition, and sheer volume of coverage involved in reporting on AIDS is unprecedented (could anyone ever have foreseen the extent to which we would be inundated with descriptions of sexual practices, homo- and heterosexual, including detailed instructions for proper condom use and affidavits concerning the sexual habits of prominent people?). Gallo has found himself working in a goldfish bowl.

It is not necessarily an enviable position. Gallo and Gallo's discoveries are at the heart of human retrovirology. That has prompted some to praise him as "a modern microbe hunter, of the same ilk as those of the past." Others, however, have reviled him as "the most dishonest scientist in the world . . . a big crook." Life ain't easy in a goldfish bowl.

"I really never thought that I'd do anything that anybody would find that interesting," Gallo says, somewhat disingenuously. Then he catches himself and laughs. "Honest, honest, this is the truth. I'm not being falsely modest, because I'm not, as everyone knows. I was busy, and I never took time to take myself so seriously. I was doing it because it was fun. I was doing it because I had an interest in the science of it, the biology of it, the competitiveness of it."

So in 1970, there he was, thirty-three years old, having fun by going against the grain and looking for a way to spot human retroviruses. He wasn't the only one, but his was a scoffed-at minority. "In the midsixties and early seventies, when every animal species that was looked at carefully finally was found to have its own retroviruses, mostly these chronic leukemia viruses, the question became, 'Why is there nothing in man?' " So says Werner Kirsten, a National Cancer Institute associate director who in the 1960s discovered a retrovirus that causes leukemia in mice, since called the Kirsten leukemia virus. "People started to work on human retroviruses, including my own lab, but probably more papers were written on the subject than ever saw publication because there was a certain resistance to the idea. The best minds looked into it and thought it was crap."

Robert Gallo, however, thought it was anything but. And one of the reasons for his interest was, ironically, Huebner and Todaro's theory of endogenous viruses. "I was very much influenced by Robert Huebner's writing," Gallo says. "I began thinking that it didn't much matter if a retrovirus causes human leukemia or not — here's a field that may tell us something about the nature of cells and of life." At the same time, Gallo was influenced by the work of Max Essex, who with William Jarrett, and William Hardy of the Memorial Sloan-Kettering Cancer Center in New York City, was suggesting just the opposite of Huebner's theory. Essex, as noted, said that retroviruses, *infectious* retroviruses, cause leukemia in cats, and not just in the lab. Feline leukemia, he demonstrated, was a naturally occurring contagious disease amongst cats, in normal, household situations. The feline leukemia virus resides in the cats' saliva, and because cats are fighters that tend to bite or lick each other frequently, they transmit the virus one to another. Furthermore, Essex and his colleagues discovered that cats make antibodies against the virus. That was yet another indication that feline leukemia was an infectious disease, transmitted by a foreign agent.

Gallo found the evidence compelling, a likely analogue for what was going on in human beings. But he was an exception. Essex and his colleagues were having a tough time persuading the scientific world of the truth of their feline findings.

"When we tried to convince others," Essex remembers, "people usually didn't listen. And when they did listen, they didn't want to be bothered because they thought it was somehow the exception." It had been only a very few years since the Epstein-Barr virus had been implicated as the cause of Burkitt's lymphoma and nasopharyngeal carcinoma, and the results of studies confirming the hepatitis B virus as the cause of liver cancer were still almost a decade away. "The epidemiologists weren't interested in looking at an infectious origin of cancer," says Essex. "They were looking at tobacco smoke and diet and lifestyle, at risk factors of umpteen different types at once. Forget about specific infectious agents. . . . The only one totally outside the feline-leukemia and comparative-animal-virus area who really grasped it may have been Gallo."

Gallo's problem was how to show evidence of human-retrovirus infection. In cats the viruses showed up in huge numbers; in humans, if the electron microscope could be believed, there weren't any at all. How could a virus infect and perhaps even cause a disease as horrible as leukemia and remain invisible? Once again, a hint of the answer and further encouragement toward Gallo's goal of finding a human retrovirus came from the animal kingdom.

In 1969, Janice Miller, a virologist with the United States Department of Agriculture, had found that cattle were yet another species of animal that suffered cancers as the result of infection by retroviruses. This "bovine leukemia virus" produced huge tumors that could kill the animals. The discovery of the virus solved a mystery that had puzzled ranchers and field veterinarians for years, for they had noticed that these tumors seemed to occur within one man's herd but not his neighbor's, as though the cancer were a communicable disease, a herd disease. Many had even come to the conclusion that the tumors were the result of an infectious agent, but nobody had ever seen an agent. Electron micrographs of tumors and blood cells had shown nothing. But here was confirmation that, yes, the vets and ranchers were right — bovine leukemia was indeed an infectious disease.

Janice Miller didn't find the virus in tumor cells. In fact, she discovered it quite without intending to, in the process of looking for something else. Using an electron microscope, she was studying abnormalities in leukemic blood cells of cattle. To facilitate her examination, she grew the cells in culture for a day or two, at the end of which time she realized that the cells were full of viruses. But when she went back to the cattle to look at fresh cells, she could find no virus particles. No virus in the cattle, but take the cells outside the animals and culture them for twenty-four to forty-eight hours, and there's plenty of virus. What in the world could be going on?

The answer was provided by an expansive Belgian named Arsene Burny. During this time Burny was a researcher in the Columbia University laboratory of Sol Spiegelman. Although Spiegelman was one of the few scientists also interested in finding a human retrovirus, and therefore Gallo's competitor, Burny and Gallo became friends. Soon thereafter Burny left Speigelman to go home to Brussels and devote his time to the bovine leukemia virus.

"These viruses are more and more sophisticated — they don't want to show themselves," Burny says in French-accented English. "A bovine leukemia virus gets into a target cell, pushes that cell to the threshold of transformation into a cancer cell, and when the step is done the virus is quiet. That cell will become a fully transformed cell, the first cell of a gigantic tumor, *pounds* of meat, that will kill the animal."

Apparently, the virus continues replicating as long as the cell remains untransformed. Once it becomes transformed, however, the virus shuts down. It may have infected thousands and thousands of cells, but it only need transform one. Just one. This one transformed cell then begins to replicate at full speed and never stops, eventually producing its huge, grotesque tumor. But by that time the virus itself has gone far underground, into the transformed cell's genes. "You don't find virus in these cells," says Burny, "but if you use a biochemical tool — namely, a probe for DNA — the provirus is there. And it is absolutely silent — shut off, quiet, but still there."

So the virus has done its job. It no longer needs to replicate because it has integrated its DNA into the DNA of the host cell — becoming what Temin called a provirus. In effect, it lets the cell do its replicating on its behalf, for as the transformed cell multiplies out of control, it produces hundreds of thousands of copies of the viral DNA. Meanwhile, another animal may become infected, eventually to undergo the same process. And so it goes, animal after animal.

But though a tumor grows quickly, the process leading up to that event moves slowly, ever so slowly. The beauty of these bovine retroviruses is that, as a part of the cell's genes, they become latent within the animal, lurking there, waiting there, for months or years, until the moment when the cell erupts and begins dividing uncontrollably. So by the time the disease is apparent, the virus is long gone, invisible, traceable only by the presence of DNA. What could be slicker than that? No wonder no virus showed up under the electron microscope; there was none to see. It's not hard to understand why so many people were convinced that no such virus existed. And it's not hard to understand why Bob Gallo found his determination to find a human retrovirus strengthened by Burny's work.

"Burny really brought the bovine leukemia system to life," Gallo says. "He convinced me that there was a good chance that the same kind of thing was going on in humans."

The pieces were falling into place for Gallo. The example of cat leukemia showed a naturally occurring infectious animal cancer that was caused by a retrovirus; the bovine leukemia virus provided an example of a retrovirus that caused cancer but was next to impossible to detect, as it only replicated *before* the onset of disease. Afterward it became virtually invisible. Gallo was more than ever convinced that there must be a human retrovirus lurking about somewhere. He just had to find it.

Coming up with a way to detect the presence of a virus that replicated too scantily for an electron microscope was a daunting task. Fortunately, however, the very nature of these viruses that work backward provided a possibility. Since, by transcribing their RNA to DNA and then back to RNA again, these retroviruses operated differently from any other viruses (or, for that matter, any cells in the body), and since that singular process was catalyzed by reverse transcriptase, an enzyme unique to the viruses, if Gallo was able to detect the presence of the enzyme in cancer tumors, he could be certain that there was a retrovirus there as well. In effect the virus left a footprint.

But while the concept was sound, and certainly had worked for Temin and Baltimore, putting it into practice in a situation in which there were hardly any viruses to deal with was difficult. "My postdocs spent night and day working out this technology so that we could assay reverse transcriptase from cells and know what the hell we were doing," Gallo recalls. "It took us years to work out a way to be sure you had real reverse transcriptase."

Five years to be exact, from 1970 to 1975. During that time Gallo and his colleagues labored to refine the assay, making it more sensitive and specific. At the same time, they realized that if finally they were to pin down this elusive human retrovirus, they'd need a much greater supply of it to work on. The problem was that white blood cells, the cells that go awry in leukemia, do not grow well in the lab. And if Gallo couldn't grow the cells, he couldn't grow the virus. It was a predicament similar to that encountered by John Enders and the other virologists who labored to overcome the simple fact that viruses were tough to cultivate. But Gallo had a leg up on his predecessors, because by the early 1970s there had been discovered naturally occurring proteins called *growth factors* that functioned in the body as catalysts to assist the growth of cells. If Gallo could find a growth factor that boosted the growth of white blood cells, he might be in business. So he began to search for the right one.

What happened next was that Gallo found his human retrovirus. That is, he thought he did. In January of 1975, he and colleague Robert Gallagher published an article in *Science* announcing that they had located an RNA tumor virus in a sixty-one-year-old Houston, Texas, woman who was suffering from leukemia. He called the virus HL23V-1, from the shorthand designation of the Houston patient as HL23, and wrote in a subsequent *Nature* article that it "may be useful in detection and propagation of similar viruses from freshly-isolated leukaemic and solid tumour tissues."

It was a modest enough statement, considering the vast implications of the find. "May be useful," indeed! If correct, his discovery blazed a trail into a new world of viruses and a new world of possibilities for fighting disease. If some human leukemias were caused by viruses, then there was hope that by preventing or controlling infection the cancer might be thwarted (all this at the same time that Maurice Hilleman was testing his newly made hepatitis B vaccine, another attempt to control viral-caused cancer). Gallo was understandably elated. He had gotten to the finish line first, outdistancing Sol Speigelman, among others. If Gallo was relatively obscure before, he would be obscure no longer.

Obscure no longer, yes — but not in the way he envisioned. In the *Nature* paper, Gallo mentioned an interesting feature of the virus: "The virus is closely related to the simian sarcoma-associated virus isolated from a woolly monkey." The resemblance to this monkey virus, SSV, for "simian sarcoma virus," was striking. The two viruses were similar on the molecular level, they induced a similar immune response, and their structure was similar — so similar, in fact, that Gallo and his colleagues felt the need to make doubly sure that they weren't identical. Their concern was that the monkey virus might somehow have infiltrated his lab, contaminated his culture, and masqueraded as the human virus he thought he had discovered. "Many precautions were taken to ensure against the possibility of a laboratory contamination," he wrote; so absolutely no monkey viruses were grown or handled in the restricted area in which the new virus was isolated. In fact, Gallo went on, if there was a similarity between the human and monkey viruses, it's probably because the monkey, a pet that had close contact with humans, was at some point infected by the human virus. No, he wrote, "though closely related, the HL23V-1 and SSV are not identical."

He was wrong. Rather than resembling each other because the monkey was infected by the human virus, the true explanation turned out to be just the opposite: the monkey virus *had* infected human cells in his lab, contaminating his experiment. What Gallo thought was a newly discovered human virus was actually a slightly modified monkey virus. And in another *Nature* article, written six months later, he had to eat his words. "We find that the particles ... are a mixture of two viruses that are indistinguishable immunologically and in nucleotide sequence from two known non-human primate viruses. . . . it must be accepted that the identification of unknown tissue culture viruses with known laboratory agents raises the obvious possibility that they arose by contamination."

It was a embarrassing confession, to say the least. "I felt terrible when it came out," Gallo recalls. "The contaminations occurred when we put the virus into a cell line. For the virus to survive, we had to get it into a new cell line; we couldn't keep the original infected cells growing anymore."

The eternal problem: how do you cultivate viruses in the lab? In trying to grow a human retrovirus, Gallo had to transfer it from the original infected cells to a new culture of cells. In so doing, he exposed his experiment to contamination from a monkey virus that inexplicably had made its way into the new tissue culture. He was then able to grow a virus, all right, but it was a monkey virus rather than the human retrovirus Gallo thought he had. He was humiliated.

"It was very demoralizing to him," remembers Max Essex. "Lots of people made jokes about it. Lots of people thought that that was the final blow for human retrovirology."

For someone like Robert Gallo, the impact of such a mistake was especially devastating. He was, and is, characteristically restless, impatient to move on quickly, impatient with being bored. And never a shy, retiring sort, Gallo was at this early stage of his career particularly flamboyant and expansive. He was verbal and outspoken, neither hesitating to say what he thought nor worrying overmuch about the effects of his opinions. He was, in a way, a latter-day Félix d'Hérelle, who through the sheer vigor and damn-the-torpedoes recklessness of his personality and the unorthodoxy of his declarations raised as many hackles as he soothed and made as many enemies as he did supporters.

No wonder people scoffed at Gallo's determination to find a human retrovirus, and no wonder people laughed when he blundered. And when he returned to the lab, no wonder he found himself virtually alone. "In the next five or six years almost nobody kept looking for human retroviruses except Gallo," says Essex. Some have attributed this determination to the fact that early in his life Gallo lost his own sister to leukemia; finding an infectious cause for the disease might be a way to appease that demon. Perhaps as likely an explanation is found in his own personality.

"I was startled at first," Gallo says. "But later I got angry. It's a typical response that I tend to get: you're knocked down, lose confidence, you're depressed for a while, then you get angry. How in hell did this happen?"

Moving beyond the debacle would take four more years. And the first step was a discovery little heard of outside scientific circles, but one that not only would make possible finding a human retrovirus, but subsequently would lead to the detection of the virus that causes AIDS.

How to grow viruses in the lab — it's a recurrent refrain in the story of viruses. Ernest Goodpasture in the 1930s showed that it was possible, and John Enders in the 1940s showed that it was feasible, but neither said it would be easy. As over the years more sophisticated ways of detecting and analyzing viruses came into play, more sophisticated ways of cultivating viruses were devised; they had to be if experiments were to be successful.

But Gallo faced a double whammy: the need to grow a virus that replicated only sparingly in a culture that itself didn't like to grow in the lab. For as we've seen, white blood cells, the immune-system cells he thought were infected, didn't grow well in tissue culture. If he was to have any chance of growing enough virus to identify it positively and describe it, he'd have to find a way to cultivate these uncooperative cells. And to do that, he needed a booster, something that would convince the cells to flourish in their laboratory home.

That's where growth factors came in. Gallo set out to find a growth factor that would boost the growth of white blood cells. He already had a lead, for in the 1960s a University of Pennsylvania researcher named Peter Nowell had discovered that a plant protein called *phytohemagglutinin,* or PHA, when applied to cells could induce them to grow larger and divide a couple of times — in effect, it could act as a growth factor on a limited scale. Gallo and his colleagues Doris Morgan and Francis Ruscetti tried PHA in bone marrow, where a number of blood cells are produced, and found that the protein did indeed stimulate growth. When the team analyzed the quickly growing cells, they discovered that they were none other than T cells, the same white blood cells that go awry in leukemia. These cells were not only growing quickly, they were releasing a growth factor. The researchers named it TCGF, for "T-cell growth factor."

It remained to see how much of a boost TCGF imparted to T cells. A great deal, as it turned out, for not only did the PHA-stimulated cells secrete the growth factor, but the cells developed receptors on their surface for TCGF. When the growth factor reunited with the cells by attaching to their new receptors, the T cells began to divide merrily, taking to their petri dish as though it were a familiar body. The year was now 1976.

Without planning to, Gallo and his colleagues had also discovered an important process of the human immune system. For the T cells' secreting and receiving of TCGF parallels what happens when the immune system gears up to fight infection. When confronted by an invader — a virus, for example — the T cells in the body react in precisely the same way, producing the identical growth factor, which then binds to the newly developed receptors and stimulates the T cells as well as the B cells, another part of the immune system armamentarium, to replicate rapidly. These activated cells then attack the infection.

At the time, Gallo and his co-researchers knew little of this parallel function — it came to light years later. Now the growth factor, christened TCGF by Gallo, is better known by the name interleukin-2, one of a series of interleukins that promote and regulate the functioning of the immune system. Its detection is yet another example of the serendipity of science. "We certainly didn't plan it," says Gallo. "It's one of the little-known ways in which retroviruses have had an impact."

So now Gallo could cultivate T cells in tissue culture, hundreds and hundreds of thousands of T cells, which meant that if T cells were infected by a retrovirus, he would be able to grow enough virus particles that he could analyze and characterize the organism definitively. He was determined that no matter what happened, he was not going to experience again the kind of humiliation he had felt as a result of his mistake with HL23V-1, the monkey virus that masqueraded as a human virus.

"He was possessed, absolutely possessed," recalls Werner Kirsten. "He knew his scientific career was on the line. If this failed, *this was it*. He was on the ropes."

On Christmas Eve, 1978, Arsene Burny in Brussels received a telephone call from Washington. "It was eight o'clock at night," Burny remembers. "I was still in the lab, the only one left in the building. It was Gallo on the phone. He says, 'Burny, we are following something. I think we have something good. But I am afraid that we might be dealing with another contaminant.' "

One possible origin for contaminants, Gallo feared, might be blood from fetal calves that he used as a source of nutrient serum in which to grow his cultures. If, unbeknownst to him, this calf blood harbored the bovine leukemia virus, it might be contaminating his cultures. "We were working on bovine leukemia virus," Burny recalls. "We had DNA, we had antibodies, we had serum from infected animals. Gallo said, 'Send me all of that, so that I can check my cultures. I'll tell you more after, but first I want to make sure we are not dealing *again* with a contaminant.'

"So we sent him all that," Burny says. "Now we know, retrospectively, that if he had not used the most stringent conditions, he would have concluded that it *was* a contaminant. Because bovine leukemia virus and the human retrovirus he was about to find belong to the same branch in evolution. If you look in the genes, there are regions of close similarity. He had to do very good molecular biology."

A year later, in 1979, Gallo knew, once and for all, that he had found his human retrovirus. From the malignant T cells of a twenty-eight-year-old black Alabama man, Ruscetti and postdoctoral fellow Bernard Poiesz isolated a medium-sized RNA virus, roughly 1/250,000 of an inch in diameter. In the heart of the virus were two strands of RNA, accompanied by several molecules of reverse transcriptase, all enclosed in a shell made of protein, which was itself enveloped by a double layer of lipid torn from the membrane of the cells in which the virus replicated. And in a somewhat subdued echo of the influenza virus, this lipid envelope was studded with projectiles, stubby knobs of protein and sugar attached to the virus with protein rivets.

There was no doubt that it was a retrovirus, and no doubt that it was a newly discovered virus, a uniquely human virus. For Gallo and his colleagues not only tested it against antibodies to the bovine leukemia virus, but against a wide variety of antibodies that normally protected against other known animal retroviruses, including the monkey virus, SSV, that had tricked him last time. But none of the antibodies could recognize the antigens of this virus; they left it alone. It was something new. Gallo named the virus HTLV, for "human T-cell lymphoma virus."

This time when Gallo announced his team's discovery, he made certain that there was no chance of another blunder. Hard upon the heels of the first paper describing the findings, in the *Proceedings of the National Academy of Science* in December of 1980, came four others, published in various journals so as to reach virtually everyone, that continued to clarify and expand upon the discovery. One characterized the nucleotide sequences of the virus; another described its reverse transcriptase; another announced that human antibodies, in contrast to antibodies from animals, reacted against the virus ("The antibodies against HTLV are thus the first evidence for a specific immune response in humans against a retrovirus"); another announced the detection of the virus from other patients with T-cell malignancies. By the time Gallo was done, no one, anywhere, would be able to raise an unanswered question about this virus; no one, anywhere, would be able to doubt the authenticity of this discovery.

"It was calculated that way," Gallo recalls. "We wouldn't publish anything until there were five papers in hand, and we released them when we thought it was appropriate, step by step, to properly convince the field by anticipating each question. 'Let's anticipate all of them on a blackboard,' we said, 'and let's have the papers so that they respond to everything.'

"So, fire one, fire two, fire three . . . It was like that." He laughs. "Not to play games, but because I felt that if we released it all in one big bullet, and if somebody didn't read it all carefully, who knows what the reaction would be? This way people will read the first one, and say, 'Ahhh, yuck.' Then the next paper will come, they'll read that, and say, 'Well, you know, maybe . . .' Then by the third one their eyeballs are going to come out — *that's* the way we'll do it."

Gallo laughs again, relishing the memory. "So, yes, it was done as carefully as I'm capable of doing something and holding my tongue at the same time."

Once again, however, the reaction wasn't quite as he expected. It was a reception characterized more nearly by "Ahhh, yuck" than by eyeballs popping out.

"The first time he presented this, people jumped all over him," Werner Kirsten says. "He was simply not credible, that's what it amounted to. People thought, 'This big style, lots of money, NIH . . .' They all pooh-poohed it. But he was right, he was convinced, he believed the data, and he was utterly frustrated. I remember an early meeting in Germany, an idyllic place, south of Hamburg, called Welsede. After the meeting, on a Saturday afternoon, we were in our hotel in Hamburg, and it was an awful, rainy day. He was very depressed because of the reaction he had gotten. We got awfully drunk that evening."

Max Essex describes another meeting in which Gallo could not convince people that his findings were either interesting or important. "Gallo had HTLV, but most people didn't believe him or take it seriously or pay much attention — it was either unimportant or wrong. I remember a meeting in Seattle in late 1981 that Gallo organized, but there were not enough people working with human retroviruses to come close to filling it up. He couldn't even convince people in his own lab."

One reason may have been that HTLV was a virus without a disease. Gallo and his team had found the virus in no more than a handful of people, after all, and the disease it was associated with, lymphoma — cancer of the lymph nodes — is relatively rare. On the other side of the world in Japan, however, researchers were investigating a far more common cancer, leukemia. There had been an outbreak of a newly recognized form of the disease, which they called "adult T-cell leukemia," or ATL. This was a particularly aggressive leukemia that generally killed people within three or four months from the time it was diagnosed. It was heavily concentrated in Kyushu and Shikoku, the southernmost islands of Japan.

The Japanese researchers had discovered a retrovirus in the infected T cells; it was found to be identical to HTLV. Soon thereafter, under the leadership of NCI epidemiologist William Blattner, the virus was found throughout most of the Caribbean, the northern part of South America, and much of Africa, where it was endemic. Gallo hypothesized that it originated in Africa, where it infected monkeys as well as people, and that it made its way to Japan and the Americas as a stowaway in slave ships.

So now HTLV had its disease. And if there was still widespread skepticism concerning the value of investigating human retroviruses, all that was about to change for good. In fact, the field quickly became the predominant one in virology. But this ascendance was due to nothing Robert Gallo or any other virologist did. What was already beginning to happen was, some say, due to an act of God. Others attribute it to the changing social and economic conditions of the times, the enormous increase in air travel, the revolution in sexual mores. It was to cause, and is still causing, the most massive marshaling of medical science in history. But like many similar upheavals, it began quietly, with no more than a hint, a puzzled whisper, that something very odd was going on.

The reports began filtering out in the late 1970s. Young, white males, many of them homosexual, were suffering a rare skin cancer, called Kaposi's sarcoma, that previously had been found mainly among elderly Jewish and Italian men and in Africa. At the same time, other young homosexuals were turning up with another rare disease, this one a type of pneumonia caused by a usually harmless single-celled animal called *Pneumocystis carinii*. Still other homosexual men were complaining of similarly uncommon diseases such as enlarged lymph nodes and a cancer called non-Hodgkin's lymphoma. The only common denominator in all these various ailments, other than the fact that most of these people were young gay males, was that every case displayed an impaired immune system — in particular, a depletion of T cells.

What in the world could be causing this strange outbreak? Early on, the suggestions ranged from ridiculous to bizarre. Perhaps the outbreak was the result of exposure to too much sperm during sexual contact, perhaps it came from the overuse of the drug amyl nitrate, "poppers," known to be widespread in certain gay communities. Perhaps it had no one cause — these men's immune systems had simply broken down because of a variety of influences. Perhaps, said some, it was the wrath of God, vented upon those engaging in a lifestyle that was excessive at the least, immoral in any case. But when soon it was found that these afflictions were also spreading among intravenous-drug users, people who had had frequent transfusions, and, most peculiar of all, Haitians, who were neither exclusively male nor homosexual; and when it was documented that women partners of male IV-drug users were suffering these same ailments; and when suspicion grew to certainty that in central Africa this disease — for a singular disease it was — was growing into epidemic proportions, irrespective of sex; and when people began *dying* from their disorders — all previous bets as to the cause were suddenly off. Rather than a function of a particular lifestyle or sexual orientation, this disease must be the result of an infection, perhaps of a virus. The disease was named *acquired immune deficiency syndrome* — AIDS.

Bob Gallo, having just discovered a human retrovirus, almost immediately began to think that this new disease must be the result of a new agent — and, given the various cancer associations, what more likely agent than a retrovirus? Maybe HTLV itself, or, if not, perhaps something much like it. And having been urged by Gallo to

start investigating HTLV, and more than familiar with feline leukemia virus, a retrovirus that caused immune suppression in cats,
Max Essex began thinking much the same thing. "I told Gallo that
I would get involved, and the first thing I was going to address was
whether HTLV caused immune suppression as the cat viruses did.
We found that it did. The suppression it causes isn't nearly as severe
as AIDS, but it was clear that a retrovirus could suppress the immune system." Impressed by these findings, and much influenced
by the parallel Essex was drawing between this new disease and
retrovirus-caused immune suppression in cats, Gallo began stating
publicly in early 1982 that AIDS might very well turn out to be the
result of infection by a human retrovirus.

Across the Atlantic in Paris, Jacques Leibowitch was beginning
to suspect much the same thing. Leibowitch, a youthful forty-year-
old physician and head of a Parisian working group on AIDS, had
read an article in which Gallo suggested that an HTLV-like virus
might be the cause of the disease. Primarily a clinician, Leibowitch
was also something of a medical detective and had been wondering
what kind of agent might attack T cells, but he had not thought of
a retrovirus. Now a light bulb went off in his head. "I said, 'Jesus,
of course! Why didn't I think of that?'" For one of his AIDS patients, a heterosexual Frenchman, had received a blood transfusion
while in the civil service in Haiti. Haiti was one of the hotbeds of
AIDS, and it was also a place where HTLV was endemic. Could the
association be something more than a coincidence? Leibowitch
wanted to find out, and he also began to wonder about Africa. He
was aware that Africa was a focus for AIDS, but it had not yet been
shown to harbor high concentrations of HTLV. Yet Leibowitch had
a leukemia patient from Zaire. Could there be an AIDS-HTLV connection in Africa as well?

"AIDS is in Africa," Leibowitch thought. "If HTLV is the cause
of AIDS, it too must be in Africa. So I called my sister, who is a
pathologist and professor of dermatology. She was working on a
kind of skin disease called Sézary T-cell leukemia that sometimes is
a manifestation of a T-cell lymphoma. I had worked on Sézarys with
my sister, so I knew that they could be caused by HTLV. I said,
'Look for anybody from Africa that has a Sézary-type syndrome and

call me.' Three days later she had one, and since then she's never had any more of them." Leibowitch laughs. "It was the strangest thing.

"She said, 'I got one. He has a very unusual Sézary — he has these crazy cells on the skin, and he too is from Zaire.'

"So I went there, took his cells, separated them, looked in a microscope, and they were typical HTLV-type Sézary. I said, 'That's it, he's got HTLV.' "

Leibowitch, an unconventional, excitable type of fellow, was definitely excited. Armed with biopsies from his Haitian and African patients, he began spreading the notion around Paris that HTLV might be the cause of AIDS. He gave public talks with the title "Africa and HTLV: A Retrovirus for AIDS." Looking for a French retrovirologist to investigate the possibility, he visited the Institut Pasteur, among other places, but with little success. And he called Robert Gallo.

"I didn't know Gallo," Leibowitch says, "and I got his secretary. I said, 'I'm calling from Paris. Please convey the message: AIDS, Haiti, transfusion, heterosexual — African, lymphoma, leukemia.' " Leibowitch laughs. "I swear that's what I said. I can't believe that he called me the following day. 'What is it you got?' he asked. And I told him."

Gallo invited Leibowitch to come to Bethesda, bring his cell samples, and state his case — but on his own money. So, with no travel budget, Leibowitch again hit the streets of Paris trying to drum up support for his theory, and a plane ticket.

Meanwhile, Leibowitch's AIDS working group had become "inseminated," as he puts it, with his idea. One of the group, a clinician named Willy Rozenbaum, went back to the Pasteur, asking again if anyone there was interested in looking for a retrovirus in AIDS. Yes, someone was. His name was Luc Montagnier. The director of the viral-oncology unit at Pasteur, Montagnier already was embarked on a search for retroviruses, but not in AIDS — rather, in a hepatitis B vaccine that the institute was helping to produce. As the vaccine, which was similar in conception to Maurice Hilleman's hepatitis B vaccine developed at Merck Sharp & Dohme, was made from blood plasma, Pasteur wanted to be certain that it was not contaminated by the newly discovered HTLV. Montagnier was looking for evidence of reverse transcriptase in the vaccine when a specimen of the swollen lymph node of a young French homosexual arrived at his lab, compliments of Rozenbaum. This "lymphadenopathy" could be an early sign of the eventual onset of AIDS, and the thought was that it might be more worthwhile to look for a virus early on rather than later, when AIDS patients became plagued by a number of opportunistic infections. It was now the first week of January, 1983.

Montagnier is short and dour-looking, with small eyes that dart warily above half-glasses and a subdued, thoughtful manner. "I got the biopsy in the afternoon but couldn't work on it until later," he remembers. "I was director of a course in biology at Pasteur, so science had to wait. After two weeks we found a small indication of reverse-transcriptase activity. This was the beginning of excitement for us. Now there were two things to do: one was to maintain the virus, the other was to characterize it to see if it was an HTLV-like virus. I had been in touch with Gallo and decided to ask him to send me reagents for HTLV."

Montagnier needed reagents — antibodies that were known to react against HTLV — in order to find out just what kind of virus he had. If the antibodies attacked the virus, it was indeed HTLV. If there was no reaction, Montagnier was dealing with something else. And it just so happened that an acquaintance of his was about to step onto a plane to Bethesda, on his way to visit Gallo's lab. Montagnier asked him to deliver a letter to Gallo personally, requesting the reagents. It was Jacques Leibowitch.

Leibowitch was by this time more than a little miffed at the Institut Pasteur, as it had rebuffed his early approaches and then had decided to work with someone else from his French AIDS group. But now, belatedly, Pasteur had decided to sponsor his visit to Gallo and had come up with money for airplane fare. Of course he'd deliver the letter.

Leibowitch was fast becoming the gadfly in this particular melodrama ("It's as though the epidemic has been created to fulfill my education," he says, ruefully). He winged his way across the Atlantic bearing a letter from Montagnier and his specimens from the French civil servant who had become infected with AIDS in Haiti and from the Africans with leukemia. In Bethesda he presented his theory that AIDS was caused by HTLV to an audience comprised of Gallo and his lab associates, and while they were already moving in the direction Leibowitch outlined, his presentation reinforced their resolve. "We were only putting about fifteen percent effort into it," Gallo says. "Jacques was instrumental in provoking us to work more on the problem." Yet, successful as his visit was, through it all Leibowitch began to suspect — at least, in retrospect, he says he did — that he was involved in a situation that could easily deteriorate into something very unpleasant.

"It was clear there was going to be a crunch — fights, egos, business interests, and it was going to be terrible," Leibowitch says. "I was trying to forget about the characters in the history and focus on the problem. But I knew it was going to be a miserable story. We should have the best people involved so we could get rid of AIDS as soon as possible. That was a fantasy."

A fantasy, indeed. Soon enough those involved in searching for the AIDS virus would be hard-pressed to feel successful about much of anything. For Bob Gallo, in particular, the coming achievements and difficulties would prove bittersweet. Finally, with the onset of AIDS, people were taking his discovery of the first human retrovirus seriously. Gallo was flying high, having recouped all he lost with his mistaken work on HL23V-1 and more. He was now seen as a man who had shown the insight and determination, established necessary techniques, and assembled the finest team. As such, many people felt, it was only a matter of time before he found the virus that caused AIDS as well — if, indeed, it was a virus-caused disease. And they were right: it was only a matter of time before he did. But someone beat him to the punch. Luc Montagnier, testing his virus with the reagents Gallo had sent him, found that the HTLV antibodies could not recognize his virus. He had something new on his hands.

In May of 1983 Montagnier and his colleagues Françoise Barré-Sinoussi and Jean-Claude Chermann published in the journal *Science* an account of what they had found. It was, they wrote, "a retrovirus belonging to the family of recently discovered human T-cell leukemia viruses (HTLV), but clearly distinct from each previous isolate," and it had been "isolated from a Caucasian patient with signs and symptoms that often precede the acquired immune deficiency syndrome (AIDS)." Just what this virus was and what it might have to do with AIDS, however, Montagnier and his colleagues, more cautious than Gallo had been in his introduction of HTLV, would not say. They had isolated it from one patient only, hardly enough evidence to provide a convincing conclusion one way or another: "We tentatively conclude that this virus, as well as all previous HTLV isolates, belong to a family of T-lymphotropic retroviruses that . . . may be involved in several pathological syndromes, including AIDS."

Montagnier was excited; so was Gallo. In the very same issue of *Science,* Gallo had two papers announcing that he had found HTLV in AIDS patients; two papers from Max Essex offered similar results, as well as information about how feline leukemia virus suppressed the immune system. In their eyes, Montagnier simply offered corroboration that an HTLV-like virus was involved in AIDS. "It seemed to us," Essex says, "that Montagnier was confirming what we had, and we were confirming what he had."

The French paper was greeted with skepticism. Not only was the evidence for this new virus slim, but the electron micrographs of the virus were not clear enough to convince some people that it was a retrovirus at all. And Montagnier and his colleagues were having a hard time culturing enough virus to analyze usefully. It kept killing the cells it was growing in, repeatedly forcing Chermann and Barré-Sinoussi to transfer it to new T cells. Still, the Pasteur group was convinced they had something, and Montagnier asked Gallo to send him some interleukin-2 so that he could more successfully grow his virus. Gallo, for his part, asked to see some of the virus.

By the summer of 1983, Gallo's own work began to stall. His researchers were picking up reverse-transcriptase activity in infected T cells from yet another virus that they simply couldn't grow. The techniques that worked so well with HTLV, allowing them to use IL-2 to keep T cells and virus growing indefinitely, weren't working with this unknown virus. Nor were the techniques working with the sample of the French virus that Montagnier had given Gallo. It was the same problem faced by the Pasteur group — the virus was killing off the cells it infected — but, used to HTLV, which caused cells to proliferate, not die, Gallo and his colleagues didn't fully appreciate the difference. They put the Pasteur virus and their own unidentified virus in deep freeze and went on looking for a variant of HTLV. "It is certainly true that in that period of time in summer and certainly by early fall, Chermann had recognized the cytopathic effect of that virus and I had not," Gallo told *Science*. "As I look back now, I could bang my head against a wall that we were so stubborn in trying to grow those cells long term in IL-2. . . . We went through loss of months with that problem."

Such was the situation in the fall of 1983 when Gallo hosted a meeting on human T-cell leukemia viruses at the Cold Spring Harbor laboratories on Long Island. Here was the first chance for everyone involved in the now thriving field of human retroviruses to get together and personally thrash out just exactly what was happening. Reading one another's papers was one thing — actually talking in person was quite another. It was here that the first signs of strain began to show.

Montagnier took the podium to announce, among other things, that he and his colleagues had named his unknown virus LAV, for "lymphadenopathy-associated virus," and that they considered it the most likely cause of AIDS. The Pasteur group was also developing a test to detect antibodies to LAV in blood, a test similar to one Essex had already devised to look for an HTLV-like virus. And Montagnier also announced that the same AIDS patients who had developed antibodies to LAV had tested negative for HTLV. The two viruses, therefore, were "clearly distinct."

Those who were there remember it as a dramatic moment. Montagnier was in effect telling Gallo, in his own backyard, that he had been wrong about HTLV and AIDS. The French virus wasn't simply a variant of HTLV — it was a distinct virus, and the likely cause of AIDS.

In the question-and-answer period that followed, Gallo took the floor. He challenged Montagnier's findings. "He told him it was terrible science," an observer later informed a journalist, "that there was no way it could be true."

"I didn't do anything but ask him questions," Gallo recalls, "but

I asked him seven in a row. He took it as a tremendous criticism. He thought I was really hard on him."

But for Jacques Leibowitch, always the iconoclast, there was something more going on than simply tough scientific give-and-take. "Gallo should've been more diplomatic," he says. "He treated Montagnier roughly, just because he presented data that Gallo didn't want to recognize. It was a fight for power more than a fight for truth. It was a situation where obviously the *ambience* was going wrong. We were going into fighting a universal battle, a disaster for humanity, something that surpassed individuals, and here it was falling into two guys fighting. Montagnier was playing the knight in shining armor with his virus that the other one didn't have, and the other one got crazy. But Montagnier represented France, and he had things Gallo didn't. Gallo should have recognized that, but he didn't."

For his part, Gallo wishes the incident had never happened. "I used to be much more blunt and direct, but I don't do those things anymore. Today I don't ask any questions like that — I keep them to myself, unless it's my own livelihood, my own staff. It looks too much like I'm throwing my weight around. I don't have the same testosterone I did then."

The meeting at Cold Spring Harbor only exacerbated Montagnier's difficulty, on both sides of the Atlantic, in having his conclusions accepted. "Nobody took Montagnier at all seriously during that year," remembers Essex, "in part because his data wasn't any more positive than ours with HTLV. We were finding HTLV in twenty, thirty, forty percent of AIDS patients, and that's all he was getting with what we now know is the right virus. It was in part maybe because we were using very polished, well-controlled assays, and he didn't have good reagents and his virus didn't grow well. But he couldn't convince anybody otherwise. They believed that he simply didn't have anything even closely resembling a complete scientific story." Essex shakes his head at the irony of it. "Little did anybody know that Montagnier's would turn out to be the right isolate and that the others would be dropped as unimportant or irrelevant or wrong." And Pasteur applied for a patent on its blood test, despite the fact that the test could only detect antibodies in 20 percent of AIDS patients.

Gallo came away from Cold Spring Harbor even more deter-mined to move ahead with his own approach. And in November he hit pay dirt. His colleague Mikulas Popovic, a cell biologist, found that when he cloned T cells from a patient with leukemia, he could infect those cells with AIDS viruses, in the form of blood serum from patients, and the cells would continue to live and produce virus in-definitely. It was precisely the break the Gallo group had been look-ing for. Now they could grow the virus, and grow it in quantity.

Then Popovic took an unconventional step. In order to increase the chances of obtaining a productive infection in the new cloned cell line, he pooled together blood serum from ten AIDS patients and used that mixture to inoculate the cells. (At the same time, in a separate experiment, he cultured serum taken from a single infected patient.) The more strains of virus attacking the cells, he reasoned, the more likely that he would get a substantial, lasting infection. The ploy worked. By December Gallo's lab was mass-producing virus.

It was the first time the AIDS virus was grown successfully, any-where, but pooling the blood would come back to haunt Gallo. Frus-trated by months of not being able to cultivate the virus, he had been desperate. Time was slipping away — time in which the disease continued to kill in ever-increasing numbers and time in which Mon-tagnier and his colleagues continued to conduct their own research. "Pasteur said it was impossible to grow the virus in cells, *impossi-ble*," Gallo remembers. "We said nothing's impossible — but we can't solve the problem without growing it in a cell line. We'll find it by trial and error. So Mika just pooled the soup of cultured cells from ten people. We were trying to get high-multiplicity infection. We were *dying* to get there."

By January of 1984, Gallo was confident that he had found the cause of AIDS. And in the May 4, 1984, *Science,* in typical flam-boyant style, he published four papers, the first time the journal had ever run so many by one group in the same issue, to describe his results. In announcing this new virus, which he called HTLV-III (having earlier discovered HTLV-II, another variant of the original human retrovirus), he and his colleagues showed that they had the means to culture huge amounts of it, continually, in their new line of cloned T cells. They had detected the virus not in just one person but in forty-eight different patients with AIDS or pre-AIDS symptoms. They had developed a technique to identify antibodies to the virus in the blood of these people, which they had done in up to 88 percent of the cases; it would become the basis for a screening test to monitor blood donations. And they had undertaken an analysis of the antigens of the virus, finding them related to the HTLVs that came before.

Finally, in a coup de grace that caught the attention of virtually everyone, Margaret Heckler, then the secretary of the US Department of Health and Human Services, called a press conference to proclaim Gallo's findings to the world. "Today we add another miracle to the long honor roll of American medicine and science," she declared, adding that the discovery should lead to a blood test for AIDS. The reaction, as with much involving Gallo, was predictable: delight, admiration, outrage, jealousy. "When it was orchestrated that Margaret Heckler made the announcement," remembers Max Essex, "that caused incredible jealousies. But people damn well had to look at the data. When they did, it was very clear that the whole story was here and he had now *proven* the cause of AIDS. Nobody even remotely up to that time — not I, not him, not Montagnier — said that they had proven the cause. What they said was that they had the leading hypothesis with the most tantalizing evidence and the strongest candidate. What slapped everybody in the face was that he had it *proven*."

Accordingly, the US government filed patents on the method of mass-producing the virus as well as on the antibody test. Samuel Broder, then the chief of clinical oncology at NCI, later told *Science* that in the war on AIDS, "the spring of 1984 was the functional equivalent of Stalingrad or El Alamein. The war is not over by any means and it is still possible that we will not have a happy outcome, but a turning point occurred."

But where did that leave Luc Montagnier? Didn't he have a virus also, and didn't he consider it to be the cause of AIDS? "Before the press conference we were informed by Gallo," Montagnier told an interviewer. "I remember well the day he came to my office in April 1984. He sat at this table, in the chair you're sitting in now, and told us he had discovered the virus that causes AIDS, which he was calling HTLV-III. It was obvious his virus was close, if not identical to ours. My reaction was altogether positive. He was confirming our work. Afterward the debate became polemical, but my first reaction was, 'Good, I'm pleased Gallo has rediscovered what we've already found.' "

So history was repeating itself. As Gallo had initially looked upon Montagnier's 1983 isolation of LAV as reinforcement of his earlier hypothesis that AIDS was the work of HTLV, so Montagnier regarded Gallo's discovery of HTLV-III as confirmation of his earlier detection of LAV. But quickly that apparent accord and good cheer turned into a rancorous confrontation, one that Pasteur researchers quickly dubbed "the war." The first thrust, whether intended or not, had actually come the *day before* the Margaret Heckler press conference describing Gallo's discovery, when the *New York Times* ran a front-page story stating that the Pasteur group had discovered the cause of AIDS. "I believe we have the cause of AIDS, and it is an exciting discovery," said Dr. James O. Mason, the head of the Centers for Disease Control. "The public needs to know that this is a breakthrough and that it is significant." The CDC had a close working relationship with the French group, and even though the Institut Pasteur denied having anything to do with the article, some in the Gallo camp saw it as a deliberate attempt to upstage their coming announcement. The second thrust was provided by Gallo's quartet of *Science* papers. Now it was the Pasteur group's turn to feel annoyed, as Gallo pointed out the "insufficient characterization of LAV because the virus had not yet been transmitted to a permanently growing cell line for true isolation and therefore has been difficult to obtain in quantity." All true, of course, but Pasteur took it as a slight to their research. "I was shocked by the way he presented our data," Montagnier said.

So, battle lines were drawn. At stake: personal reputation, national pride — and economics. Who should get credit for discovering the virus that causes AIDS?

"A Viral Competition Over AIDS," headlined a *New York Times* editorial.

What causes acquired immune deficiency syndrome, the deadly disease that has terrorized the homosexual community in America in recent years? Within the week, two officials have declared that the guilty virus has at last been found. Less reassuringly, each named a different candidate.

Some kind of progress is surely being made. The commotion indicates a fierce — and premature — fight for credit between scientists and bureaucratic sponsors of research. Certainly no one deserves the Nobel Peace Prize.

And who should own patent rights to an AIDS blood test, a privilege claimed by both sides? Ownership of such rights would mean millions of dollars of royalties. The search for the AIDS virus, which had involved brilliant research, exemplary and innovative scientific techniques, quickly turned into low farce — which might actually have been funny had it not been so petty and mean-spirited. For example, the CDC deferred to Montagnier's 1983 paper as the earliest mention of an AIDS virus, and used the term LAV in its publications. The NCI then petitioned the government to order federal agencies to use the Gallo designation, HTLV-III. The CDC responded by calling the virus LAV/HTLV-III, a label that in its acronymic clumsiness symbolized how far away the battle had gone from the reality that out there in the world people were dying of a mysterious and barely understood disease. When in May of 1985 a blood test patent was awarded to the United States, not France, which had made its application months earlier, and when in December of the same year the Institut Pasteur brought suit against the US government contesting the action, the rift was complete.

Gallo was often cast as the villain in the piece, in France and at home as well. The French accused him of trying to steal Montagnier's thunder, and that of the Institut Pasteur, by claiming, fraudulently, that *his* virus was the one true cause of AIDS. It was a matter of ego, some said, perpetrated by the person with the biggest ego in microbiology. Then two surprising events transpired to fuel the accusations, raising speculation that it wasn't only Montagnier's thunder that Gallo had stolen.

The first was the revelation that when it came down to it, there was very little difference between HTLV-III and LAV. Information that included an analysis of another AIDS virus, ARV, which had been discovered in early 1985 by Jay Levy of the University of California at San Francisco, indicated that whereas ARV differed from both other viruses by almost 600 out of the 10,000 nucleotides in the viruses' genes, HTLV-III and LAV were almost identical. The difference between them was only about 150 nucleotides, virtually nothing. But how could that be? How could a virus discovered in Paris be almost identical to a virus isolated in Bethesda, thousands of miles away? To some, there was only one explanation: Gallo had stolen Montagnier's virus and claimed it as his own. He was not only a usurper, he was a thief.

It was a charge that was never made publicly by anyone in the scientific community, although innuendos circulated under the breath. The media, however, was not so circumspect. The *San Francisco Chronicle* noted that "there was speculation that Gallo's lab may have taken the French viral isolate and claimed it as its own discovery. Montagnier stopped short of delivering the charge publicly. Other Pasteur researchers said privately that Gallo had 'stolen' the virus." Gallo himself fended off such attacks with typically blunt language (again, from the *San Francisco Chronicle*): "I'm the father of human retrovirology. Even without HTLV-III, my place in the history of science was secure. I don't even need to answer such charges, they're so absurd."

If so, why are the two viruses virtually identical? One explanation is that they actually are strains of the same virus that happened to show up in different patients. For although Gallo and Montagnier isolated their viruses thousands of miles away from one another, and at different times, the people who provided those viruses had something very much in common: gay life in New York City in the 1970s.

Through it all, Gallo and Montagnier stayed in contact. Gallo remembers a telephone call during this period. "I called Montagnier and said, 'You know something, HTLV-III and LAV are so close.'

"He said: 'They should be, it's the same virus. They come from New York. The guys were infected at the same time, in New York in the late seventies.'

" 'Yeah, that's probably it,' I said."

"My virus and HTLV-III are probably the same virus," Montagnier says now. Then he adds: "They came from the same patient, or from patients that had sexual relations. My patient was French, but he went several times to New York. He could have picked up the virus there."

When you acknowledge that in the middle to late 1970s, when the AIDS virus was just beginning to make inroads in this country, there certainly were not as many different strains as later on, it may seem possible that sexually active gays in the same place at the same time might have been infected by identical viruses. And the possibility that this common virus might have turned up in both researchers' labs heightens because of the fact that one of the ploys Gallo and Popovic used to grow their AIDS viruses was to pool together viruses from ten AIDS patients. Perhaps an identical virus to LAV was in this mix. "Is it possible that there's a cross-mix?" says Gallo. "Yes, it's possible."

But, "They came from the same patient . . ." What is Montagnier implying? Could it be that the viruses are so close because they actually are one and the same virus — that is, Montagnier's virus? Could the explanation involve the dirty word of virology, one that Bob Gallo in particular has heard much too often already: contamination? "Since the two isolates are so close," explains Montagnier, "it's not impossible that one is the contamination of the other. Because Gallo and his collaborator also grew my virus, not only his own isolate, but grew my virus in the same lab."

Perhaps LAV contaminated Gallo's cultures, just as the monkey virus had contaminated his culture back in 1975. For it's true that Gallo did have LAV in his lab, and it's also true that contaminations, even in this era of disposable plastic flasks and other modern precautions, are always a very real possibility. Gallo himself admits to the possibility. "The truth is that either a contamination occurred, which is possible, or it *is* the fact that these infections are early, mid to late seventies, and there were not so many viruses around then. But we have other isolates — who cares?"

Then surfaced yet another unexpected twist in a tale filled with them. In one of his four *Science* articles, Gallo, it turned out, had inadvertently included in a composite illustration showing similarities and differences among human retroviruses an electron micrograph of none other than Montagnier's retrovirus, LAV. What was identifed as HTLV-III in the illustration was actually LAV. And in an April 1986 letter to *Science,* Gallo and his associates admitted as much.

How embarrassing. Especially when you're in the midst of a court battle over who discovered the virus and who has patent rights to a blood test. Accusations aside from Gallo's detractors, who saw in the mistake another indication that he had stolen Montagnier's virus, this was an opportunity for Pasteur's lawyers to contend that the mislabeled micrograph "adds to the circumstantial evidence" that Gallo used Montagnier's virus to gain essential information. It was another indication, in other words, that credit, and patent rights, should go to the Institut Pasteur.

Nonsense, retaliated Gallo. He had ordered electron micrographs of LAV, which Montagnier had freely given him, simply to confirm that the virus was indeed a retrovirus, something that had been a matter of contention. Besides, his was not an electron-microscopy lab. He had to send his samples out to NCI's Frederick Cancer Center to be micrographed. The mix-up happened there, not in Bethesda. "It was my idea to make a composite," he says, "but the composite was done at Frederick, not by me or anybody in my lab."

Gallo shrugs his shoulders, exasperated. "Let's look at it philosophically for a moment: *nobody did anything wrong!* Sure Montagnier and his people gave us virus, but so what? We gave them IL-2, we gave them reagents to HTLV-I, we gave them reagents to HTLV-II, we talked to them on the phone, gave them our protocols. It's a two-way street.

"What if we *had* taken their virus and grown it in mass culture, which they said was impossible. What if we *had* made a blood test with it and solved the etiology with it. I would say, hurrah! Mika Popovic goes crazy. He says, 'You know, if that's all we ever did in our whole life, you'd think there would be people patting us on the back, not throwing things at us.' "

Yet you must wonder at the sheer volume of confusion surrounding the events involved in the discovery of the AIDS virus. And as for Bob Gallo, who seems to find himself at the center of one whirlwind after another, he certainly might wish that at the least he had handled the situation somewhat differently.

"Gallo made a mistake in not saying that this was the line his virus could have come from," says Werner Kirsten. "Why didn't he say, 'This is wonderful that Montagnier has a virus. That's a European isolate, we have an American isolate.' I think he should've been more generous. If I would have any criticism of Bob, it's that."

Jacques Leibowitch concurs: "Gallo made a mistake. He showed himself as very strong but arrogant. He didn't realize that he had to play politically with Institut Pasteur. He could not slap at Pasteur — he was offending a whole nation. He was touching our pride. Touching the Pasteur was like touching the French flag. He should have understood the stakes. He eventually understood this, but it took too long."

So as 1987 approached, AIDS had provoked two full-scale wars: one against the disease and one between the Institut Pasteur and the NCI. But the war was ending. In the fall of 1986, Prime Minister Jacques Chirac of France sent a message to President Ronald Reagan indicating that the time had come to make amends; it seemed that to bring about this armistice, nothing would do except the involvement of the highest political power in both countries. Both sides were ready, well nigh exhausted. With the help of Jonas Salk, who acted both as intermediary and messenger, an agreement was reached in March of 1987. It included official recognition of the Montagnier and Gallo groups as codiscoverers of the AIDS virus, which was now to be called HIV, for "human immunodeficiency virus" — no more of this unwieldy LAV/HTLV-III business. All pending legal suits were to be dropped, and the US patent for the antibody test was to be amended to include the names of the Pasteur scientists, just as the French patent application would contain the names of the NCI researchers. The agreement stipulated the creation of an AIDS research foundation financed by 80 percent of total royalties received, with the remainder to be split between the two sides. And, finally, Gallo and Montagnier were to agree on an official scientific chronology of research on AIDS. It was published, under both their names, in an April issue of *Nature*.

The agreement was announced at the White House by Reagan and Chirac on March 31, 1987 — a historic date, for there had never been anything resembling it before, nor has there been since. "This agreement opens a new era in Franco-American cooperation," said Reagan, "allowing France and the United States to join their efforts to control this terrible disease in the hopes of speeding the development of an AIDS vaccine or cure."

"We will now work together . . . to fight against AIDS," said Chirac. "I think it's a great step to be successful in this important battle."

As for the principals, they seemed content with the accord, or at least resigned to it, and they hailed it primarily as an opportunity to get back to work. "I think the agreement is fair, and I feel good about it," said Gallo. "There is the recognition that there was extensive scientific cooperation from the beginning — and throughout. Now, instead of being distracted by all the legal business, I'll be able to return full-time to trying to do something about this disease."

"I cannot help thinking, deep inside me, that it was a surrender," said the Pasteur's Jean-Claude Chermann. "With time everyone would have seen that we were right. But, with that said, I am happy for the Institut Pasteur, which can thus continue to work effectively in the battle against AIDS."

And today, Gallo and Montagnier more or less echo those sentiments. "The joint agreement is okay with me," says Montagnier, "because it recognizes our contribution. There are two distinct parts: scientific and legal. The scientific recognizes what we have really done; the legal part is the consequence that we share the discovery. I think the war is over. It was a truce. Now back to a good relationship."

"We know what we did," says Gallo. "Montagnier knows what we did, and Montagnier has always asked me if when we made the agreement on the history I would settle for roughly a fifty/fifty contribution. I don't mean to human retrovirology, but to AIDS. And I said, yes."

POSTSCRIPT: In November of 1989, the *Chicago Tribune* published a lengthy, critical article examining the hunt for the AIDS virus, in particular the confusing circumstances surrounding Gallo's discovery of a virus virtually identical to that already found in Luc Montagnier's lab.

In the wake of the *Tribune* story, the NIH was contacted by Representative John Dingell of Michigan, the House watchdog concerning matters of science. As a result, the agency has taken the unprecedented step of enlisting the help of the National Academy of Sciences and the Institute of Medicine to conduct a review of Gallo's actions. And there are rumblings that in response to this new criticism, the French are considering trying to break the agreement stipulating that royalties for the AIDS blood test be shared equally by both sides. Perhaps the war is not over after all.

As of this writing, the results of the inquiry are not yet in. The melodrama continues.

Meanwhile, human retroviruses have more than assumed the significance Gallo first saw in them and have attracted some of the best minds in basic and applied science. This is fortunate, given the difficulties retroviruses pose. The leukemia viruses, HTLV-I and its brethren, aside, HIV is challenge enough. Montagnier has said: "If I were a devil creating a malicious virus to cause the most problems for the human race, the virus would be AIDS. The virus has found the Achilles' heel of the immune system."

Why is that so? Well, to begin with, HIV primarily attacks T cells, white blood cells that play a major role in the body's immune response to invasion by viruses and other infectious agents. It is the T cells that directly attack infected or malignant body cells, while at the same time helping to orchestrate deployment of other immune-system tactics, such as the production of antibodies. And HIV specifically goes after the very T cells — called T_4, or *helper T, cells* — that do the most to facilitate these functions and activate as well another kind of white blood cells called *killer T cells*. These cells patrol the body, killing infected cells on contact. The killer cells bind fast to their targets and deliver a lethal dose of toxic chemicals that bore holes in the infected cells' membranes. Thus perforated, the cells leak out their fluids and soon burst. So knock out T_4 cells, and you knock out a big chunk of the body's immune response. It's a masterful and insidious strategy.

But that's not all. For HIV also attacks macrophages, another one of the indispensable characters of the immune system. Macrophages are scavenger white cells that roam the body engulfing foreign matter and cellular debris. They also play an important part in initiating immune response in the first place, as they alert the T cells that there is an invasion to be fought. So by infecting macrophages, HIV further debilitates the immune system. At the same time, infected macrophages spread the virus throughout the body, as these scavenger cells don't quickly die of infection, as do T cells. They continue traveling the body, now bearing an unwelcome burden of HIV. In fact, these circulating, infected macrophages may be the vehicle by which HIV causes the neurological difficulties often seen in AIDS. Loss of memory, failing judgment, and declining bodily coordination during the course of the disease may be the results of macrophages carrying their deadly cargo to the brain.

So HIV's attack strategy is far more sophisticated and clever than that of any other human virus: it debilitates the immune system itself. It disarms the body, not by overwhelming it from without but by infiltrating from within. Yet, crippled as it is, the immune system still has resources to fight the virus — if it could find the invader. For not content with simply disabling the immune system, HIV is also a master at hiding from it. Other than the capricious influenza virus, there is no virus as adept as HIV at changing the configuration of its surface so as to elude immune detection. HIV mutates incredibly rapidly, often within the very person it infects; there have been instances of viruses in the same individual varying by as much as 30 percent. The idea here is as old, and effective, as espionage: if you can't find me, you won't stop me.

Yet surface disguise may be the lesser of HIV's wiles. Its other formidable ploy hearkens back to the characteristic that gives it and its brethren their family name, the feature that Temin and Baltimore discovered twenty years ago: the ability of these retroviruses to transcribe their RNA genes into DNA. In so doing, HIV, and other retroviruses, jettison their outer membrane and protein shell and the rest of their viral accoutrements and insert their newly formed DNA directly into the DNA genes of the cell they infect. And there they stay, no more than a blip along the cell's chromosomes, *absolutely invisible* to immune-system surveillance.

It's this modus operandi that imparts to HIV infection its particular insidiousness, for while the virus hides out, the infection may not produce obvious symptoms for some time, perhaps up to ten years or more. There in the heart of the cell the virus sits, silently, patiently, while the infected person, unaware, goes about his or her business — a business that may involve spreading the virus to others. Only later does the virus again rouse itself, begin producing more viruses, and renew in earnest the process of killing off T cells, with the result that the individual starts suffering the symptoms that eventually lead to AIDS.

What instigates HIV's return to active duty? Why, after hiding out so quietly for so long, does it explode into action once again? No one knows for sure, but increasingly it seems as though other infections may spark the virus's return. People can be coinfected with HIV and other viruses. HTLV and HIV, for example, can infect the very same cell in the very same person at the same time. In such cases, the development of AIDS can be considerably enhanced. So, besides causing leukemia, the other human retroviruses may play an important role in AIDS as well.

The same role may be played by another virus, HHV-6, the herpesvirus that Gallo and his colleagues discovered almost five years ago. HHV-6, like Epstein-Barr, another herpesvirus, is ubiquitous. In the United States at least 60 percent of people are infected by the virus during their first two years of life, and 80 percent of adults are chronically infected. Ordinarily we handle the virus well, keeping it under contol to the extent that we're almost never aware of its presence (although, as we've seen, the role, if any, of HHV-6 in chronic fatigue syndrome has yet to be nailed down). But when chronic HHV-6 infection is joined by an attack of HIV, the delicate balance may be upset. With the immune system at least marginally depressed by HIV, HHV-6, no longer under control, may begin replicating, setting into motion a vicious cycle. HHV-6, too, can infect and kill T cells, and in so doing it may awaken HIV from its long sleep. HIV may then begin reproducing, further depressing the immune system, which allows yet more HHV-6 production, which activates more HIV — and so it goes. The result: an accelerated onset of AIDS.

The role of such cofactors in AIDS — a role that is just beginning to be uncovered — may turn out to be more important than anyone suspected. "Cofactors are producing a dramatic augmentation of AIDS," says Gallo. "When I look back at our first cases, we thought AIDS occurred like *that!* One of the first people we saw in 1980 died quickly after blood transfusions; it turns out he was doubly infected with HIV and HTLV-I. Now we understand these things so much better. We're now seeing a new field of virus-virus copathogenesis. It doesn't mean that I don't think HIV is the cause of the disease — it *will* bring it on eventually. Cofactors just make it go faster."

It's a scary story: a lethal disease caused by a sophisticated and insidious virus, the debilitating effects of which can be accelerated by coinfection by other viruses, one of which, HHV-6, already resides in most of us. No wonder AIDS is leaving an indelible mark on our times, both in the frightful number of people infected and dead, and in terms of its harrowing impact on our consciousness. Here's a disease, and a virus, that upon gaining a toehold simply can't be stopped.

At least such has been the history of AIDS so far. But today, finally, scientists are beginning to see light at the end of this particularly dark tunnel. For the first time, people are envisioning a world that, while not without AIDS, may be one in which the disease is brought under control. "The beauty of where we are right now," says Samuel Broder, now the director of the National Cancer Institute, "is that it's possible that HIV infection will become the functional equivalent of high blood pressure. What we'll do is identify people who are infected and then, as we do with people who have hypertension, give them treatment that will forestall or prevent the disease, just as we can forestall or prevent a stroke."

The reasons for that optimism rest for the most part upon three emerging areas of treatment: antiviral drugs, immunotherapy, and, finally, vaccines, which may prevent the disease before, or even after, infection. Of these approaches, antiviral drugs are most well known, primarily because of AZT. The first successful anti-AIDS agent, AZT (for "azidothymidine") works by preventing HIV's reverse transcriptase from making DNA out of its RNA genes. By knocking out that ability, AZT knocks out the virus. Because it's unable to make DNA to integrate into the target cell's DNA, the virus simply can't infect. But AZT, and other drugs that stop the virus's replicative process, simply aren't specific enough to shut down the virus only. They interfere with the cell's ability to reproduce, as well. The result is a wide range of side effects that make the drugs helpful and dangerous at the same time. The success of these drugs will depend on using them in just the right doses, in combination with other drugs, so that the disease may be slowed or eliminated without too severely compromising the health of the patient. It's a tricky business, a kind of high-wire balancing act, but the future of anti-AIDS drugs looks promising. And that's saying a great deal, because antivirals in general have enjoyed very little success up to now. It's really the pressures of AIDS that have sparked this heretofore moribund field.

Immunotherapy is another avenue that presents hopeful possibilities. The idea here is to provide a potion for already infected people that will boost their immune systems so that they can give the virus a good fight. Like antiviral drugs, it's an approach that has been infused with life because of AIDS.

Daniel Zagury of Pierre and Marie Curie University in Paris may be the most well known practitioner of this art, if for no other reason than that he has inoculated himself with his own immunotherapeutic vaccines. Zagury pinpointed the precise moment during an attack of HIV when the cell signaled by a protein marker on its surface that it was being infected by the virus. It's that signal that prompts the immune system to unleash T cells and natural killer cells and the rest of its infection-fighting armamentarium. Zagury took these infected cells and arrested them at that moment, killed and fixed them with formaldehyde so that they would always display the sign of infection but no longer contained living virus. They were thus immunogenic but not infectious. He then returned the cells to the patient by slow-drip intravenous injection, with the result that the immune system was indeed boosted, by the production of antibodies and white blood cells.

But when it comes to HIV, working with whole virus, killed or no, is risky. So after that initial success, in November of 1986, Zagury scrapped the virus per se and turned to recombinant methods in which he used vaccinia virus that has been genetically altered to show HIV proteins on its surface. The result is a virus that looks to the immune system as though it's HIV but is actually the old standby vaccinia, which has been used safely in smallpox vaccines for two hundred years. Building on the methods he had used before, Zagury took cells from an infected patient, inoculated them with his new vaccine, fixed the cells as before so that they would permanently display the signs of infection on their surface, and let them drip back into the patient. And, as before, he let the immunogenic cells drip into his own veins as well. Again, the results were hopeful, to the point that Zagury could write in a spring 1988 issue of *Nature,* "Our results show for the first time that an immune state against HIV can be obtained in man."

Today Zagury is continuing to test his approach, which while extremely promising is also extremely cumbersome and time-consuming. Still, no technique is too cumbersome if it works, and at the least Zagury's methods may provide a basis for procedures that simplify the mechanics of immunotherapy while retaining and increasing its effectiveness. Already his approach, and that of others working to provide successful immunotherapy, has supplied a measure of hope and optimism so far missing from the fight against AIDS.

The same optimism is slowly creeping into the vaccine field as well. It's more than welcome, for producing a vaccine against AIDS has long been considered the most difficult to achieve of all possible forms of treatment. As we've seen, vaccines work by causing a mild form of infection that fools the body into thinking it must gear up its defenses against the real thing. The result is immunity to the disease. With HIV, it seems, no such strategy is possible. Because the virus takes up permanent residence in the genes of infected cells, let it in at all, no matter how subdued and ineffectual the attack may be, and you run the risk of developing AIDS. To be successful, then, an AIDS vaccine may have to block the virus completely, every last particle — something that's never been achieved with any vaccine.

Making a vaccine that uses an attenuated version of HIV, therefore, is almost certainly out. Unfortunately, so is the equally venerable practice of killing the virus so that it can't infect but can still raise antibodies against a real attack — at least such is the view of most experts. The problem here is, how can you know if you've killed every last virus? The answer, as the sad, early experience with killed polio virus showed, is that sometimes you can't. So an AIDS vaccine will most likely have to be made of harmless parts of the virus that can still stimulate immunity in the body.

That's the first difficulty with a vaccine: the forbidding and complicated nature of HIV. The second is that even if these barriers could be surmounted, there would be no way to test the results of the vaccine — other than in people, that is. Except for chimpanzees, there are no animals other than humans that can become infected by HIV. And chimps, besides being large, expensive, and difficult to deal with, are on the endangered species list. Without an accessible, viable "animal model," so called, on which to test precisely how the virus works in a living system and how it might respond to a variety of vaccine strategies in a living system, the prospect of coming up with a vaccine is dim, indeed. Advances in tissue culture notwithstanding, you can only do so much in the lab.

Why, then, this current optimism? In the first place, it may be that the elusive nature of this virus has been somewhat overestimated. For example, on its knobby surface HIV displays a looplike protein that seems to be the part of the virus that is most readily recognized by the immune system. In early stages of infection, people quickly produce antibodies against the loop that are effective in stopping infection. This leads scientists to speculate that these antibodies might actually control the spread of virus during the latent period of the disease. But that same loop is one of the regions on the surface that mutates the most, which suggests that it might function as a decoy — as though the virus shouts, "Look at me," and then, having caused the body to raise antibodies in defense, changes the configuration of the loop to elude those defenses. Now it appears that the loop may not be as slippery as was once thought. It seems to have an area that stays the same throughout all its gyrations. That being the case, a vaccine that provokes the production of antibodies against the conserved region might be effective.

That kind of possibility gives researchers new hope that instead of having somehow to tailor a vaccine to hit a constantly moving and changing target, it may be possible to count on certain exposed areas of the virus being available all the time. Not quite a sitting duck, not nearly — but much more accessible, perhaps, than had ever been suspected. The virus may not be the foolproof escape artist it was once thought to be.

In the second place, it may well be that the notion "once infected is always infected" is simply not the case when it comes to HIV. Daniel Zagury's immunotherapy work has suggested that already-infected people can benefit from a vaccinelike application, and during the spring of 1989 came results of another dramatic experiment that reinforced that possibility. Jonas Salk, director of the Salk Institute, in collaboration with Clarence Gibbs of the NIH, inoculated infected chimpanzees with a preparation consisting of whole, killed HIV. It was precisely the approach many people said would be impossible because of the danger posed by the virus (and it was much the same approach that was used in Salk's killed-virus polio vaccine). It produced surprising results: by boosting the body's immune response, the vaccination actually cleared the blood of HIV infection. The chimps went from HIV-positive to HIV-negative, an outcome that had other researchers shaking their heads in disbelief. But how long this reprieve from infection may last remains to be seen, as does the effect of such an approach in human beings. Early experiments with people carrying the virus or in the early stages of disease showed that the preparation was safe — it didn't enhance the infection in these patients — and that it may have prolonged a state of good health. More extensive testing is going on now.

And what about the accepted wisdom that when it comes to HIV, you can't let even one cell become infected because once the virus takes up latent occupancy in the cells it invades, it's inaccessible to any attempt to flush it out? "I don't know if every last particle is gone, and I don't care," says Salk. "It's a state of affairs in which the augmented immune system will take care of virus in virus-expressing cells. What we have established is that resistance to reinfection is possible, and that somehow, in the course of infection, reversal from an active to inactive infection is possible. Now we'll have to wait and see what happens over time."

So experiments such as these are suggesting that the AIDS virus may not be quite as tough a nut to crack as had been thought. And there's yet another development that is making researchers feel hopeful: now, for the first time, there's the possibility of useful animal models for studying the effects of the virus. A retrovirus called SIV, for "simian immunodeficiency virus," causes a disease much like AIDS in monkeys. Researchers are finding that rhesus monkeys vaccinated with killed SIV do not, for the most part, develop disease and in many cases have actually resisted infection. Whether SIV behavior in monkeys corresponds to HIV behavior in humans is still not known, but the research is certainly a promising beginning toward finding out.

But although monkeys are more manageable than chimpanzees and in many cases are not threatened with extinction as are chimps, they're still unwieldy, difficult animals to deal with in the lab. How nice it would be to have available a tiny, docile, prolific animal — such as a mouse, for example — that could be infected by HIV and develop AIDS. A mouse with a human immune system, say.

Science fiction. But, astonishingly enough, it's an impossibility that is becoming possible. Two researchers in California are spearheading the development of just this sort of model using "SCID" mice. *SCID* stands for "severe combined immunodeficiency" and describes a mutant strain of mice that is born without an immune system (a fate shared with about two hundred children a year, including the "bubble boy"). These mice, like their human counterparts, are completely unable to defend themselves against even the most innocuous infectious invader. They are, in a sense, voids waiting to be filled. Donald Mosier and Mike McCune are busy filling those voids with *human* immune systems.

Early in 1987, while a postdoc in immunology at Stanford, Mike McCune came up with the idea of trying to produce mice that mimic the human immune system from start to finish. His thinking went like this: if these mice were to have a functioning, self-renewing human immune system, they'd have to be equipped with the organs that produce and maintain immune-system cells. So he decided to take the ambitious step of surgically implanting human organs into SCID mice. In May of 1987, McCune and three Stanford colleagues provided eleven SCID mice with human organs obtained from aborted fetuses.

The organs they transplanted into each mouse were a liver, the organ in which immature immune cells are made in fetuses; a thymus gland, the one in which T cells mature (thus their name, "T" cells); lymph nodes, to which the cells migrate and where they interact with other immune cells; and bits of skin, another producer of immune cells. McCune and his team couldn't use mature human organs — they would undoubtedly reject life inside a foreign mouse — but fetal organs could learn to regard mouse tissue as their own. That done, the researchers injected the mice with fetal immune cells, hoping they would circulate to these home organs and develop into mature, functioning human immune cells. "All the blood system is is something that ties those organs together," says McCune. "Picture the blood system as a highway. The cells that move along the highway are just trafficking from one organ to another."

The experiment was a success. The mice began showing resistance to infections that were killing ordinary SCID mice, and their blood showed evidence of T and B cells, as well as macrophages. The next step was to see if these mice could be infected by HIV. If so, they would become the only animals other than chimps susceptible to this human retrovirus. In April of 1988, McCune successfully infected his mice, the first time such a feat had been accomplished. So now these "SCID-hu" mice (*hu* for "human") not only share with us our protective immune system, they have succumbed to infection by one of our most notorious tormentors. Still to come is the answer to the question of whether the mice will actually come down with AIDS.

McCune, who has since left Stanford to start his own research firm, called Systemix, is now thick in SCID-hu mice that are ready to provide information about the course of AIDS. And because his mice show the potential of exhibiting all the functions of a human immune system, it may be that he will be able to study the entire life cycle of HIV. At present he and his associates are testing antiviral drugs on the mice, in the hopes of determining how, and how well, these drugs deal with the virus.

Mosier is an immunologist at the Medical Biology Institute in La Jolla. In early 1988 he performed a variation on McCune's theme by injecting SCID mice with mature human blood cells. Would these human cells flourish in the mice, or would they turn on the mouse cells as foreign objects to be attacked? It turned out that the cells survived beautifully, circulating in the animals' bloodstreams as though right at home. Now, in part, these mice too became armed with what they had never had before, an immune system.

Once the immune cells were established, the question became, could these mice become infected with HIV? The answer was yes. HIV infected these mice with gusto. But would they get sick? Chimps don't — they merely test positive for infection. If so, these mice would be the first animals to succumb to AIDS. Again the answer was yes. The mice are now showing signs of disease, and some are actually dying. A mouse with AIDS — it is an almost unthinkable contradiction.

Mosier's SCID mice, then, may prove ideal for the testing of all sorts of AIDS therapies and treatments — a wonderful boon. What they cannot provide so far, however, is a strong initial response to infection. "These little mice don't fully mimic every aspect of the human immune system," says Mosier, "but in large part they reflect where the cells come from." The cells Mosier has injected into the mice come from the veins of an adult human: T and B cells that recirculate through the blood and that last for ten to fifteen years, if not a lifetime. These are cells that have at some point in the past successfully beaten back infectious agents and are capable of doing it again. They're the cells that defend against familiar invaders time and time again. What they don't do as well is defend against an unfamiliar invader, an infectious agent the body may encounter for the first time. An unfamiliar agent, for example, such as HIV.

Mosier is trying to amplify the range of the immune response of his mice by injecting them with B and T cells that haven't yet encountered infection. He's using cells from the spleen, the tonsils, even from umbilical cords, so as to mimic the immune system of newborns, allowing him to investigate the onset of AIDS in children. Mosier anticipates that in the future his mice will be able to react more strongly to first infections.

So, two different approaches, two different kinds of humanlike mice. Mosier's mice are in a sense living test tubes that house human blood cells. The animals neither manufacture nor perpetuate the cells — they simply provide them a home. McCune's mice, on the other hand, more nearly resemble small, four-legged, furry humans, in that they incorporate not only cells of the human immune system but the human organs that produce and maintain them. But Mc-Cune's mice carry a heavy cost in time and effort, for whereas Mosier can produce hundreds of human SCID mice in just a few hours, simply by injecting them with human blood cells, McCune must perform laborious surgery on each mouse. And whereas the tools of Mosier's trade — human blood cells — are readily accessible, McCune must rely on the availability of aborted fetuses. It's another cumbersome process, but the early results suggest that it is well worth the effort. "The animal is one in which all of the organ systems that support human immune function are present," says McCune. "We're looking at cells actually being created by those organs, and those cells appear to function. And we've demonstrated that HIV is able to infect those organs, as it infects those organs in people."

Between McCune, Mosier, and a host of other researchers now entering the field, these mice with human immune systems may provide just the impetus AIDS research has been lacking. At least, these scientists hope so.

It is a hope shared by many researchers, part of the general upswelling of optimism that today surrounds the fight against AIDS. Dani Bolognesi of the surgical-virology laboratory at Duke University Medical Center has been working on AIDS vaccines and treatments for the past six years. As far as he is concerned, those involved in the battle have most definitely turned the corner. "I've always been hopefully optimistic," he says, "but I think now we're really moving. Admittedly, it's early. My enthusiasm shouldn't be unbounded. But I think we're looking at some exciting prospects. Enormous barriers are no longer there. I have a feeling we're going to take off from here."

Bolognesi's enthusiasm is echoed by many others, as is his sense of being involved in something important and historic. "AIDS is a tragedy," he says, "but it's wonderful to have the opportunity to work on something as important as this. We are writing medical history. That's a very strong driving force."

Retroviruses, the viruses that came to dinner — and stayed — have had an impact larger than anyone could have anticipated. They've been the prod for a new understanding of how viruses interact with one another and how they are able to live inside the cell, latent, for years. They're providing fresh insights into the nature of neurological disease and the mechanisms of how cancer is caused. In fact, retroviruses have been our entrée into the formidable and fascinating world of cancer — not only retroviral-caused cancer, but cancer in general. Retroviruses have been the instrument by which scientists have become aware of the existence of oncogenes, those normal, essential genes that mutate to cause cancer tumors. And besides that unintended service, which is allowing researchers to plumb to the private heart of the body's cells as never before and understand the workings of genes as never before, retroviruses are offering an unprecedented opportunity for us to harness the power of viruses for our own ends. Retroviruses, after all, conduct their business by infiltrating the very essence of life, our genes. They therefore allow us access to our own beginnings and our own fate. And they allow us the chance to *change* that fate.

Given all that we've been through with viruses, it only seems poetic justice that they might serve as tools to promote health rather than sickness, that they might hold the promise of better days rather than exclusively serving as the bearers of bad news. And the final irony is that not only may viruses turn out to be disease-fighting tools of unprecedented promise, they are our conduits to basic biological knowledge. It is viruses that provided us access to the submicroscopic world of genes in the first place. Were it not for viruses, we might still be on the outside looking in.

So, for the science of molecular biology and genetic engineering, thank viruses. For the structure of DNA, thank viruses. For the emerging knowledge of the origins of cancer, thank viruses. And for the promise of fighting heretofore unassailable disease, thank viruses. It is that promise that we'll explore in the next section.

IV

· · · · · · · · · · · · · · ·

The Promise of Viruses

· · · · · · · · · · · · · · ·

Fifteen

. . . .

Gods in the Laboratory

. . . .

The setting is a large meeting of microbiologists, including a number of the top AIDS researchers from around the world. The speaker is an outspoken veteran scientist, whose career has taken him from the days of pre-Enders virology in the 1940s to the intricate genetic engineering of our own time. He asks for the houselights to be raised and begins to tell a story.

It seems that a woman, marrying for the fourth time, informs her husband on their wedding night that she is still a virgin.

"A virgin?" the man replies. "You've been married three times before. How can you still be a virgin?"

"Well," the woman explains, "my first husband was an elderly man, and he died on the way out of the church. My second husband was a very nice boy, but he turned out to be more interested in the best man than he was in me."

"What about your third husband?" asks the newly married man.

"Oh, he was a molecular biologist. He spent our entire married life standing at the foot of the bed telling me how good it was going to be when it happened."

It's not a very funny story, really, but as delivered by a scientist of this stature it breaks up the house. It's his way of teasing his audience, of reminding them that while molecular biology may be the predominant biological science of our day, it is still as much potential as achievement. It may be fair to say that the great accomplishment of molecular biology to date has been in finding out, in gathering knowledge. But how that knowledge may be applied is a question just beginning to be answered.

Beginnings are always intriguing — hopeful, frightening, bewildering, stimulating. "There are just *really* exciting things ongoing," says Dennis Panicali of Applied Biotechnology in Cambridge, Massachusetts. "We're paddling as fast as we can to keep up. We're riding the crest. It's a lot of fun, a lot of tension, a lot of frustration, sometimes, but it's really exciting."

In 1937, two years after Wendell Stanley had sent shock waves through the scientific community by showing that a virus could be crystallized and so seemed to be nothing more than a tiny particle of protein, two English biochemists named Frederick Bawden and Norman Pirie discovered that yes, the virus could be crystallized, but, no, it wasn't pure protein. There was a trace of something else in there. That something was RNA. Now the question became, what in the world was this RNA doing there?

It wasn't the only question facing virologists. In fact, at that point you might have asked what questions *weren't* facing them. Viruses infected living cells — how, no one knew. They multiplied in or around those cells — how, no one knew. They went off to infect other cells — how, no one knew. And what did they look like? No one knew that, either.

So when Max Delbrück arrived at the California Institute of Technology in 1937, he had his work cut out for him. Delbrück was then thirty-one years old, a German physicist who had come to Caltech as a visiting research fellow. But he didn't come to do physics per se. For in Berlin Delbrück had become fascinated by the mysteries of heredity. And he had decided that the best way to approach those mysteries would be through the investigation of viruses. They were, after all, the simplest material known to reproduce. Perhaps investigating how viruses were able to propagate would cast some light on how human beings did the same thing. The viruses he settled upon were controversial, little-understood, and largely neglected. They were the bacteriophages.

Years before, Félix d'Hérelle had developed a technique in which he grew a thick culture of bacteria on a base of solid agar. When he spread a solution containing bacteriophages onto this bacterial "lawn," it soon developed clear spots, or "plaques." Each plaque, d'Hérelle contended, was actually a hole in the lawn of bacteria formed by bacteriophages that had infected and killed the microorganisms in that particular spot. Therefore, each plaque represented a colony of bacteriophages, all of which were descended from a single parent phage. That being so, it was possible to estimate the number of phage particles present in any particular culture simply by counting the number of plaques — the more phages, the more plaques; the fewer phages, the fewer plaques. And since the plaques were separate from one another, the experiment showed also, once and for all, that bacteriophages were indeed distinct, self-contained particles — "discontinuous," in d'Hérelle's terms. In other words, they were individual viruses.

The first thing Delbrück did was to expand upon d'Hérelle's plaque observations. In 1939, with Emory Ellis, another research fellow at Caltech, Delbrück devised an experiment that was to give the clearest picture yet available of what happens during a viral infection. For in the late thirties, it was known only that viruses *did* infect cells; how they went about it, and how quickly and to what extent they went about it — all that was unknown. Delbrück and Ellis's procedure was called *one-step growth*. They mixed bacteria and phages in a test tube filled with nutrient broth. Then, while the phages were busy infecting the bacteria and beginning the mysterious process of reproduction, Delbrück and Ellis stopped them in their tracks by taking a portion of the solution, placing it in another test tube, and diluting it a thousandfold. They then repeated the procedure in a number of other test tubes. They now had solutions that contained a relatively sparse mix of phages, uninfected bacteria, and already infected bacteria — mixtures so dilute that the phages simply could no longer find any fresh bacteria to infect. Therefore, nothing new could happen in the solutions. Except in the case of the already infected bacteria, in which the process of phage replication was beginning, they had brought things to a stop. That was the first step.

Next, every couple of minutes, Delbrück and Ellis emptied one and then another test tube onto petri dishes covered with fresh lawns of bacteria. If the sample contained any phages or infected bacteria, they would soon produce plaques in the bacterial lawn — Delbrück called them *infective centers*. By counting the number of plaques, each of which was produced by one phage or infected bacterium, the researchers were able to determine how many phages and infected bacteria there were in each portion. Now things began to get interesting. The first dilute sample they spread onto the bacterial lawn, taken from the test tube after just a couple of minutes, produced relatively few plaques. So did the next sample, spread onto a petri dish a couple of minutes later; and the next, a couple of minutes after that; and the next, another couple of minutes later. In fact, for the first twenty-four minutes' worth of samples, the number of plaques remained relatively stable. When Delbrück and Ellis made a graph of the results, it turned out to be a horizontal line.

But at twenty-four minutes, something dramatic happened: the number of plaques increased astronomically and continued increasing for the next few minutes' worth of samples. The line on the graph ascended precipitously, heading for the stratosphere. Then, about ten minutes later, the numbers stabilized and the line leveled off again. Whatever had caused the explosion of plaques had run its course. The result was a hundred times as many phages as before.

The significance was clear: rather than a phage's infecting a bacterium and producing new phages immediately and continually, as some had suspected, something had been going on within the infected bacteria for those twenty-four minutes, after which time it burst open with phages, a hundred of them in each bacterium, until after a few minutes more the process ended as abruptly as it had begun. It was, then, a three-part process: infection, then a latent period in which the virus multiplied a hundredfold *within* the unfortunate cell, followed by a release period during which the young phages exploded into the outer world.

Delbrück could now be certain that whatever it was that the virus did to reproduce, it happened *inside* the cell (it would have continually produced more and more plaques, otherwise), and it was an astonishingly efficient operation. An infection could involve an exponential increase in viruses — a disquieting thought. Yet, for all it revealed, the one-step growth experiment could not answer the essential question of *how* the phages managed to replicate within the cell.

Delbrück was a thin, angular fellow, with a serious demeanor, a dry wit, a love of practical jokes, and an ability to anticipate the value of any particular approach. With bacteriophages, he knew he was onto something good: solid information about heredity and reproduction. Delbrück also could see, however, that he was not going to be able to find everything out by himself. His one-step growth experiment had resurrected the investigation of bacteriophages; now the field must become more populated. What was needed was a school of phage study, a community of phage investigators. To that end the times cooperated, for war had recently broken out in Europe, and, like Delbrück, other scientists began leaving their homelands for the safety of America. One of them was an Italian microbiologist named Salvador Luria, who in 1940 came to Columbia University as a research assistant. Luria and Delbrück met shortly thereafter, marking the beginning of a loosely aligned but tightly focused cluster of like-minded scientists that came to be known as the "phage group." "It wasn't really a cohesive group," remembers Renato Dulbecco, a future Nobel laureate who arrived at Caltech from Italy in 1949 to become part of the effort. "It was rather a church. A phage church" — with Delbrück as "the pope."

Soon after Delbrück and Luria formed their band of two, a quiet, dedicated young RCA electron microscopist named Thomas Anderson was drawn into the circle. The electron microscope had only recently been developed in Germany, and Anderson remembered that when he first heard of it "it almost seemed to be a hoax perpetrated on the rest of the world by the Nazis." For how could anything be that good? Under the "lens" of the electron microscope, the invisible viral world could become visible. In early 1942, when Luria brought Anderson solutions chock-full of bacteriophages, phages clearly swam into view. Tails they had, and heads. Their physical characteristics caused investigators, Delbrück among them, to speculate that perhaps, like a tadpole, or sperm, a phage swam forward by means of the whiplash action of its tail. And that, again like sperm, it infected by burrowing its head into the bacterium, where it began the process of replication. Delbrück went on to suggest that, just as the human egg when fertilized by a single sperm prevents any other sperm from penetrating, a bacterium similarly shuts out other bacteriophages upon being infected by the strongest and quickest phage.

It was an appealing analogy, familiar and understandable. And it was reinforced by Anderson's electron micrographs that showed no phage particles inside an infected bacterium but a number of them loitering outside, as though having been rebuffed from entering. On the other hand, the analogy had certain conspicuous holes in it. As Anderson commented, "Of course the analogy breaks down completely when we consider the products of the two reactions for phage-infected bacteria are lysed [ruptured] with the production of more phage while the fertilized egg has never been observed to break down with the production of more sperm." Besides that, electron micrographs were ambiguous concerning which end of the phage seemed to attack the cell. Some showed phages moving in headfirst, as might be expected, but others showed phages that seemed to attach to the bacteria by their tails. By their tails! Phage researchers, it was becoming more and more clear, were playing with new rules.

So as the war ended and the decade of the forties passed, Delbrück, Luria, Anderson, and their colleagues, more numerous since 1945, when Delbrück began a yearly course in bacteriophages at Cold Spring Harbor, were confronted by a most unlikely and outlandish creature. A submicroscopic, spermlike virus that attacked bacteria perhaps headfirst, perhaps tailfirst; that somehow — for there was no trace of the infecting phage inside the cell — created within the microorganism a hundred new phages in little less than a half-hour.

In 1944 Oswald Avery published his smooth-pneumococcus / rough-pneumococcus experiment that suggested that the "transforming principle" of life just might be the nucleic acid called DNA. By 1950 it became known that bacteriophages were chock-full of

DNA. It resided, in a long strand coiled around itself like a ball of twine, in the head of the phage. When liberated by chemical means, the DNA left behind an empty, depleted head, which Anderson called a *phage ghost*. That same year, the problem of whether phages infect headfirst or tailfirst was solved also, as Anderson succeeded in making electron micrographs that clearly showed the phage particles attached to the bacterial cell by the tips of their tails. As before, none of the phages seemed to enter the cell, but now Anderson could see that a number of them displayed empty heads — they were phage ghosts, relieved of their burden of DNA. Where the DNA had gone was still a mystery, but now the researchers began entertaining an intriguing possibility. "I remember in the summer of 1950 or 1951," Anderson later recalled, "hanging over the slide projector table . . . in Blackford Hall at the Cold Spring Harbor Laboratory, discussing the wildly comical possibility that only the viral DNA finds its way into the host cell, acting there like a transforming principle in altering the synthetic processes of the cell."

It was still a "wildly comical possibility" because Avery's findings hadn't had the impact that now, in retrospect, we know they deserved. But if Avery's work didn't convince, it still served to bring to the level of speculation the possibility that DNA might be the transforming principle. "Sometimes things have to mature," remembers Dulbecco. "There was enormous discussion about Avery's experiment. Perhaps it put in the back of lots of people's minds the possibility that DNA *was* the carrier of information — although they didn't admit it. So when this other evidence came up, then it exploded, and everybody accepted it."

"This other evidence" was supplied by Alfred Hershey. Hershey had met Delbrück in 1943. In 1952 Hershey was at Cold Spring Harbor, a gaunt, shy, and withdrawn forty-four-year-old microbiologist who had been galvanized by Anderson's pictures of phages with empty, DNA-depleted heads attached to bacteria by their tails. He also recently had received from a colleague named Roger Herriott an extremely evocative letter: "I've been thinking — and perhaps you have, too — that the virus may act like a little hypodermic needle full of transforming principles; that the virus as such never enters the cell; that only the tail contacts the host and perhaps enzymatically cuts a small hole through the outer membrane and then the nucleic acid of the virus head flows in the cell." Science fiction, to be sure. Almost a cartoon image. But by this time, with phages, as with viruses in general, anything seemed possible.

The letter's seemingly outlandish suggestion — "the virus as such never enters the cell" — was corroborated by another member of the phage group, A. H. (Gus) Doermann, of Vanderbilt University. Curious as to what went on within infected bacteria during the twenty-four-minute latent period before they exploded with hundreds of new phages, Doermann decided to try to open up the bacteria at various stages along the way to see what was inside. To do so, he repeated Delbrück's "one-step growth" experiment, but instead of spreading the diluted samples directly onto bacterial lawns in petri dishes, he mixed each solution with cyanide first. The cyanide broke open the bacteria but didn't harm the phages. In effect, he had made a window into the bacteria so as to see what was inside at various stages in the process.

Doermann found that for the first ten minutes of the experiment there were no phages inside the bacteria. It was only afterward that phages began appearing, gradually increasing in numbers, until finally, twenty-four minutes later, a hundred phages burst forth. Doermann called the barren early period the *eclipse*.

But what was going on during those quiescent first ten minutes? To all intents and purposes, the bacteria then were completely empty of any trace of phages. Not only did no viruses seem to enter the cell, *nothing* seemed to enter the cell. How could new phages spring full-blown from such a vacuum?

Hershey, with the help of colleague Martha Chase, decided to find out. What followed has been called "the experiment that revolutionized biology." It certainly led directly to Watson and Crick's determination of the molecular structure of DNA just a year later. It established the essential role of DNA in heredity, its function as the carrier of genetic information. It vindicated Oswald Avery and his colleagues. Yet while it led to the high-tech world of genetic engineering, this momentous experiment was performed in an ordinary Waring blender.

Hershey and Chase wanted to find out what happened to DNA

carried in the phage heads once the virus attached itself to a bacterium. Where did it go in the process of leaving the phages behind as ghosts? Electron microscopy didn't provide the answer, as Anderson's pictures could show phages initially attaching to the cell, their heads full of DNA, and later the ghost phages depleted of DNA, but they couldn't show what happened in between. Something ingenious was needed, and, luckily for Hershey and Chase, that something was at hand: radioactive tracers.

Recall that in the experiment in which he discovered reverse transcriptase, David Baltimore tracked a particular chemical by making it radioactive. That way, in effect, he could trace it with a Geiger counter. Hershey and Chase, who preceded Baltimore by almost two decades, did much the same thing. They knew that the DNA in the phage's head contains phosphorus and that the body of the phage, which is made of protein, contains sulfur, but that the reverse is not the case — that is, DNA contains no sulfur and protein contains no phosphorus. So, in 1952, Hershey and Chase added radioactive sulfur to one set of bacteria and radioactive phosphorus to another set. Then they infected both sets with bacteriophages. As the phages multiplied within the bacteria, they incorporated the radioactive chemicals. The result was two different kinds of phage, one whose headful of DNA now contained radioactive phosphorus, and another whose protein, which comprised the body of the phage, contained radioactive sulfur. These two essential phage components had become radioactively marked. Now, when the phage infected new bacteria, Hershey and Chase would have a way of tracing what happened to its cargo of DNA as well as its protein container.

The next step was to infect fresh bacteria and then separate them from the phages so as to trace the radioactivity in both. Hershey and Chase mixed together the radioactive phages and new bacteria, waited until the phages had enough time to infect, then poured the solution into a Waring blender and let it spin. Using a blender — and a borrowed one, at that — was a last resort, for their other attempts at separating phages from bacteria simply hadn't worked.

After a few minutes, the researchers emptied the contents of the blender into a centrifuge for another brief spin. When they turned off the machine, they had two distinct parts: a solid pellet of infected bacteria and fluid containing any particles smaller than bacteria — which they hoped would include the smaller, lighter phages.

Now came the moment of truth: Hershey and Chase traced the radioactivity in both parts. What they found was that in the pellet containing bacteria there was no radioactive sulfur — in other words, no phages. In the solution, however, there was lots of radioactive sulfur — a great number of phages. The shearing forces of the blender had indeed torn loose the phage particles from the bacteria.

But what about the radioactive phosphorus — in other words, the DNA? It turned out that there was no sign of phosphorus in the solution containing phages, whereas the bacteria was loaded with it. The results could mean only one thing: the phages had deposited their DNA inside the bacteria, while retaining their protein. Science fiction had become science fact. A phage did indeed act like a hypodermic needle, neatly injecting its DNA into the bacterial cell.

So much for the question of where did the DNA go: it went directly into the bacteria. But there was more. Hershey and Chase found that the infected bacteria, with the ghost phages whisked away by the blender and therefore thoroughly out of the picture, *still produced new phages*. The protein body of the hypodermic needle, in other words, had no role in replication other than that of injecting the DNA into the cell in the first place. DNA was all that was needed for phage replication. Once the DNA was safely inside the cell, the depleted phage was expendable. "The phage protein removed from the cells by stirring consists of more-or-less intact, empty phage coats," Hershey wrote, "which may therefore be thought of as passive vehicles for the transport of DNA from cell to cell, and which, having performed that task, play no further role in phage growth." The implication couldn't be clearer: the medium of replication, the carrier of genetic instructions, the agent of heredity, the transforming principle itself, was DNA.

So *that* was the reason for the fallow period Doermann had discovered, the so-called eclipse. There was no sign of phages inside infected bacteria during the first ten minutes because the only thing inserted into the bacteria was DNA. It took those ten minutes for the DNA to instruct the bacterial cell to produce more phages.

Avery had been right. Just about everyone else had been wrong.

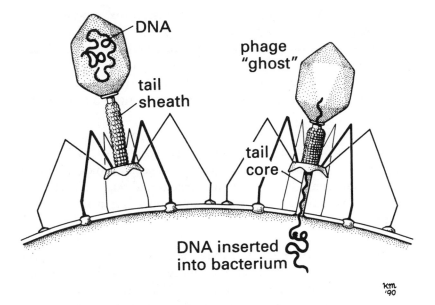

Bacteriophage and how it infects

T-4 bacteriophage has a sophisticated way of getting its DNA into the bacteria it infects. First the virus's tail fibers adhere to proteins on the bacterial surface; then the tail sheath contracts, sending the tail core plummeting into the cell wall. A series of chemical reactions help breach the cell wall, and the phage injects its DNA — hypodermic-needle style — into the bacterium. The empty phage "ghost" remains on the surface.

Rather than simply some neutral, boring, almost irrelevant substance, DNA was the giver of life, the sum and substance of genes themselves. Now it was protein that was relegated to the role of vehicle, of protective envelope for the essential nucleic acid.

It was not an event welcomed by everyone. Microbiologists had become used to dealing with protein, had built careers out of investigating protein, but there was nothing anyone could do about it. And when in 1956 Heinz Fraenkel-Conrat of Wendell Stanley's University of California virus laboratory and, independently, two German researchers, A. Gierer and J. Schramm, proved that when it came to an RNA virus such as tobacco mosaic, the same was true as for phages — nucleic acid carried the genetic instructions — the matter was settled for good. Nucleic acid, whether DNA or RNA, was the stuff of life. In 1969 Max Delbrück, Alfred Hershey, and Salvador Luria shared the Nobel Prize for their ground-breaking efforts. The lowly bacteriophage had indeed performed an invaluable service.

"The identification of DNA as *the* essential viral constituent, the determination of its structure, and the recognition of its informational content opened a new branch of molecular biology," wrote Thomas Anderson. "When we finally recognized this . . . it was like finally hearing the hilariously improbable punch line at the end of a long preposterous tale."

If Anderson had only known at the time how improbable and preposterous the tale would become. With the discovery that DNA and sometimes, in viruses at least, its sister, RNA, carry genetic instructions, and the realization that viruses offer an unmatched access to this source of life, the possibility arose not only of examining at close range the very heart of all life, but of *manipulating* it. Studying life at close range is one thing; controlling it is quite another. But controlling it is what genetic engineers set out to do. In effect, genetic engineers mimic nature. Just as nature, over eons, alters genes of organisms so as to construct life that is best able to perpetuate itself — the process of evolution — so genetic engineers, over a comparatively minuscule time period, alter genes to construct life that is best able to perform the task they have in mind. It's a kind of force-fed evolution. The results of such manipulations are already numerous despite the short history of this science, and are rapidly becoming much more so. The production of large amounts of insulin in bacterial "factories" constructed by splicing human insulin genes into the genetic material of bacteria is perhaps the best known of the products of genetic engineering. Other such mass-generated products are interferon; food-generating enzymes such as rennin, which is used in making cheese; growth hormones to increase the milk production of cows; and microorganisms that break down garbage and other wastes into alcohol or natural gas.

This ability to mimic nature by altering genes and manipulating life has also led to the development of brand-new kinds of viruses that show promise in performing the tasks *we* want them to. Turnabout is fair play. It is a delicious reversal of roles. Now viruses are being used to fight the diseases they themselves cause.

Old Edward Jenner stumbled onto a gold mine when he had the presence of mind to use the virus that causes cowpox to produce a vaccine against smallpox. It was an inspired strategy in more ways than one, because not only did it lead to the eradication of smallpox, the only disease as of yet eliminated by human hand, but it provided the scientific community with a most interesting and accommodating virus. Interesting, in that vaccinia, as we've seen, is a huge virus; it measures about 1/100,000 of an inch on a side, over twice the size of other large viruses such as EBV and influenza and almost ten times the size of small viruses such as poliovirus and rhinovirus. It is so large that it is one of the few viruses that can be seen through an ordinary light microscope — although it shows up as little more than a dot. Like our own cells, vaccinia carries genes of double-stranded DNA and enzymes capable of expressing those genes. They have a great deal to express, as the genome in vaccinia extends to approximately 187,000 nucleotides. By comparison, influenza viruses carry about 14,000 nucleotides and poliovirus 6,000. This long gene strand has the capacity to make up to two hundred different viral constituents, including an unusual growth factor that enhances vaccinia's ability to infect. So vaccinia is not only one of the largest viruses known, it's one of the most complex.

But not so complex that researchers haven't turned to vaccinia as a foremost candidate for the brave new world of disease prevention — recombinant vaccines. "There are several unique properties of vaccinia virus which make it attractive as a vaccine vehicle," says Dennis Panicali. "The first is that it has been used for almost two hundred years as a vaccine, so it's well known as to what effects it has in the human population. It's a very stable virus. Freeze-dried, vaccinia can survive for years and years, which is relatively unique for viruses. That allowed it to be used in the Third World, in Africa and Asia, because they didn't have to worry about maintaining a cold chain. It's also very easy to administer by scratching the skin. That may not be the way to do it with recombinant vaccines — we might use a vaccine gun — but either way it's certainly easy to administer. Vaccinia is a virus that grows like a weed in many cell cultures, and very quickly, so that's going to reduce the cost of manufacturing. It'll be by far the most inexpensive vaccine to produce. And it's a very big and complex virus, but not so complex that we haven't been able to figure out how to regulate certain aspects of its replication. So from a genetic-engineering standpoint, it's relatively easy to manipulate. And since it's so big, you can put lots of material into it."

After all we've been through with viruses, imagine speaking of the creatures as though they're some kind of usable commodity! Yet that's exactly what vaccinia is proving itself to be. Panicali is one of the pioneers in remodeling vaccinia virus so that it can serve in a vaccine that will protect against more than simply smallpox. It is to be a kind of all-purpose vaccine, a skeleton key vaccine that will fit whatever ails you. The idea is to add genes from other viruses to vaccinia's already abundant supply of DNA, with the expectation that vaccinia will then dutifully express those imported viral characteristics on its own surface, thereby inducing the body to prepare an immune defense against a host of possible invaders, rather than just one. Vaccinia therefore becomes a man-made hybrid virus, the subminiature equivalent of the mythological chimera. The virus is so large, in fact, it might be able to accommodate genes from as many as twenty other viruses and still have room to express those genes on its surface — if such largess is appropriate. "I know Enzo talks about putting in twenty antigens at one time," says Panicali, "but I don't think it's necessary. You can put five in one, five in another, and so on."

Enzo is Enzo Paoletti of the New York State Department of Health. He, Bernard Moss at NIH, and Panicali are spearheading the effort to develop viable multiuse vaccinia vaccines. And, like so much in science, the effort began by mistake. "In 1980 I was in Albany, working with Paoletti," explains Panicali. "I was doing some work on cloning DNA from vaccinia, studying transcription in the virus, and I found that there was a large section of DNA missing from one of our virus stocks. It was probably a spontaneous mutation. So as an exercise in recombinant techniques, Enzo and I decided to see if we could put those genes back in."

Panicali, now forty-one years old, is of middle height, with a trim mustache, curly dark hair, and an enthusiasm that intensifies the more he talks about his work. "Those missing genes didn't seem to affect how the virus replicated in tissue culture — they just weren't necessary in that stock of virus — but we put them back in anyway, just to demonstrate that it could be done in vaccinia. No one had been able to do that before."

Then, having shown that it was possible to return missing vaccinia genes to the virus, a thought came to the two researchers: what about inserting genes from other viruses? What about the possibility of making vaccinia into a vaccine vector? Just as a mosquito, for example, can transmit microorganisms such as malaria protozoa and yellow-fever viruses, so a virus itself might be able to transmit characteristics of other viruses. As the mosquito is a vector for causing disease, the virus could be a vector to prevent disease. "People were putting foreign DNA into all sorts of things," says Panicali. "Into bacteria, yeasts, even into other viruses. We said, 'Could we use vaccinia as a vector system as well?' So we decided to try and do that."

For Panicali and Paoletti it was the beginning of a direction that has dominated their research since. They took a gene from a herpesvirus and found that it readily set up shop within vaccinia's DNA and directed the virus to make a herpes enzyme, just as though it were in residence back home in the herpesvirus where it started. Now the old vaccinia virus was a vaccinia virus no longer. It was something new, a hybrid virus — a little of this, herpes, and a lot of that, vaccinia. And now it was as well a vector, a vehicle for *transporting* a part of the herpesvirus to the places vaccinia likes to go.

"From there," Panicali remembers, "we thought that if we're able to do that, vaccinia, being the benign smallpox vaccine that we've all had, might be a logical carrier for other antigens. Perhaps we could use vaccinia virus as a Trojan horse, perhaps carrying genes from herpesviruses to vaccinate against herpes simplex, or maybe carrying other genes. Then we started putting in a variety of different genes, different antigens, with the hope in mind of making a vaccine."

At about the same time, Bernard Moss began working on a similar idea. "When I initially started thinking about this type of vaccine approach, I looked at it as a people's vaccine for China or Asia." Moss, who is somewhat ill at ease talking publicly about his work, grins sheepishly from behind his beard and glasses, embarrassed at the presumption of such a scheme. "I got the idea from the smallpox vaccine. That was a vaccine manufactured in the countries in which it was used. It was a low-technology operation. In fact, they made the vaccine in the skin of live animals. They vaccinated them, essentially, but over a wide area in the skin, and after a week they removed liquid and with a few steps put it in bottles, and that was the vaccine.

"So my thought was that although it's high technology to make the recombinant virus in the first place — no question about that — it's no longer high technology just to make more of it. Once it's made, it could be propagated just like smallpox vaccine. In fact, once you've made one vial of vaccine, and you send it out, people could continue to make it. They'd simply have an animal make more of it."

For example, China has a high incidence of hepatitis B — about one in ten people suffers from the disease. It's a country that could benefit enormously from a hepatitis B vaccine. "They really need a billion doses of vaccine," says Moss, "and it's difficult to think how in the next ten years they could get that many doses either by conventional ways or other kinds of recombinant ways. But they sure could get a billion doses of this."

So with that ambitious thought in mind, Moss started working on making a vaccinia virus that would carry with it hepatitis B antigens. The technique he used was identical to that of Panicali and Paoletti in upstate New York. To alter the virus, the researchers used various kinds of animal cells as vaccine factories. First, visualize the DNA of the viruses — no matter whether of vaccinia, or of other viruses like hepatitis B or herpes — as long ribbons of genes. The scientists then locate on this DNA ribbon the gene they're interested in adding to vaccinia — let's say a gene from a hepatitis B virus.

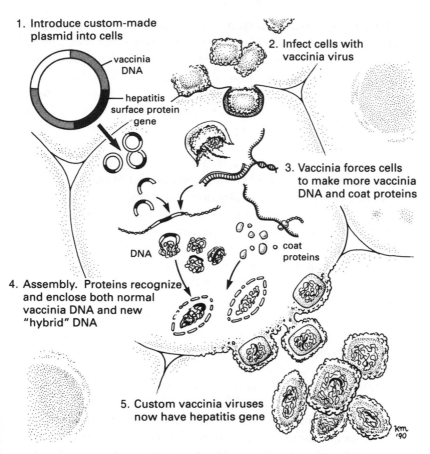

1. Introduce custom-made plasmid into cells

vaccinia DNA

hepatitis surface protein gene

2. Infect cells with vaccinia virus

3. Vaccinia forces cells to make more vaccinia DNA and coat proteins

DNA

coat proteins

4. Assembly. Proteins recognize and enclose both normal vaccinia DNA and new "hybrid" DNA

5. Custom vaccinia viruses now have hepatitis gene

How to make a hybrid vaccinia virus for use in new vaccines

Vaccinia virus has been used for many years in a vaccine for smallpox. Researchers have recently customized the virus (steps 1–5) as a vehicle for carrying genes of other viruses, in an attempt to develop safe, effective vaccines against other diseases.

Then, using special enzymes able to dissolve and reassemble the strand of DNA — in effect, a pair of scissors with which they can cut the ribbon wherever they like and tape to put it together again — they snip out the gene and splice it into another ribbon of DNA, this one a small, circular strand called a *plasmid* that's often found floating in bacteria.

Next, the scientists snip from a vaccinia DNA ribbon a number of genes, sometimes quite a lengthy strand, which they then proceed to splice into the plasmid loop *on each side* of the hepatitis B gene. So now within the plasmid loop the foreign gene is flanked at both ends by vaccinia DNA. If a vaccinia virus were to come along, it would feel right at home with this plasmid and its vaccinia DNA. In fact, it might overlook the single hepatitis B gene altogether, which is precisely the idea.

In effect, then, the scientists are trying to trick the vaccinia virus into recognizing the plasmid DNA ribbon as part of its own. In fact, they're depending on that recognition, for the next step is to insert the plasmid loop into a living cell in tissue culture, then infect the cell with a vaccinia virus. When the virus begins to replicate inside the cell, it will more likely than not enclose the plasmid in its own DNA ribbon, thereby integrating the hepatitis B gene and its vaccinia DNA escort into its own complement of DNA. When the cell begins spitting out new vaccinia viruses, now with the hepatitis B gene on board, the scientists harvest them for use in a vaccine. The entire process is accomplished in a matter of weeks. It's that simple, and that amazing.

But while vaccinia is a most amenable virus, it's still a virus. And that means that infection with vaccinia poses some risk, usually of a cowpox-like skin eruption. Those most likely to become infected are people whose immune systems are impaired. In those people, the virus can cause serious complications. For example, recently a military recruit was routinely vaccinated against smallpox with ordinary vaccinia virus. Unbeknownst to the doctors who vaccinated him, the recruit had AIDS. One of the hallmarks of AIDS, of course, is a deficient immune system. As a consequence, the recruit became seriously ill. And it's not only AIDS that can cause an impaired immune system; other circumstances — chemotherapy, even skin diseases like eczema — can open a person to complications from vaccinia infection.

"The risk of complications is generally between ten and a hundred per million people vaccinated," explains Panicali. "One hundred if you consider some of the more minor complications, fewer and fewer if you get to the more serious ones that can result in death — vaccinia encephalitis and other things. So there is a level of complication that may be greater than that of certain other live-virus vaccines, greater than that of the Sabin live-polio vaccine, for example."

One way of dealing with the problem is to confine the recombinant vaccine to diseases that are themselves so dangerous that some risk is acceptable compared to what might result from a bout with the disease. AIDS is certainly such a disease. Malaria is another; over ten million people a year die of that venerable affliction. Hepatitis B, especially in high-risk areas such as China, might be another. At the same time, however, vaccinia researchers are trying to engineer a safer vaccinia virus. In contrast to poliovirus, for example, vaccinia during all the years of its use as a vaccine has never been attenuated so as to weaken its impact. Panicali's Applied Biotechnology, as well as other labs, are attempting to do just that — this time, however, by recombinant methods. By figuring out the function of a number of vaccinia's genes and snipping away or altering those that are involved in causing complications, they may be able to fashion an even more cooperative virus.

"There are certain functions of the virus that need to remain," Panicali says. "You can build a million different types of houses and buildings, but you need a solid foundation, you need certain structural members, it's nice to have a window here and there. So certain core components will remain within the virus, but we hope to make it safer and more efficient."

To that end Bernard Moss and his colleagues at the NIH have taken yet another tack. They've transplanted into vaccinia another set of genes, those that produce interleukin-2, the immune-system component and growth factor. When they vaccinated SCID mice with this altered vaccinia virus, they found that the ordinarily defenseless mice were protected against any possible side effects.

So the outlook for vaccinia-based vaccines is bright. Already genetically manipulated vaccinia viruses have produced immune protection in animals against a number of diseases, including hepatitis B, rabies, herpes, influenza, and measles. In humans, a vaccinia-based vaccine against AIDS produced by Oncogen, a Seattle subsidiary of Bristol-Myers, has been tested in over twenty volunteers, none of whom suffered any ill effects from the vaccine, while at the same time the majority have developed signs of immunity to HIV. However, it's as yet too early to know if the immunity is strong enough to prevent the disease — a familiar story at this stage of AIDS-vaccine development.

While the vaccinia in Oncogen's AIDS vaccine expresses proteins from HIV on its own surface, thereby inducing the body to build up an immune defense against the real virus, Applied Biotechnology is using vaccinia to make whole HIV that lacks only its core of nucleic acid. They do so by infecting cells with specially altered vaccinia that express all the structural and envelope proteins of HIV — but *not* its genes. The cell, tricked into reacting as though it were an authentic HIV invasion, dutifully constructs what look to be real HIV particles. However, the particles are noninfectious because they carry no nucleic acid. They're simply empty shells, albeit convincing ones. "We hope to stimulate an especially effective immune response," says Panicali, "because we have the whole antigen, the whole virus there, without any potential of reverting to virulence." Other projects under way in various labs include vaccinia vaccines against melanoma, tuberculosis, and leprosy.

Yet as promising as vaccinia vectors may be in the fight against specific diseases, they may be even more valuable for their role in disclosing the workings of the immune system. The control offered by recombinant vaccinia makes the virus a finely tuned tool for exploring the wilds of the immune response. For example, researchers at Applied Biotechnology have made a series of genetically engineered vaccinia viruses that each express specific aspects of HIV, as though the virus, like a jigsaw puzzle, had been broken apart into its separate constituents. Into one vaccinia virus they've transplanted the gene that produces protein from the core of HIV; into another has gone the gene expressing an HIV envelope protein; into yet another, the gene that makes HIV's reverse transcriptase. In collaboration with the University of Massachusetts Medical School, the researchers then infect cultures of human immune cells, B and T lymphocytes, with each virus and examine the cells' response. And working with the New England Regional Primate Research Center, they do much the same thing with vaccinia-based parts of SIV, simian immunodeficiency virus, which they inject into macaques. In this way they can monitor the immune system's response to specific components of the virus.

"We can measure what types of immune responses a particular protein is eliciting," explains Panicali. "When only one protein is expressed, does it elicit a cellular immune response — natural-killer response, cytotoxic-lymphocyte response — or does it elicit the production of antibodies? We can really pinpoint who does what to whom. We're using the vaccinia virus to explore the workings of this inner world. The human immune system is extremely complex, and we only understand a fraction of it."

The benefits of this knowledge are wide-ranging and difficult to pin down. In the short run, researchers may be able to use a heightened awareness of the impact of various viral proteins to design the most effective possible vaccines. And in the long run, the advantages of this kind of research may include a fuller understanding of the way our bodies work in general. It's in that larger realm that viruses may offer their greatest boon.

"Viruses are helping to unfold basic life processes," says Panicali. "We're learning more and more about the normal functioning of cells, primarily by monitoring what goes *wrong* with those cells, and in particular how viruses affect what goes on in those cells. It's hard to tell what's going on in a cell when everything is going perfectly fine, but when something starts going wrong you start picking out what's going on and seeing where that fits in in the overall scheme of things. Viruses allow us to do that.

"Viruses are just exquisite parasites," he adds. "They get in there and start modifying and manipulating the machinery of the cell. They're very, very valuable tools for understanding the biology of cellular regulation and growth."

Yet it may be that in this fantastic world of viral manipulation and recombination, viruses themselves — that is, naturally existing viruses — are on the way out. Vaccinia, no matter how valuable a tool for vaccine delivery and cellular probing, has the aforementioned drawback of being an infectious organism. Researchers may be able to attenuate it and modify it and reduce its ability to cause woe (nominal though it may be), but still vaccinia is a virus. Its job is to infect. So if the techniques are at hand, the skill is there, and the motivation is high, it's a natural step for these manipulators of the genes of viruses to manipulate them right out of the picture. The next step may be man-made, synthetic viruses that pose absolutely no risk at all.

"I think that the vaccinia virus will have an immediate role, maybe from now to fifteen years from now," says Bernard Moss. "It's not the ultimate. The ultimate is when we learn enough, we'll do everything ourselves. It'll all be synthetic. We won't have to use a live virus, we'll construct it ourselves. The ultimate will be beyond nature."

For W. French Anderson, going beyond nature has become a way of life. Anderson, chief of the molecular-hematology lab at the National Heart, Lung, and Blood Institute, and his colleagues are involved in an activity that has enormous potential to alleviate human suffering. Yet it is easily as controversial as it is promising: gene therapy.

For many people, those two words are enough said. The idea of mucking about in our genes, altering the way nature, or God, has constructed us, is not only dangerous but unethical and immoral. It's an objection that Anderson is well aware of. A trim fifty-three-year-old in his white lab coat, with a full head of graying hair and a direct, businesslike mien that attests to an astonishingly active life and an abhorrence of wasting time, Anderson has been fighting the gene-therapy battle ever since he wrote in his application to Harvard that he wanted to understand genetic disease at a molecular level. That was in 1953, just one year after Hershey and Chase had demonstrated conclusively that DNA is the carrier of genetic information, and the same year that Watson and Crick gave the world the double helix as the molecular structure of DNA. It was a bit early in the game to be interested in the molecular properties of genetic disease, especially for a sixteen-year-old from Tulsa, Oklahoma (even if this particular sixteen-year-old had been reading college science books for fun since the age of nine). And even a few years later, when Anderson was about to graduate from Harvard, it was still early enough that DNA was not yet fully recognized as the stuff of genes. "The Watson-Crick discovery was still very controversial," Anderson remembers. "My undergraduate thesis at Harvard dealt with genes in bacteria, but the term I used was *transforming factor*. It was DNA, but it was still called 'transforming factor.' "

By this time Anderson had homed in on the possibilities of gene therapy, in no small part influenced by taking the first course offered by a microbiologist also interested in genes who had just come to Harvard from the Cavendish Laboratory at Cambridge University in England. It was James Watson. After graduation in 1958, Anderson reversed Watson's journey and took up residence in the Cavendish, where he worked with another microbiologist interested in genes, Francis Crick. "It was an exciting time in the fifties," Anderson remembers. Suddenly he laughs. "Probably the reason I don't bother with large offices, which I could have if I wanted, is that when I first joined the Cavendish I was given one square foot of space. Crick and Sydney Brenner shared an office that was so tight that when Crick wanted in Brenner had to leave the room — there simply wasn't enough space. When Crick came, Brenner got out and Crick went in, then Brenner came back in.

"Space was an absolute. But I was delighted. We were given one pipette of each size, and if you broke it you had to go check out another pipette. Keeps you humble. Nowadays I find it easy to stay conservative in terms of what equipment we buy. I don't care about the latest versions of things with the latest bells and whistles. You not only don't need it, it distracts you."

By 1965, after having returned to Harvard, where he trained as a pediatrician, he was at the NIH to stay, working in the lab of Nobel laureate Marshall Nirenberg, whose group was the first to crack the genetic code. It was Anderson who rose before two thousand people in April of 1966 to announce the findings. By 1968 he had his own lab devoted to gene therapy, and by 1980 he was successfully employing existing techniques to inject — or "transduce" — genes into animal cells.

Coincidentally, 1980 was also the year that Richard Mulligan finished up his PhD at Stanford and returned to MIT, where he had done his undergraduate work. Coincidentally as well, Mulligan's dissertation was entitled "Transduction of Genes into Animal

Cells." And not so coincidentally — for, as he says, "that's my business, retroviruses" — Mulligan began to work on developing a means by which retroviruses might be used as vehicles to transfer genes into cells. For although it worked well, the existing technique of injecting genes into individual cells by hand was simply too slow and laborious a task. It took about one minute to insert a gene into any one cell, a lapse of time that seems quite short until you consider that the body is comprised of billions of cells. If you wanted to transfer genes manually into 100 million cells, say, not an unreasonable number at all as experiments go . . . well, the limitations were obvious. Retroviruses, on the other hand, offered the chance of making such a chore not only possible but reasonable.

"What fascinated me about viruses," says Mulligan, "was the possibility of manipulating them." Now thirty-six years old, long and tall, with reddish hair and beard, Mulligan cuts a raffish figure in his flannel shirt and blue jeans as he leans far back in his chair at the Whitehead Institute near MIT. He is the consummate molecular biologist — young, inordinately bright (he has been the recipient of a MacArthur Foundation "genius" award), and drawn to viruses as finely tuned molecular machines there for the tinkering. "Viruses have very sophisticated and unique programs for entering cells and multiplying. For example, there's the monkey virus SV-40, which goes into a cell, replicates, makes a million copies of its genome, in the course of infection kills the cell, and then comes out for the next infection. That's a virus that provides a good way to make lots of product in a short time. Lots of DNA, lots of RNA, lots of protein.

"Then you go to vaccinia, which is a totally different virus. It's obvious for a vaccine. And then to a retrovirus. Obvious for gene transfer. It does exactly what you want. You couldn't possibly have conceived of a better life cycle for effecting gene transfer."

For, as we've seen, in contrast to most other viruses, which simply commandeer the reproductive machinery of the cell to make new viruses and let it go at that, a retrovirus actually slips its genes inside those of the cell it invades. After infection by a retrovirus, then, the very genetic material of the cell now says virus as well as cell. What could be a more perfect mechanism to achieve the purpose of ferrying designated genes into designated cells? All that was necessary

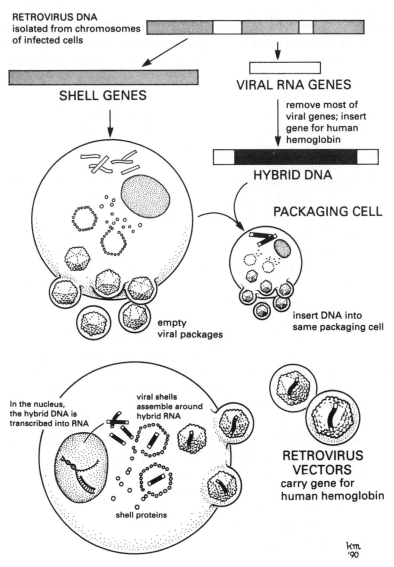

RETROVIRUS DNA
isolated from chromosomes
of infected cells

SHELL GENES

VIRAL RNA GENES

remove most of
viral genes; insert
gene for human
hemoglobin

HYBRID DNA

PACKAGING CELL

empty
viral packages

insert DNA into
same packaging cell

In the nucleus,
the hybrid DNA is
transcribed into RNA

viral shells
assemble around
hybrid RNA

RETROVIRUS
VECTORS
carry gene for
human hemoglobin

shell proteins

km
'90

Creating retrovirus vectors for use in gene therapy

Viruses may someday help us cure certain diseases. Since they are
adept at inserting genetic material into cells, they may be utilized to
ferry desirable genes into those cells. Richard Mulligan has succeeded
in devising such a process using retroviruses.

was that scientists find a retrovirus that naturally invaded the cells they targeted and then devise a way to have the virus insert the genes *they* wanted to introduce into the cell rather than the virus's own genetic material. In effect, they would have to hijack these viruses and force the microorganisms to carry new cargo.

In 1983 Mulligan, David Baltimore, and Baltimore's graduate student Richard Mann came up with an ingenious way of doing just that. They boiled down the task to essentially two steps: make a viral shell scoured of its own genes that will serve as a carrier, and bundle inside that shell the foreign genes that are to be ferried into the cell. To do all that, they transformed ordinary animal cells into gene-therapy factories. They called them *packaging cells*. The process goes like this.

First, the researchers allow a retrovirus to infect a cell and go about its natural routine of changing its RNA to DNA by means of reverse transcriptase and then inserting that DNA into the genes of the host cell. The scientists then go into the cell, locate that strand of viral DNA among the cell's chromosomes, and snip it out. Now they have in hand the virus's naked genes, the secret blueprints of how to make a retrovirus. "So what we do is to play a trick on the virus," says Mulligan, "and get rid of everything in the DNA that we don't absolutely need" (that is, need for their own purposes).

The DNA includes plans for an entire new retrovirus — plans for proteins for the viral shell and the reverse-transcriptase enzyme, plans for new RNA that will become the genes of the new virus. Mulligan and his colleagues separate away the genes that code for the shell of the virus and the viral enzymes from the genes that produce the viral RNA. Now they have two different vectors of retroviral DNA: one that will make an empty viral shell and one that will make the genes that go within that shell. Then, using the familiar manual method, they insert one of those vectors into a new cell.

"So we take the genetic information for making the shell and the enzymes and put them into the packaging cell," explains Mulligan. "That packaging cell is a normal animal cell, yet now what it makes is virus proteins. It doesn't make the viral RNA." Not yet, that is. Lacking plans for the retrovirus's RNA, but abundantly supplied with blueprints for the viral shell, the cell starts spitting out empty viruses. Half of the task is done.

The second step is to supply genes for the empty virus. But not the virus's own genes — rather, genes that the researchers want the virus to ferry into the target of their choice. So they go back to the original viral DNA, which now lacks those genes encoding the virus's structural proteins, but still retains the viral-RNA-producing genes. Snipping away those genes, the researchers replace them with genes of their own choosing — a human hemoglobin gene, for example. Then the scientists insert this new gene strand into the packaging cell. "So what you do is to go into that cell with this DNA from which you've cut out all those other sequences," Mulligan says, "and that cell becomes a factory for making virus particles that now just carry the gene of interest. That's the basis for the system."

Pretty nifty. Now fully programmed, the packaging cell will churn out viral shells that contain the new, human-inserted genes — which become retrovirus vectors — virtually forever. Mann, Mulligan, and Baltimore had supplied the methods to play God in the laboratory, by tinkering with the mechanisms of life to produce a brand-new virus. "Now that particle is a very interesting particle," says Mulligan. "It's infectious in the sense that it'll go into another cell at high efficiency, it'll undergo reverse transcription, and integrate its DNA, but since you've now chopped out all of the virus bits, all that's really left is a hemoglobin gene. It doesn't go any further. It's pretty cool."

They had come up with a retrovirus vector that infected like a normal retrovirus, integrated its genes into the host cell's chromosomes like a normal retrovirus, but then, instead of making more viruses, instructed the cell to make whatever it was that the new, human-selected gene coded for. Once it had deposited its genetic cargo into the cell, the virus itself fell away. And these manufactured viruses inserted their genes into *every* target cell. They were not only faster than the old, manual method of transplanting genes, they were much more efficient.

For French Anderson, it was just what the doctor ordered. "It started to become clear by '83, and very clear by '84, that retroviral vectors were the way to go because you could go into a hundred million cells in just two hours." With the promise of this kind of speed and effectiveness, it might be possible for Anderson actually to carry out meaningful gene-therapy tests with living humans, rather than being relegated to laboratory and animal experiments. In 1984 he wrote for *Science* an enthusiastic article called "Prospects for Human Gene Therapy," in which he presented the possibilities and problems as he saw them. Yes, techniques were now at hand to do the job, primarily that of using retroviral vectors. These viral vectors had been successful in transferring genes into mice, and the procedures were becoming increasingly efficient. They boded well for human gene therapy. All that remained was to establish the safety of the procedures by further animal experiments. "Once these criteria are met," Anderson wrote, "I believe that it would be unethical to delay human trials."

Ah, yes — the question of ethics. Having the technical means to proceed was one thing; having an ethical go-ahead was quite another. For the prospect of gene therapy, no matter its benefits, raises questions that transcend scientific and medical concerns. Treating disease is one thing; altering the very heart of life, our genetic makeup, is quite another.

"At the core of society's concern about beginning the human application of genetic engineering may be our sense that we are developing a capability to *change* who and what we are," Anderson wrote. "Might the procedure be able to produce alterations in the fundamental structure of our existence — our humanness? We do not really understand what makes up our humanness, nor do we know precisely what role genes may play. But whatever our humanness is, we fear the possibility of its being tampered with." "Change . . . alterations . . . tampered with" — the words themselves produce uneasiness. I may not know who I am, or like who I am, but I *am* who I am, and that's the way I'm going to stay.

Anderson confronted the controversy head on once again when, in 1985, he published in the *Journal of Medicine and Philosophy* an article entitled "Human Gene Therapy: Scientific and Ethical Considerations." And he continued thereafter to write about the problem and voice his concerns. Part of his reason, of course, was practical: before he could proceed with his plans to ferry genes into humans, Anderson would have to elicit the formal approval of the NIH's Recombinant DNA Advisory Committee, a task that would not be easy, as most members of the committee had no gene-therapy experience. They would have to be convinced of the project's safety as well as its ethical appropriateness before bestowing their blessing. Accordingly, Anderson touched upon every possible objection.

He differentiates four categories of human gene therapy: *somatic gene therapy,* in which he would introduce normal genes into a person suffering from a genetic defect or other disease; *germ line gene therapy,* which would involve inserting a gene into the reproductive tissue of a patient so that a disorder might be corrected in the patient's offspring; *enhancement genetic engineering,* in which introduced genes might "enhance" a known characteristic — for example, placing an additional growth hormone gene into a normal, although undersized, child; and finally, *eugenic genetic engineering,* in which gene therapy would be used to improve complex human traits such as intelligence and personality.

The last of these, using gene therapy to improve a person's intelligence, or strength, or character, Anderson rejects out of hand. Not only is such a process impossible at our present level of skill and technology, but it's simply unethical. So those people interested in breeding a passel of seven-footers for basketball or blond, blue-eyed babies for the movies will have to wait a while longer and, when and if the time comes, work with someone other than Anderson. Another of the possibilities, gene therapy that would persist through generations, is also far from being feasible given our present state of accomplishment — and Anderson defers that judgment until later on.

Remaining are the two types of therapy that are technically feasible now: somatic therapy and enhancement therapy. And between these two therapies Anderson draws the line. For him, gene therapy to treat disease is ethical; therapy to enhance normal conditions is not. For enhancement engineering could be medically hazardous. Correcting a disease-producing defect by inserting a gene would be one thing; adding a gene in the hopes that it might produce a desired characteristic is quite another. As an illustration, Anderson points to a TV set. "If the set were to stop working and we saw a broken wire inside," he says, "we might think that if we replaced the wire the set would be fixed. That could be a pretty good assumption. But what if the set worked fine but didn't have as sharp a picture as our neighbor's set? And what if we noticed that his set had an extra part inside? Would our picture become as sharp as his if we inserted his extra part into our set? Chances are we'd do more harm than good. We might have to take the part right out before it caused real damage. But once you insert a gene into a person's cells, you can't take it back out. The fact is, we simply don't know enough to understand the effects of an attempt to alter, rather than just correct, our genetic machinery."

So enhancement therapy could be dangerous. The second objection is that the practice would be morally precarious, and perhaps could lead to increases in discrimination. For, assuming we completely understood the effects of the process and that there were no risks at all, who should decide which people receive the therapy and which don't? What should the criteria be? Should a person be allowed to receive an extra gene upon request? If you come from an overly short family or an overly tall family, should you routinely receive therapy to normalize your height? What if you're at risk of dangerous radiation at your job — should you be given a gene that would protect you? What if you didn't want such a gene but your employer demanded it? And, finally, what about those, for whatever reason, who didn't receive such enhancement therapy? Might they be discriminated against? No, says Anderson, the problems, both real and possible, both technical and ethical, are too great. Gene transfer should not be used for enhancement therapy.

That leaves therapy for the treatment of disease. Anderson not only approves of this use of gene therapy, he's absolutely enthusiastic about it. This is therapy that is ethical "because it can be supported by the fundamental moral principle of beneficence: it would relieve human suffering." This is what he has been pointing to since he was a teenager in Tulsa dreaming about Harvard University and the molecular characteristics of genes. And this is what, since May 22 of 1989, he has actually been engaged in doing. On that date, duly approved by authorizing commissions from the NIH to the

FDA, Anderson and colleagues Steven Rosenberg and Michael Blaese of the National Cancer Institute began the world's first use of gene therapy in human beings. Accordingly, a sign appeared at Anderson's lab: "That was a small step for a gene, but a giant step for genetics."

Actually, these first experiments, performed on five patients with advanced cases of melanoma, a lethal type of skin-related cancer, were less an attempt at gene therapy than they were an opening test to see if the researchers' techniques were safe and effective. At the same time, however, the experiments offered some hope of improvement for these terminally ill patients, as they wedded Anderson's genetic engineering expertise with Rosenberg's ongoing work in cancer therapy. For some time, Rosenberg had been taking cancer-fighting cells called TIL, for "tumor infiltrating lymphocytes," from the diseased organs of his patients, growing them in the lab to increase their numbers vastly, and then reintroducing them into the patients in the hope that this augmented defense force might slow or stop the progression of the cancer. And so it did, but only in about half the patients. Rosenberg wanted to know why the other half were receiving no benefits from what seemed as though it might be very effective therapy.

That's where Michael Blaese came in. Blaese, an expert in the rare condition called ADA deficiency, or SCID, the lack of immune-system functioning, had already been working with Anderson for five years, as ADA deficiency is one of the genetic diseases targeted for gene therapy. Blaese realized that his NCI colleague Rosenberg's TIL therapy would be an ideal mate for Anderson's genetic-engineering methods, and he got the two men together. It has been a fruitful collaboration, as the combination of the two approaches has provided Anderson a vehicle for testing the effectiveness of his techniques and Rosenberg a means of figuring out why his TIL therapy has not been more widely successful. For by tracking the whereabouts in the patient's body of the genes Anderson inserts into the TIL cells, the researchers can follow the path of the reintroduced cells.

"You want the TIL cells to go where they'll be effective in fighting the cancer," says Anderson, "and you assume that's the tumor. But it might not be. They might be going to lymph nodes or someplace else and producing an immune response, and it's the immune response that's killing the cancer. Whatever it is that kills the cancer, that's what we want to learn. How the cells do it and how to make them do it better. We want to correlate where they go and what they do when a patient's tumor disappears verses when a patient's tumor gets worse."

So what's important in these experiments is effectively tracking the reintroduced TIL cells within the body. To do that, the researchers take a sample of cancerous tumor from the patient's body and treat it in the lab in such a way that the tumor dies but the TIL cells that are circulating in the tumor, so far unsuccessfully fighting its spread, continue to live. They then cultivate the cells, bathing them in the growth factor interleukin-2, to increase their number from perhaps 10 million cells to 300 million. It's then that Anderson turns to his viral vector, which is made from a retrovirus that primarily infects mice but also has the capability of infecting human cells with no ill effects. The added gene these altered viruses carry is an easily traced bacterial gene, also with no potential of doing harm. He takes about a third of the huge accumulation of cells and transduces them with the engineered virus and its gene cargo.

Anderson emphasizes that the proper term is *transduce* rather than *infect*. "It's not really an infection because there are no viral genes," he explains. "It produces no new viruses." He laughs. "In fact we joke with the patients about it. They certainly like the idea that their cells are transduced with a vector rather than infected with a virus."

Once they've transduced the cells, Anderson and his colleagues grow them and the nontransduced TIL cells until their combined total reaches a whopping 200 billion and then give them back to the patient by slow-drip IV. If the engineered retroviruses have indeed ferried their new genes into the heart of the cells, the researchers will be able to track the route of those billions of gene-marked cells and determine where they go. And while these particular genes have no therapeutic value for the patients — they're just markers — perhaps the massive infusion of TIL cells will be of some help. At the least, the experiment will do the patients no harm and the researchers great good. At the most, it will be a boon to both parties.

That was the plan as of May 22, 1989. It wasn't a plan everyone in the scientific community approved of. The familiar ethical, moral, and just plain irrational objections aside, some people felt that the technology of retroviral vectors simply wasn't established well enough to risk using it on humans. The effort, some said, was well intentioned but simply premature. One of those was Richard Mulligan. "I think it's absolutely crazy to take the technology we have and think you can do an intelligent experiment on a human," he said in a scientific newsletter. "I can guarantee you it won't be successful." Mulligan's stance typified that of some other scientists, who felt that Anderson was jumping the gun. In a way, the debate galvanized along the lines of PhD versus MD. Pure researchers, like Mulligan, who were not medical doctors and had never treated patients, preferred to go slowly, testing in the lab and in animals, to squeeze out every last bit of information before subjecting humans to an untried procedure. "If you jump to the patient at this point you preempt a series of experiments that could yield information that will eventually be useful for the patient," declared Mulligan.

Anderson, however, and other MDs like him, saw no reason to wait. "My orientation is and has always been to try to treat patients," says Anderson. "If it works, I don't care if I understand it. That's the way clinical medicine is. We have to be sure we don't harm the patient, so we have to understand it well enough to know what the risks are, but to do all the experiments to know exactly what's happening before ever going into a patient — there would be lots of people dying or dead by the time you did.

"One of the basic scientists at the NIH review actually said, if we designed our experiments well enough and worked hard enough to develop animal models, we'd *never* have to go into patients," Anderson recalls. "That's a direct quote. Steve Rosenberg and Mike Blaese and I looked at each other and said, 'No wonder we're having a hard time getting this approved!' "

They did get it approved, however, although Anderson and the others would have to wait until fall, months after the initial May 22 therapy, to know if any of it would pan out. In order to be absolutely precise in analyzing the information they were accumulating, the researchers took periodic blood samples from their patients, gave the test tubes a number code, and packed them away in the deep freeze, awaiting the day when enough samples from all the patients would be available so as to provide results comprehensive enough to make some meaningful judgments. As the summer months of 1989 rolled by, Anderson, Rosenberg, and Blaese took blood and tumor biopsies from their patients, filed them away in the freezer, and waited.

That's not to say that they didn't have anything to do, however. In fact, Anderson, who for years had been wedded to his work, now found himself spending even more time at the lab, to the point that he found it hard to get worked up about the tantalizing prospects before him. "Exciting?" he asks. "In one sense it really is, but I've been working on this for twenty years and it's so exhausting now because I've been working around the clock and I've just sort of . . ." He sighs. "I'm sure that in retrospect it will be very exciting. It's just that I'm exhausted right now."

On tap also are lab experiments in which Anderson and his colleagues are transducing TIL cells with different vectors that contain various anticancer genes, among them genes that direct production of interleukin-2 and interferon. He plans next to try out the vectors in animals and, if successful, to move on to human beings. In that case, the TIL cells will not only in themselves pose a threat to cancer tumors, but they'll carry genes that, once ferried by the cells into the heart of tumors (or wherever the cells go), will begin producing their own anticancer substances.

Also ongoing, in collaboration with Robert Gallo, are experiments designed to halt the spread of HIV, the AIDS virus. In this case, Anderson has taken a gene that makes the T-cell receptor to which HIV binds, called CD4, and has inserted it into ordinary animal cells, in effect producing a host of CD4 factories. The hope is that once he implants these cellular factories in an animal and they begin to secrete huge amounts of the CD4 protein, HIV will attack these decoy receptors rather than those that actually permit entrance to T cells. Again, if successful, the cache of factory cells might be able to provide a constant supply of CD4 into the bloodstream. Richard Morgan of Anderson's lab is now involved in testing the procedure on Donald Mosier's SCID mice in La Jolla.

"We've done it in rats, mice, rabbits, and monkeys," says Anderson. "The difficulty is that we might be able to make enough CD4 in mice to do something, but to make enough in humans — we're not efficient enough yet to be able to make a large enough amount of material. We can make a little factory about the size of a dime, and that works fine in a mouse. But it wouldn't work in a human. We have to learn how to scale up. That's what we're trying to do now in monkeys, to see how many factories we can put in."

Finally, Anderson is continuing to work toward his initial goal of treating genetic disease. One focus is hemophilia, which is currently treated by injections of various substances that clot the blood. But these clotting factors, which are derived from human blood, run the risk of being contaminated. That risk might be reduced by instead giving patients the genes that produce the clotting factors, so that they can in effect make their own.

But hemophilia is relatively rare, as is another focus of Anderson's work, the even more uncommon problem of SCID. Says Anderson: "One of the problems is, here you're trying to do something for no more than a dozen patients in the world that are candidates, and they're all infants. So if you actually do cure them, they've got seventy years for something else to go wrong. That's why we backed up and instead of doing a therapy in an infant for a rare disease, we went into a study of an adult with a common disease, namely cancer."

Cancer, AIDS, and genetic disease. "All three are being done concurrently," Anderson says, "and whichever one gets the animal data that is sufficiently strong to put in a protocol, and we get approval, that's what will go into humans first."

All of this makes the TIL-cell experiments even more crucial, for if they're successful, a precedent will have been set, and the door to human gene therapy, so long locked, will have been opened, perhaps for good. If they're not successful, it'll be back to the drawing boards for Anderson and his fellow pioneers. They won't stop fighting, certainly, but they'll be facing a battle with suddenly longer odds.

"I'm the eternal optimist," Anderson says. "I think when we break the code and have the data, they will be encouraging. We're moving ahead as though they will be. We've made the vectors for the next step, they've all gone into cells, and they've gone into animals. Safety studies have started, and as quickly as we're satisfied that we have adequate safety data to say that the risk-benefit ratio is appropriate, we'll put in a protocol. So we're charging down the road, and we'll continue to charge."

In the fall of 1989, the first results came in. Anderson, Rosenberg, and Blaese translated the blood sample codes of three of the five patients in their study and reported to the NCI on what they had found. In the first place, none of the patients had suffered any side effects or other problems from the infusion of gene-marked TIL cells. That alone was cause for celebration among the researchers, but there was more. In the first patient they analyzed, a twenty-six-year-old woman with extensive melanoma, Anderson and his colleagues were able to detect the gene-marked TIL cells for up to 19 days after they had been introduced; in the second patient, for 51 days in the bloodstream and 64 days in the tumor; and in the third patient, for 20 days. Those results indicated that the marker genes were indeed doing what they were designed to do — allow the researchers to track the cells through the body — and that, as they had anticipated, a goodly amount of the TIL cells made a beeline for the tumor. And in the spring of 1990, results from the rest of the patients in the study reinforced the earlier findings.

"We're starting to see a pattern now," Anderson says. "It seems like the cells are in the bloodstream for three weeks or so, then break down and are only produced again if needed, if the tumor begins to grow again. In fact, the second patient we looked at didn't have the cells in his blood from days twenty-five to fifty. It's a real tentative speculation — we've only looked at a handful of patients — but that's what things look like as of now." Furthermore, the patients have begun to respond to the TIL therapy: all of them have remained alive longer than they might have been expected to otherwise; and a large tumor on the soft palate inside the mouth of the first patient, which had begun to obstruct her ability to swallow, was growing smaller, as were many cancerous nodules on her skin. "We're very pleased," says Anderson. "What we're getting is turning out better than we might have hoped."

Emboldened by these results and by their ongoing animal experiments, the researchers have submitted requests for, and have received, preliminary approval for human trials involving TIL cells transduced with genes that manufacture cancer-fighting substances. For it seems as though lymphocytes may be the ideal vehicle for delivering these therapeutic genes to people who can use them. And they've received a provisional okay to provide gene therapy for victims of ADA deficiency, as well. "A small step for a gene, but a giant leap for genetics."

Another person pleased with the progress of the experiment is Richard Mulligan. "I want this to go right," he says. "I'm responsible for some of the basic techniques; therefore I feel responsibility for it going to the clinic. I'm happy now that the experiment is going forward. People will gain confidence in the concept, and that's probably one of the most important things, that people don't look on it like something from science fiction. It's a legitimate medical procedure." Meanwhile, Mulligan is working to refine further the viral-vector techniques he helped to pioneer, as well as apply them to the treatment of other diseases. For example, he and his group are investigating a defect in a gene in liver cells that causes the overproduction of LDL-type cholesterol. High levels of LDL cholesterol, in contrast to HDL, have been linked to atherosclerosis and heart disease. In fact, says Mulligan, people with this particular gene defect "have a predisposition to heart attacks in their childhood. The frequency of the problem is one in a million, and there's absolutely no cure except a liver transplant."

He has been working on removing liver cells from rabbits who also exhibit that particular liver defect and inserting healthy genes by means of his retroviral vector. In the lab, the gene transfer has been successful in causing the cells to reduce the amount of LDL-type cholesterol in circulation. The next step is to return the cells to the rabbit's liver to see if the same results apply in the living animal. And Mulligan plans soon to try to bypass completely the need to remove cells at all, thereby dispensing with the cumbersome tasks involved in transducing them in the lab and returning them to their original home in the body. How much more efficient — and more comfortable for the patient — would it be simply to inject the genes and their viral carriers directly into the body and trust them to find their target on their own.

"We're going to try to take virus and do direct injections," says

Mulligan. "Just inject virus into the circulation of the living animal, so it will infect liver cells. It would be best if we could do that, it would be really great. I think it'll work."

The possibilities for the use of retroviral vectors seem just about endless. For example, why not use them to uncover how specific genes function in the body? If you want to see what a particular oncogene does, put the gene into a retrovirus, let the virus plant the gene in a laboratory mouse, and track its effects. If you want to see how the blood system functions — discover the precise relationship among blood cells — use a retrovirus to insert genes into bone marrow cells and then track their destination in the body.

And gene vectors needn't be limited to retroviruses. "What you'll see in the future is the evolution of more gene-transfer systems that employ different viruses," says Mulligan. "You have at your disposal all these different viruses, and you can pick and choose aspects of them to make them useful. For example, you may not want to integrate genes into a cell's DNA. For safety's sake, you may want to stay with a replicating virus like EBV and papilloma. You might want to use herpesviruses to get you into nerve cells and fight disease there. There are so many ways that you can really manipulate these things."

And beyond these applications, Mulligan has grand plans indeed: actually to tinker with the makeup of viruses so as to construct custom-made versions. Rather than depending on a particular virus's ability to invade particular cells, why not instruct it to target the cells *we* want it to? "You could give a virus a very specific envelope protein to target it just to this or that cell," Mulligan says. "We're just getting cooking to try to put a protein on the surface of the virus particle for which there's a surface receptor on some target cell that we want to transduce. You could take a liver cell, say, find out what is on liver cells that might be a good docking mechanism for a virus, and then use the protein that reacts with this receptor to target viruses specifically to that [type of] cell."

Mulligan laughs, delighted by the prospect. "We're just putting our foot in the water here. We're touching everything. If you're into viruses, there are many things to think about. There's a lot of room for creativity. It's an exciting place to be."

*　　*　　*

"It's important to give people the idea that viruses can be harnessed," Mulligan says. It's a revealing remark, for it acknowledges both the power of viruses and our growing ability to use that power for our own good. A fundamental question remains, however. It is one many of us ask each time we come down with a runny nose and sore throat, or are laid up with some more serious viral ailment, or, even worse, stand by helplessly while a loved one dies: why *are* there such things as viruses? Do they serve any purpose other than that of tormenting us? In a universe that exists through the interrelated functioning of its constituent parts, do viruses play an important role?

For a number of people, perhaps the majority, the answer is simple: no. "Maybe they're just here, like the rest of us," says David Baltimore. "There's no reason anything has to have a function." Moreover, says Baltimore, if there were no viruses, we wouldn't miss them one iota. "Are viruses good for anything? I don't think so. If you subtracted them from the world, where would we be? I think we'd be better off."

But not everyone feels that way. "If viruses were nothing but bad news," says Michael Syvanen, a molecular biologist at the University of California at Davis, "you would expect cells to be taking great pains to evolve resistance to them. And they do express resistance in some ways, but they also seem awfully accommodating in others." For we've seen how cells invite viruses inside, obediently loan their own reproductive mechanisms for the making of new viruses, in some cases offer up pieces of their own outer membrane to serve as viral overcoats, and often expire for their trouble. You might think that if viruses posed a threat and nothing but, we might have evolved more effective ways of keeping the organisms at bay.

In the next chapter, we'll explore the ramifications of such thinking as well as various theories as to where viruses came from. It seems a fitting farewell to these invisible invaders — to consider their origins and larger purposes, if any.

Sixteen

· · · · · · · · · · · · · ·

Co-Travelers through Time

· · · · · · · · · · · · · ·

In October of 1989, Michael Bishop and Harold Varmus of the University of California at San Francisco won the Nobel Prize in Physiology or Medicine for discovering a genetic source for cancer. In so doing, they convicted the Rous sarcoma virus of petty theft.

It had already been discovered that this most venerable of retroviruses provokes cancer by inserting an oncogene, a cancer-causing gene, into the DNA of the chickens it infects. In trying to figure out where that devastating gene came from, Bishop and Varmus came up with an astonishing answer: it's part of the chicken itself. While screening healthy fowls, they found that the very same gene the retrovirus had used to cause tumors is a regular resident of the birds. How, then, did it get into the virus? The only explanation was that at some point in its journey through the animal's cells, the Rous sarcoma virus snatched the gene from the chicken and packaged it alongside viral genes in new, emerging viruses. Then, when the viruses went on to infect a new bird, returning the gene to its original home, its behavior was changed by association with viral genes — it was now capable of causing cancer.

The news, published in 1976, was stunning. Here was evidence that cancer is the work of normal genes from normal cells that somehow have undergone a metamorphosis into a new, dangerous form. Therefore the chicken carried the seeds of the disease within it all the time. It needed only the virus to alter the gene to the point that it became cancer-causing, or oncogenic.

Next Bishop and Varmus discovered something else equally as astounding: the gene didn't show up in chickens only. It appeared in a host of animals, from birds to mice to cows to fish. In fact, just about any animal they looked at carried the same gene, and the list also included human beings.

Bishop and Varmus's discovery triggered a rush into the field of oncogene research, with the result that today over forty oncogenes have been discovered. Scientists hope that by screening for the presence of these genes they might be able to detect the likelihood of cancer before the disease actually erupts. Less dramatic, but no less momentous to those interested in viruses, the discovery raised a question: how did this gene get around so widely in the first place? Fish to humans, birds to rabbits — that's a pretty wide range. Is this dispersal a result of heredity? Is there a likelihood that all these animals, ourselves included, have an ancestor in common whose lineage has provided us a gene in common? Or is there another answer, one that combines heredity with another evolutionary mechanism? Could it be that at some point in the distant past the gene was the victim of the same kind of viral burglary as that performed by the Rous sarcoma virus? Did it become inserted into genome after genome as the pilfering viruses infected animal after animal? Could it be, in other words, that viruses are the reason that the gene has been swapped around among so many different animals? Besides conferring disease among the organisms they infect, then, could it be that viruses play a role in the development of species? Could viruses be agents of evolution?

It's an evocative question, an important question, and an unanswered question. It may be that no one will ever know for sure, but that hasn't stopped people from speculating. "Viral transmission of DNA molecules may be an important source of genetic variability," says Michael Syvanen. "What the ability to move genes around reflects is the fact that in the course of evolution an important source of genetic variation has been foreign genes. Animals you see today, the successes in the great evolutionary race, are animals that have the ability to read this foreign information. If that's true, then how is the DNA moved around? Viruses. Viruses pick up foreign DNA."

The prime suspect in this viral-evolutionary service would seem to be retroviruses. As in the case of the Rous sarcoma virus, retroviruses are known to snatch segments of DNA from the humans they infect and take those pieces along with them, perhaps to deposit them in the genes of the next organism along the line. But other viruses, too, carry foreign genes, and a number of organisms have been found to include viral DNA in their own genomes. In this case, in contrast to the situation with retroviruses, such integration may not be part of the virus's survival plan but rather a result of sloppy execution — the virus snitching the host's genes by mistake and then mistakenly inserting this new, viral DNA into a newly infected animal — but it happens nonetheless. And, as we've seen, mistakes can play as central a role in the form and behavior of viruses and other organisms as any "planned" behavior.

So viruses *do* swap genes around in the organisms they infect, snatching from here and depositing there. That much is demonstrable and undeniable. But coming up with a reason for this gene swapping, this viral pack-ratting behavior, leads to speculation, and speculation is not a game for everyone. "Viruses may serve some kind of evolutionary function, carrying genes around," says David Baltimore, "but the evidence for that is very slim. It's speculative. I don't know of any evidence that any gene has been benefited by a retrovirus or actually moved around by a retrovirus. You could imagine it happens, but I don't know of any evidence. Except cancer-inducing genes, which are clearly bad."

Yet the more Michael Syvanen thinks about it, the more he becomes intrigued with the idea that perhaps viruses do have a grand purpose after all. He remembers when the notion first came to him, sometime in 1983. "I was in an unusually contemplative mood. Everybody spends time doing research on very specific things, but I was just pondering the question of why the concept of coadaptation was empirically turning out not to be true."

Coadaptation is a concept that was preferred by biologists of a couple decades ago: over time an animal becomes used to its constituent parts — coadapted to them — and those parts, including genes, can't move around to other animals. What you have, in other words, is uniquely yours. But in the late 1970s, with the advent of genetic engineering, a time when scientists began moving genes and genetic traits all over the place, the concept of coadaptation started to unravel. Today the idea is not as widespread. As it turned out, the development of these recombinant-DNA techniques catalyzed Syvanen's thinking as well. "It occurred to me," he says, "that maybe these gene movements aren't accidents. Maybe they've been part of evolution all along."

Intrigued, Syvanen, who spends his professional time researching bacterial resistance to antibiotics and insect resistance to insecticides, found himself looking into the literature of paleontology. "I went to paleontology because I was thinking that if my speculations were correct, the big evolutionary picture would have to be influenced," he says.

What he found was that about seventy-five million years ago, when the dinosaurs suddenly vanished from the earth, there remained only two or three orders of mammals. But in the next ten million years, little more than an instant in geological time, those few orders exploded into the sixteen mammalian orders we know today. "It's not possible to track the ancestry in a treelike pattern," Syvanen explains. "They just radiate from a point, as far as the fossil record is concerned. So all the changes that occurred to create the whales, the horses, the bats, the primates, occurred in this very short period of time."

How? How could so great an assortment of animals emerge from so few varieties of ancestors in so brief a time? For Syvanen the answer is obvious: the slow process of heredity and natural selection was augmented by viruses that transmitted genes from species to species. And, once given a jump start, so to speak, the various species went on from there to evolve into modern mammals, following a parallel, simultaneous pattern of development. "If you start asking, what did those early mammals look like? Well, they were very primitive," says Syvanen. "They were mammals, but they didn't look like modern mammals. As time moves on, over the next thirty million years, all the mammalian orders became modern to our eyes." And they became modern in concert. For example, the grinding molar evolved independently in a number of different lines. So, there are records to show species evolving in a parallel fashion — they evolved together.

The continuing viral transmission of genes from species to species not only could have contributed to their rapid development, but also may have ensured that they evolved in parallel fashion, developing similar features virtually simultaneously. And because the genes at the heart of this evolution initially came from a common source — those relatively few species that survived at the time of the dinosaur's disappearance — they used a genetic scheme accessible to all the different emerging species. These new animals could make use of the intrusion of foreign genes, readily translating the scheme into development and growth. All courtesy of the DNA messenger service run by viruses.

Why not? Certainly no one would be bold enough to make the claim that we know all there is to know about viruses. As we've seen, we've been aware of the creatures for less than a hundred years, and that awareness has primarily been in the context of disease. And as Syvanen points out, "You can be infected by a much larger number of viruses than cause disease. Viruses go in, they uncoat, they deposit their nucleic acid inside the cell, they even start to replicate, but something goes wrong and the infectious cycle is aborted. A lot of your resistance to viruses occurs at that level. The virus just doesn't get started, so you don't even have to have an immune reaction to it." Who's to say, therefore, that viruses aren't busy doing all sorts of things the likes of which we can hardly imagine? Viruses as agents of evolution — it has an appealing ring to it, even if it hasn't been proved, and even if relatively few people seem interested in tackling the possibility.

"The idea is formulated at a level that's so general that it really does scare away a lot of biologists," says Syvanen. "Everybody's a specialist nowadays. When you start dealing with a problem that touches so many different areas, there's no single specialist who can critically evaluate all of its parts. Still," he adds, "it's been more fun to ponder than anything I've done over the last few years."

*   *   *

People are also having fun trying to figure out where viruses came from in the first place. For while we often think of these invisible invaders as alien objects that attack us from afar, it may be that they're not so foreign after all. In fact, just as viruses may play a role in the evolution of living organisms, they may have sprung from those same living organisms, then escaped to an independent — and infectious — existence away from their original homes in living cells. It may be that viruses are actually like long-lost cousins, by now many times removed; in their cycle of infection, replication, dispersal, and infection again, they perhaps are simply coming home. With viruses, therefore, we may be dealing with returning bits of ourselves.

What would be required for a virus to spring from a living cell? First, there would have to be a supply of basic viral ingredients. Second, there would have to be a way that a piece of DNA or RNA could break loose from a cell and assume an independent existence. And finally, there would have to be a mechanism by which these newly escaped genes could undergo replication into a self-contained particle. As it turns out, cells are abundantly supplied with all these necessities. Viruses are bits of nucleic acid packaged in shells of protein — cells contain both. Cells also contain enzymes whose job it is to catalyze such processes as genes replicating — precisely what an emerging virus would need. And cells contain plenty of candidates for viral precursors in the form of bits of DNA or RNA that threaten to run amok. Rather than being staid, settled organisms, living cells are continually in flux. When it comes to life, the only constant is change.

"There are two or three different potential origins for viruses, depending on whether you're talking about DNA viruses, or RNA viruses, or retroviruses, which probably did have independent origins," says James Strauss, a molecular biologist at the California Institute of Technology who does research on the evolution of viruses. "One popular theory is that RNA viruses arose from messenger RNAs that escaped the cell and acquired the ability to be replicated and packaged." For messenger RNA, as we've seen, routinely ventures out from the nucleus of a cell to its cytoplasmic hinterlands, carrying genetic instructions to the ribosomes, the cell's protein-making factories. As an inveterate traveler, then, perhaps mRNA has at some point continued its adventuring all the way out of its own cell and into another and then another, using the new cells' supply of enzymes to self-replicate and eventually acquiring the ability to provide for itself a protective shell of protein.

So mRNA is a likely suspect to escape and become a virus. Another is an *intron*. Introns are sections of RNA that perform no known function. As far as anyone knows, they're simply there, hogging space along the RNA strand, doing nothing — they're sometimes called "junk genes." At the last minute, however, before mRNA travels out to a ribosome to direct the production of protein, these seemingly useless introns cooperate by cutting themselves out of the RNA strand. Might one of these little scraps of RNA, unrestrained and aimless in the cytoplasm, spring loose to an independent existence? It certainly seems possible.

Another possibility is that RNA viruses may be a link to the very beginnings of life, the primordial RNA. "Some people would argue that RNA viruses are very ancient," says Strauss. "They discuss the possibility that parts of the genes of these viruses might be related to the very earliest RNA molecules that were the basis of living systems. I don't know how to handle something like that because RNA viruses are changing so rapidly, there's no way of knowing how they were in the beginning. One can only speculate."

So much, then, for RNA viruses, which, mercurial as always, continue to raise as many questions as they answer. What about DNA viruses? As we've seen, DNA viruses tend to be more stable than RNA viruses, more reliable in terms of performing as expected. That doesn't mean, however, that their possible cellular precursors are any more responsible than those of RNA viruses. For cells contain yet another kind of material that might be the stuff of emerging viruses: *transposons*. Transposons, as their name suggests, are unique sections of DNA that tend to break loose from the cell's genes, move about the nucleus as though involved in a game of musical chairs, and transpose themselves back into the genome in a different spot. As such, they're sometimes called "jumping genes." "I think some people would argue that transposons have served a function in the evolution of higher organisms, and that's why they're there," Strauss says, "but no one really knows what they're doing."

No matter what their function is, however, transposons are prime suspects as the forerunner of DNA viruses. It may be that some of these jumping genes have jumped clear out of the cell and gone on to independent lives as infectious viruses. Perhaps some are doing it still. When it comes to viruses, the process of creation is certainly far from over.

Yet transposons, mysterious as they are, are not implicated in the origin of all DNA viruses. Vaccinia and other pox viruses may have evolved separately. They are different enough from other DNA viruses — much larger and more complex, and more nearly brick-shaped than geometrically symmetrical — that some feel they're actually degenerate bacteria. "The idea would be that you have a parasitic bacterium," Strauss says, "that devolved to the point at which it gives up its independence and ultimately gives up almost everything else. It gives up all of its ribosomes and its ability to generate energy and just becomes totally dependent on the host."

The theory has some precedent in a similar hunch concerning the origin of the cellular structures called *mitochondria*. These cucumber-shaped "organelles" situated in the cytoplasm function as the cell's power packs, producing through various chemical means energy the cell needs to function. Some people suspect that they, too, are degenerate bacteria that have found their way into living cells, for like bacteria mitochondria carry their own genes and are able to self-replicate. Pox viruses might be another instance of bacteria operating in a severely minimalist mode.

Finally, there's the question of the origin of retroviruses. As might have been expected when it comes to these notorious creatures, the possibilities are especially intriguing. For there exist in cells special kinds of transposons called *retro-transposons*. These are much like the jumping genes that may give rise to DNA viruses, as they pop away from their place in the cell's genes and bounce along until they find a new spot more suited to them. In this case, however, the wandering fragments of DNA contain instructions for producing reverse transcriptase, and this causes them to act suspiciously like retroviruses. While retroviruses transcribe their RNA genes into DNA, which then integrates into the cell's DNA genes, retro-transposons transform themselves from DNA to RNA and then back to DNA while jumping about in the cell's nucleus; then they finally slip into the genome at a spot different from that in which they started. Furthermore, the DNA itself of retro-transposons is remarkably similar to that of retroviruses. In fact, if retro-transposons had a protective protein shell, as do retroviruses, it would be hard to tell the two apart.

"They're very much like retroviruses," says Strauss. "Once you have retro-transposons, it's easy to envisage them becoming retroviruses. On the other hand, one could argue that the process goes the other way and they're defective retroviruses. But I think most people would say that they have served a function in evolution, that they're there for some reason, and that retroviruses are escaped transposons."

So it just may be that most viruses are no more or less than minute, wayward, and unruly parts of ourselves — something like adventurous teenagers who have fled the nest but just can't resist coming back home at every opportunity, sometimes to overstay their welcome, sometimes to wreak absolute havoc, sometimes to make us better through their mere presence. And, like loving parents, for better or for worse, we almost always leave them a key to the front door.

If these speculations are correct (and no one knows for sure whether they are or not), the ramifications are inescapable: perhaps viruses are our co-travelers through time — there with us from the beginning, evolving as we've evolved, perhaps even augmenting that evolution, and continuing to accompany us through our best days, our worst days, all our days. They are certainly our intimate companions, sharing as they do our very genes and subsisting as they do in the very heart of our cells. And for all the sorrow they can cause, they offer us a great boon: as they may arise from ourselves, they can provide knowledge of ourselves.

In essence, this story of viruses has been a story of ourselves. It has chronicled over two centuries of exploration, the result of which has been nothing less than to discover the basis of all life. The purpose has been, and continues to be, to discover precisely what makes us tick. Viruses offer an unmatched entrée into these essential mysteries, which, as we've seen, are rapidly becoming less and less baffling. For those who have explored the enigmatic world of these invisible invaders have been remarkably successful. Viruses, our cotravelers, our submicroscopic kin, have generously given up their secrets. As counterbalance to the devastation and sorrow they can bring, it is this, the unique access to basic knowledge that viruses offer, that defines their great boon.

Where we go from here is anyone's guess. This chronicle of viruses and virus hunters has been as much prologue as history. And the final irony is that the virus-associated insights and techniques that have given us access to this knowledge are themselves being supplanted by even more subtle methods. Molecular biologists are now able to do with cellular genes what they could previously do only with viral genes. No longer must viruses be intermediaries between us and ourselves. Having guided us to our own central workings, viruses are no longer necessary for us to explore them.

Having opened the door to the gene and been appropriated by molecular biologists as tools to investigate and manipulate genetic secrets, viruses are now passing back into the hands that first found them, the medical detectives. Of course, they've never been abandoned by our modern-day Jenners and Pasteurs — this book is testament to that. Yet now these intrepid medical explorers are doubly fortified, for they're able to use the techniques of molecular biology in their quest to understand, cure, and finally prevent disease.

So, for better or for worse, we're stuck with viruses. Sometimes it seems as though we won't be able to survive the tiny creatures; yet it's become increasingly likely that we won't be able to survive without them, either. In any case, our destiny seems inextricably bound to these invisible invaders. We're all in this together.

Epilogue

. . . .

AIDS in 1994 — The Virus That Can't (Yet) Be Stopped

. . . .

The statement from the Department of Health and Human Services was as curt as its content.

> November 12, 1993 . . .
> RE: In the Matter of Robert C. Gallo, M.D.; Case No. 93-91 . . .
> In light of recent Panel decisions, the Office of Research Integrity ("ORI") now feels compelled to withdraw its determination set forth in the findings of scientific misconduct at issue in this matter. This letter serves as notice of the withdrawal. ORI therefore considers these proceedings concluded.

For Bob Gallo this terse pronouncement might as well have been beautiful music. During the previous four years, ever since the appearance in November 1989 of a book-length article in the *Chicago Tribune* accusing him of stealing credit from French researchers for the discovery of the virus that causes AIDS, if not stealing the virus itself, Gallo had been a man under siege. From the media, from his own parent organization, the National Institutes of Health, from the United States Congress. Now the Office of Research Integrity, the government agency charged with investigating the matter, was calling a halt to the whole business.

In the view of many, it was about time. The puzzle at the heart of the inquiry had been why Gallo's AIDS virus proved to be virtually identical to the virus discovered a year earlier and thousands of miles away by Luc Montagnier of the Pasteur Institute in Paris. Families of viruses are like families of people — alike in general but different in particulars. It was highly unlikely that two viruses, separated in time and place, could be as alike as identical twins.

The answer to the puzzle came in 1991. In an irony transcending the acrimony between the French and American camps, it turned out that Montagnier's LAV and Gallo's HTLV-III were indeed the same virus — an aggressive third strain of AIDS virus that, unbeknownst to either researcher, had contaminated first the Pasteur lab and then Gallo's lab at the National Cancer Institute. Without realizing it, both men were working with the same organism. No wonder the viruses appeared to be identical — they were.

The debate might have screeched to a halt then and there. "Why does this not end it?" commented the British journal *Nature*. "When contamination is proved to have occurred on both sides of the Atlantic, which group can claim to be scientifically cleaner than the other?" Apparently not the Gallo group, as the government investigation into his behavior trudged on. The result was that a year later, in December of 1992, Gallo (as well as a member of his lab team, Mikulas Popovic) was charged with "scientific misconduct," a devastating censure for a scientist of his stature.

The accusation focused on one sentence from one of the four seminal papers Gallo's group had published in *Science* in 1984. In it, as quoted earlier in this book, Gallo stated that apparent differences between his and the French virus might be due to "insufficient characterization of LAV because the virus has not yet been transmitted to a permanently growing cell line for true isolation and therefore has been difficult to obtain in quantity." Not so, contended the ORI. Although it was true that the Pasteur group was still struggling to grow the virus, Gallo's lab had succeeded in doing so. How then could he say that LAV "has not yet been transmitted to a permanently growing cell line"? The statement was simply false, and intended to mislead the scientific community, argued the ORI. Thus the accusation of scientific misconduct.

Nonsense, countered Gallo. He meant the statement to refer only to the unsuccessful French attempts to grow the virus. Taken in context, he thought the meaning was clear. He did not intend to misrepresent anything or mislead anyone. He regretted any lack of clarity; he absolutely denied the intent to deceive. He would appeal the censure. In an open hearing he would have a chance to state his case.

The hearing was scheduled for the end of 1993, but before it could take place the ORI case against Gallo unraveled. First, in November of 1993, a Department of Health and Human Services (HHS) appeals board cleared Gallo's colleague Mikulas Popovic. Popovic had been censured because of charges similar to those leveled against Gallo — alleged misrepresentation in the same *Science* papers (for which Popovic had done the bulk of the research). For example, in the papers appear two charts describing the Gallo laboratory's success in isolating its virus in patients and growing it in cells in lab dishes. In those charts Popovic uses the notation "ND" to refer to experiments which, as explained in the legend, were "not done." But when the ORI investigating panel checked the original research records, they found that the experiments in question had indeed been done.

Popovic explained that to him "ND" meant "not determinable." Yes, he had done the experiments, but the cells infected were so few and the infections were so problematic that it was hard to come up with useful data. Moreover, as a recent Czech refugee, his English wasn't good enough for him to edit the papers. The discrepancies, he declared, had just passed him by.

That didn't matter, concluded the panel. Popovic was responsible for the mistakes, and for two other similarly nominal discrepancies. He too was guilty of misconduct.

But after Popovic's hearing, in which he and his lawyers presented his case, the appeals board simply obliterated the government's charge. "One might anticipate that from all this evidence, after all the sound and fury, there would be at least a residue of palpable wrongdoing. That is not the case." The board went on, in seventy-nine pages of strongly worded opinion, to clear Popovic of all charges. "How could it happen that such a massive effort produced no substantial evidence of its premise?" the board asked. Its answer was that once the major question — did the Gallo lab appropriate the French virus? — had been settled, there was nothing left to contest. "This dispute is largely vestigial."

And if the Popovic dispute was vestigial, and so forcefully rejected, what about the Gallo case? For the ORI, the handwriting was on the wall. If the agency was unable to bear the burden of proof against Popovic, a comparatively small fish to fry, what hope did it have against Gallo? "ORI Faces High Hurdle in Gallo Case," headlined *Science*. So high that little more than a week later the agency gave up the attempt.

But it did so neither gladly nor graciously. In fact, the ORI blamed "a new definition of scientific misconduct as well as a new and extremely difficult standard for proving misconduct" for its failure. The agency stood by its condemnation of Gallo — it simply felt it could not prove it in an appeals hearing.

Reaction was swift and strong. The *Chicago Tribune,* whose relentlessly critical articles about Gallo had fueled the inquest, declared: "When scientists, lawyers argue, justice is the loser." Other publications saw the result differently. "Method and Madness, the Vindication of Robert Gallo," read a headline in the *New York Times. Time* declared, "Victory at Last for a Besieged Virus Hunter." And AIDS activist Martin Delaney, director of Project Inform in San Francisco, accused the ORI of attempting to save face by blaming its failure on a new definition of scientific fraud. "That is completely untrue, as even a casual reading of the appeals board ruling will demonstrate," Delaney stated. "The ruling explicitly shows that it tested these cases against the existing 1989 HHS definition of misconduct. These people have clearly lost their case on the basis of the evidence, yet are now pretending otherwise. This is shameful."

And as for Gallo himself? His relief was mixed with indignation. "I don't feel good about the way the ORI went about it," he says.

"But the press has said I'm vindicated, and you can't ask more than that."

Still, "vindicated" or no, it is something of a Pyrrhic victory for Gallo. While no longer having to face official censure, he still must contend with a damaged image forged by the protracted investigation and by popular media presentations such as the recent HBO television movie *And the Band Played On*. In this account of the onset of the AIDS epidemic, Gallo is portrayed by Alan Alda as a scheming, self-serving glory hunter who stole his ideas and methods from the French and others, while withholding credit where it was due — precisely the charges, and more, that the ORI admitted it could not prove.

And, at this writing (fall of 1994), his ordeal may not be over yet. Still to come is one final official thrust, a report from Michigan congressman John Dingell, who set the whole Gallo investigation in motion some five years ago. The *Chicago Tribune* expects that it will contain "much new information — and even some answers," an assessment which, for Gallo, means, better be prepared for the worst. Nevertheless, he is now gearing up a renewed effort to stop the AIDS virus.

"I'm determined to make up for what has been lost," Gallo says. "What else can I do? It's best for me not to talk to the media about what happened — it won't help anything. My biology is the only way I have of making up for lost time."

That's good news, because the fight against AIDS can use all the help it can get. Since the publication of the hardcover edition of this book in 1991, when a hopeful optimism pervaded the field, the mood has become decidedly gloomy. Nothing seems to work against this insidious virus. Although we have learned more about HIV, and in a shorter time, than any microbe in history, it still refuses to give up essential secrets — for example, how does it bring about disease, and what chink in its armor can be exploited to stop it? None of the approaches discussed in this book — antiviral drugs, immunotherapy, or vaccines — has yet fulfilled its early promise.

Consider, for instance, the antiviral drug AZT. In 1991 it seemed a promising way to keep the virus under a modicum of control and delay the onset of AIDS. Today, although a recent study has shown that the drug reduces the transmission of HIV from infected pregnant women to their unborn infants and so is a promising tool for preventing pediatric AIDS, there is debate as to whether AZT provides any real benefits at all. What is certain, however, is that AZT can cause all sorts of debilitating side effects. In the absence of proof that the drug offers offsetting benefits, physicians are beginning to wonder whether to prescribe it. And other antivirals seem to offer similarly indeterminate blessings. It may be that an ingenious combination of these drugs is necessary to do the trick. That, too, remains to be seen.

As for immunotherapy, the attempt to boost a person's immunity to HIV after infection, the news appears to be just as bleak. Although experiments have revealed some signs of an increased immune response after the administration of a variety of boosters, there have been no reports of reversing or even slowing the course of disease. Nor is it at all clear what kind of immune response might actually do the job. It's a major frustration involved with HIV infection: if you don't know how the virus causes disease, how can you know what might stop it?

Vaccines have been similarly disappointing. Although a number of trials are now going on, no vaccines have shown any real usefulness. And the primary insight offered by animal studies, particularly those involving rhesus monkeys, which can be infected and killed by a cousin of HIV, a retrovirus called SIV (for "simian immunodeficiency virus"), is how difficult the problem continues to be. Not so long ago it looked as though these monkeys could be protected from infection by SIV vaccines. Now it's clear that those early successes were due to experimental quirks rather than anything real. For most SIV researchers, it's back to the old drawing board. (A conspicuous exception is Ron Desrosiers of the New England Primate Center, near Boston. For over two years his monkeys have resisted infection after immunization by a vaccine made of live, weakened SIV. But because researchers are wary of a live retrovirus vaccine, fearing that the virus might revert to virulence or cause other unforeseen problems down the road, Desrosiers's approach has not yet been tested in people.)

All of which is not to say that a solution to the problem of HIV and AIDS won't be found. Some of science's greatest achievements have arisen not from a base of understanding but from intuition, or accident, or just plain luck. Two primary examples are discussed earlier in this book — Edward Jenner happening upon the world's first vaccine, against smallpox, by noting that the mild affliction called cowpox seemed to confer immunity to the more dangerous disease, and Pasteur devising a chicken-cholera vaccine, and later a rabies vaccine, by making use of the serendipitous weakening of pathogens in his lab. Neither man understood the mechanism of the disease in question; neither man had any idea of the existence of anything so bizarre as a virus (viruses were not discovered until after both were dead); neither man understood the intricacies of the immune system. Yet they prevailed. Perhaps something similarly unexpected will happen with AIDS. As the saying goes, the darkness is deepest just before the dawn.

So AIDS researchers continue to push on. And one of their more exciting new approaches harkens specifically back to Pasteur and Jenner: trying to develop a strategy for combating disease from evidence of resistance to it. For, common perceptions to the contrary, some HIV-infected people *do* resist contracting AIDS. For example, Jay Levy of the University of California at San Francisco is keeping track of some fifty people who've lived more than eight years (some as long as fifteen years) since being infected, while retaining essentially normal immune systems.

Some people even resist becoming infected at all. Immunologist Gene Shearer of the NCI is following thirty-five high-risk gay men who have been exposed to HIV numerous times but haven't become infected. And Francis Plummer, an epidemiologist associated with both the University of Manitoba and the University of Nairobi, is studying perhaps the most provocative group of all. These are some 1,600 female prostitutes in one of Nairobi's oldest and worst slums. Although most of these women are infected with HIV, a consistent 3 percent of the group remains free of the virus. Plummer has followed some of them for nine years. They've had hundreds, even thousands of opportunities to become infected — yet they're not.

Why? What is protecting these people? The answer could have

enormous ramifications. If scientists can discover the source of their resistance, perhaps it could be the basis of a new vaccine against or treatment for AIDS.

Easier said than done, of course, but there are a few hints as to why these people are able to fend off the virus and why the infected survivors have thus far warded off disease. Perhaps the infecting virus itself is less virulent than other varieties, in effect immunizing these people in much the same way as a live virus vaccine confers protection. Perhaps the virus carries with it stolen bits of protein from other infected people, thereby providing more targets for the body's immune defenses. Or perhaps the secret of resistance involves a relatively little-known arm of the immune system called CMI. Everyone hears about antibodies, those molecular scouts that travel the bloodstream seeking out viruses and other intruders and marking them for destruction by other immune forces before they have a chance to infect any cells. This antibody-dominated response is called humoral immunity. Less well known is the fact that cells, too, can mount an immune response. Known as cell-mediated immunity, or CMI, this immune defense actually responds before antibodies arrive on the scene. Its heavy hitters are white blood cells called killer T cells. Whereas antibodies attack free-floating viruses in the bloodstream, killer T cells destroy cells already infected with viruses. And because the primary way HIV enters the body is not directly into the bloodstream (during IV drug use, for example) but through infected cells (during sex), it may be that CMI is the primary means of defense against the virus.

At least such is the evidence being accumulated by researchers working with long-term survivors of HIV and people who resist becoming infected. Accordingly, they are exploring ways to induce a protective ratio of humoral to CMI response, or CMI alone, that mimics the way these people successfully thwart the virus. And even more fundamental is the attempt to discover why these people are armed with such an effective mechanism of protection in the first place. What in their makeup confers this immunity? The answer may be in their genes.

With that, AIDS research in particular, and viral disease research in general, enters the labyrinth of the human genome. And labyrinth it is. Each of our trillions of cells encloses twenty-three pairs of chromosomes, within which are twisted and packed the strands of DNA that contain our genes. If uncoiled, each cell's DNA strands would stretch six feet in length. Connected like minuscule chain links along this virtually invisible thread are somewhere between 50,000 and 100,000 genes, each of which is comprised of smaller chemical compounds called nucleotides. It's in this molecular maze that scientists are now searching for the roots of resistance to disease. For it's been obvious for years, if not much investigated, that different people react differently when exposed to the same germs. Some get very sick, some just a little sick, some not sick at all. Why? The devastation of AIDS, and onrushing advances in techniques of molecular biology and genetics, have provided scientists the motivation and ability to explore the question. The story of two researchers, pathologist Murray Gardner of the University of California at Davis and geneticist Stephen O'Brien of the NCI, illustrates the promise of this approach, the often unexpected origins of scientific insights, and the ever surprising nature of viruses.

In 1968, as part of President Nixon's highly publicized War on Cancer, Gardner, who was then based at the University of Southern California, began investigating the possible link between viruses and cancer in animals. In particular, he set his sights on retroviruses, and soon focused on mice. It had been known for some time that retroviruses readily cause cancer in lab mice, which are inbred, but whether such viruses infected their wild cousins, much less what the effect might be if they did, was a mystery. To find out, Gardner and his crew went mouse hunting where huge numbers of the animals tended to show up — in dairies and egg farms, duck and squab farms scattered throughout several hundred square miles in and around Los Angeles.

"We probably collected twenty thousand mice over the next decade," says Gardner. But of all those, the only signs of infection by retroviruses were in mice from a squab farm in a canyon near Lake Casitas, north of Los Angeles. These mice were suffering cancer in huge numbers. Twenty percent of them were infected, and 18 percent of those were plagued by lymphomas, cancerous tumors of the lymph nodes. It was an astonishingly high proportion for a wild animal — "Nature almost never lets that many tumors occur in a species," says Gardner. Not only that, 12 percent of the infected mice suffered an often fatal paralysis of their hind legs as well. When Gardner injected the virus into laboratory mice, who, being inbred, would be expected to succumb somewhat more readily, fully 90 percent of the animals promptly became sick and paralyzed. This was one virulent virus.

Gardner was puzzled. Faced with such an efficient killer, the Lake Casitas mice should have been wiped out. Yet despite the overwhelmingly high incidence of disease, the colony as a whole was flourishing among the grain bins and hay beds of the squab farm. Why? Could it be, he wondered, that some of the animals were genetically resistant to infection?

At the time, it was an amazing notion, but there was some precedent for it. Genes that in test tubes protected cells from viral invasion had been extracted from chickens and laboratory mice. But no one had ever discovered genetic protection from disease in a free-living, wild animal. If he could prove that such was the case in the Lake Casitas mice, Gardner would be breaking new ground.

In 1977, Gardner decided to mate healthy Lake Casitas mice with infected lab mice, so as to monitor their offspring. If the healthy wild mice did indeed enjoy some kind of genetic resistance to the devastating virus, perhaps they would pass on those genes to their newborns. Sure enough, the mice produced uninfected, healthy progeny — there *was* a protective gene involved after all. But they also gave birth to infected offspring. Genetic inheritance isn't a simple matter. Due to the fact that all animals, including humans, receive two copies of every gene, one from each parent, and that some genes are dominant and others recessive, the permutations of inheritance can become complicated. Gardner would have to resort to statistical analysis. "The pattern didn't unfold for about two years," he recalls. "It took that long to breed enough progeny to be sure."

By 1979 the results were clear: a strong, dominant gene was protecting the mice. Gardner was excited. "I got right on the phone to Steve O'Brien. I asked him if he wanted to help me characterize the gene."

A colleague of Gardner's in the virus cancer program, NCI geneticist O'Brien was also intrigued with the idea of a gene that might confer resistance to disease. His own studies of mice had convinced him that such genes could exist. In fact, he welcomed the possibility, as it illustrated his view of the process of evolution.

"Darwin talked about the genetic response to predators, climate, ecological habitat, and all this kind of stuff," says O'Brien. "But what about the biological environment? What about all these nasty bugs that are always coming in? As far as I'm concerned, the most important selective pressure in evolution involves infectious diseases and the genes that respond to them. There's a dance going on between pathogens and the species they infect."

Far from being an isolated phenomenon, therefore, disease-resistant genes might be a common evolutionary response to millennia of battles with infectious viruses. They might be a living legacy of our past — in O'Brien's term, "footprints of historic epidemics."

Sometime in the distant past — so went his scenario — ancestors of the Lake Casitas mice had been beset by an identical retroviral epidemic. Over generations and generations they had evolved genetic resistance against the viral invader. With the battle won, the gene conferring this resistance had remained in the animals' genetic arsenal, to be passed down through subsequent generations of mice to the present day. Evolution had selected the gene as a way of allowing the mice to protect themselves against this particular virus.

It was a pretty slick premise. Now all Gardner and O'Brien had to do was back it up with concrete evidence. That came a year later, in 1980, when they pinpointed their gene. They dubbed it a "restriction" gene, because it restricted infection. But not until 1983 was it determined just how this intriguing gene worked. And that turned out to be as amazing as the fact that there was such a gene in the first place. Its ability to protect was made possible by the ancient invading retrovirus itself.

Viruses, of course, are little more than a few coiled strands of genes enclosed by a protective shell made of protein — "bad news wrapped in protein." Viruses infect by latching on to specific sites on a cell's surface. These "receptors" allow the virus access to the interior of the cell, at which point the virus shucks its shell and releases its genes to commandeer the cell's reproductive machinery and force it to churn out new viruses instead of new cells.

Retroviruses, as we know, go one better. They not only may take over the infected cell's reproductive capability, they actually deposit their genes inside the cell's much larger supply. There the viral genes hide out, invisible to the body's immune surveillance. Some retroviruses may lie low for years, a ticking time bomb. Others continually produce viruses, constantly infecting from within.

At some point during that ancient epidemic something interesting had happened. An invading retrovirus had failed to deposit its entire complement of genes inside an infected mouse. Instead, only one of the virus's genes made its way into the cell's DNA. "Sometimes when retroviruses integrate their genes into cells they get screwed up and only deposit some of them," says O'Brien.

As it happened, the function of this particular gene was to make the protein shell of the virus. And that's exactly what it continued to do inside the cell, make viral shells. But because the rest of the virus's genes were absent, the shells had nothing to enclose. They remained formless and impotent, nothing more than noninfectious fragments of viruses — in effect, no news wrapped in protein.

In this case, no news was good news. Doing what comes naturally, the protein fragments made their way to the surface of the cells, burst forth, and, as though they were whole and infectious viruses, adhered to receptors on the cell's surface. In the process, however, they blocked the entrée of related viruses invading from the outside. Says O'Brien: "It's as though they sat on the surface of the cell and said to their foreign cousins, 'Go ahead and try to infect. You can't do it, because we've got the receptors occupied.' "

If such accidental interference is to protect an entire organism, however, it must take place not just in one cell but in every cell in the body. For that to have occurred, the misfiring retrovirus must have invaded a female mouse on the verge of becoming pregnant. The virus then passed into the nascent embryo, with the result that all the multiplying infant cells incorporated the viral gene. Says O'Brien: "It's a grand example of how evolution can come up with solutions to the problem of infectious disease."

So this restriction gene in the Lake Casitas mice was nothing more nor less than an inadvertently useful remnant of a past viral invasion. It was the result of an accident, really, a viral mistake that the mice had retained for their own benefit. All of which got Gardner and O'Brien to thinking: if evolution can come up with such an ingenious method of preventing infection, why, in this new day of molecular biology, can't we? "This research has been a springboard," says Gardner. "It's led to the idea that maybe we can engineer protective genes like that into cells, such that those cells would be resistant to infection."

Japanese researchers have recently done just that. By injecting cells containing the restriction gene into the bone marrow of mice infected with the deadly leukemia virus, Yoshiaki Nakagawa and his colleagues at Kyoto University and Kansai Medical University have protected the animals against disease, an affliction they call murine AIDS, or MAIDS. And a National Institute of Allergy and Infectious Diseases team lead by Teresa Limjoco and Jonathan Silver has produced lab mice that carry the gene, which, generation after generation, protects them against infection.

These efforts bode well for people — assuming, that is, that such genes can be found, or made to work, in human DNA. O'Brien is optimistic about the possibility, so much so that he has devoted a major portion of his lab's efforts to locating human restriction genes. His enthusiasm is largely based on the startling fact, known since the mid-1970s, that we humans carry around inside our genes vestiges of hundreds of ancient retroviral infections. These genetic remnants, "endogenous" retroviruses, are similar in origin and character to the mouse restriction gene. But they differ in one important sense: human endogenous retroviruses offer no protection against viral invasions — no protection that we're aware of, that is.

"The reason we don't know of any protection in humans," says O'Brien, "is that we haven't had a decent gene map and we haven't had a decent retroviral epidemic. Now we have both."

The emergence of ever more sophisticated techniques for mapping human genes — an effort now formalized as the Human Genome Project — is providing geneticists like O'Brien with unprecedented knowledge about our genes and the ability to gather more information quickly. But without an event against which to measure such possibilities, they may mean little. For example, if viral cancer and paralysis had not plagued so many Lake Casitas mice, Gardner never would have been able to find his restriction gene — there would have been nothing for it to restrict. Until recently there has been no comparable human retroviral epidemic. Now, of course, because of HIV and AIDS, all that has changed.

Accordingly, it is among long-term survivors of HIV infection that O'Brien is hunting for human restriction genes. "We've embarked on a fishing expedition," he says. "We're trolling through human DNA to find genes that influence the outcome of exposure to HIV. The ultimate hope is that if we can discover human genes that confer resistance to HIV, we can use them in a new kind of vaccine against AIDS, a genetically engineered delivery of resistant genes."

Such an approach would first target already infected people. By supplying the protective gene to their uninfected cells, it might be possible to build up resistance to further HIV invasion. If that strategy works, the next step would involve introducing the gene to uninfected people who have a high risk of contracting AIDS.

O'Brien's fishing expedition is coming up with some bites. "This is the year we're going to announce that we've found some genes," he says. O'Brien anticipates that about twenty genes, acting separately and in concert, are involved in resistance to HIV. The interrelationships among genes and virus are going to be complicated and involve intricate feats of sorting out, but as far as O'Brien is concerned, the problem is solvable.

And not only may the problem of HIV infection be solvable — all viral diseases may lend themselves to such genetic investigation. "This avenue is defining a new way of dealing with disease," says O'Brien. "Whether it be polio, flu, the common cold, AIDS — the next generation of geneticists is going to see it all unravel."

If so, the answers will arrive not a moment too soon. Viruses, and the problems they create, are not going away. Nor, in all likelihood, will they be confined to the types with which we've become familiar. AIDS itself is a prime example of a new disease caused by a newly emerged virus. Estimates are that HIV may have made its appearance no more than fifty years ago. Another powerful reminder that the natural world is fluid and ever changing is the frightening spring 1993 disease outbreak in the Four Corners region of the southwestern United States. By late 1994, forty-seven people had died from what was determined to be infection by a deadly strain of hantavirus, an organism previously thought to be confined to Asia and US port cities. The virus was harbored by tiny deer mice, excreted in their urine, and spread through the air. It took nothing more than living near the rodents, or just inadvertently stirring up dust that had previously been soaked in mouse urine, for people to become infected.

Such episodes make it all too clear that the invisible invaders are, now and always, very much with us. It's anyone's guess as to what their next surprise might be. Virus hunters have their work cut out for them.

· · · · · · · · · · · · · · ·

Sources

· · · · · · · · · · · · · ·

Most of the information in this book was drawn from personal interviews, historical texts, and technical publications. This brief listing of sources includes only that material best suited to give the reader a general glimpse into the world of viruses, as well as to illuminate further some of the historical personages mentioned in the text.

Paul De Kruif, *Microbe Hunters* (New York: Harcourt, Brace & World, 1926, 1953).
> Still highly readable and interesting some sixty-five years after its original publication. A number of the scientists I talked with pointed to this romantic book as their inspiration for pursuing a career in science.

Greer Williams, *Virus Hunters* (New York: Alfred A. Knopf, 1960).
> Williams does for the early virus hunters what De Kruif did for the early hunters of bacteria.

David M. Locke, *Viruses: The Smallest Enemy* (New York: Crown Publishers, 1974).
> A thorough and readable introduction to virology — although now dated.

Andrew Scott, *Pirates of the Cell* (Oxford, New York: Basil Blackwell, 1985, 1987).
A more contemporary look at virology than Locke's. Although somewhat too technical for the average reader, it contains a wealth of information.

Index